Nuclear Power Plant Systems and Equipment

Nuclear Power Plant Systems and Equipment

KENNETH C. LISH, P.E.
Chief Quality Assurance Engineer, Burns & Roe, Inc.

INDUSTRIAL PRESS INC., 200 Madison Avenue, New York, N.Y. 10016

Library of Congress Cataloging in Publication Data

Lish, Kenneth C. 1919–
Nuclear power plant systems and equipment.

1. Atomic power-plants—Equipment and supplies.
I. Title.
TK1078.L57 621.48'3 77-185989
ISBN 0-8311-1078-3

FOURTH PRINTING

NUCLEAR POWER PLANT SYSTEMS AND EQUIPMENT

Contents

Preface

The contents of this book originated in oral orientations given to young engineers assigned to work on nuclear projects. These men were already familiar with heat transfer, fluid flow, stress analysis and the other quantitative disciplines required on the design team of a nuclear facility. Similarly, students of power engineering and nuclear engineering have many sources available to them dealing with the analytical aspects of the commercial application of nuclear power. Few sources, however, cover the hardware and the systems needed to assemble a nuclear power plant. This book is written to fill that gap. The book is qualitative in character and is designed to answer the question "What do I need?" rather than "How big should it be?"

It describes the various systems required by each of the three reactor types available commercially today in the United States: the boiling water, pressurized water, and gas-cooled reactors. The equipment requirements for the light-water reactors are further analyzed so that having selected the quantitative rating of the hardware, the mechanical requirements: material, codes, tests, etc., may be determined. The gas-cooled reactor has only been generically described because it would be premature to go further than that at this time. There is no detailed description of gas-cooled reactor equipment because nothing has been built as yet, in a commercial size.

Nuclear power plant design is still in the process of evolution. The design information given in the book is the accepted practice today. Modifications of present designs can be expected to take place as experience accumulates and more convenient systems are designed. Wherever possible, definitive recommendations have been given; where practice still offers choices, guidelines have been given so that the engineer can make an informed decision.

Industry in the United States today has evolved the concept of the "nuclear steam supply system" (NSSS), in which the reactor manufacturer furnishes the major portion of the material required in a nuclear island. The engineering organization designing the complete facility must then integrate this NSSS supplied material into the rest of the facility and produce a complete design. The starting point of this integration is an evaluation of the proposal of each NSSS bidder. A conceptual design of a complete plant is needed, upon which can be superimposed the scope of supply of the NSSS proposal. This book furnishes a basis for the conceptual design.

The book would not have been possible without the wholehearted cooperation of many industrial organizations. The American power reactor manufacturers, Babcock and Wilcox Co.; Combustion Engineering Inc.; General Electric Co.; Gulf General Atomic Inc.; and Westinghouse Electric Corp.; all contributed without hesitation.

I want to thank my own company, Burns and Roe Inc., and Mr. Al Gick for their assistance with the illustrations, and to Mr. George Nugent and others who have read all or portions of the manuscript and have given me many constructive comments.

Finally, my deepest thanks to my wife, Norma, for her encouragement and patience.

CHAPTER 1

Introduction

1.1 Theoretical Basis

The basic concept of nuclear energy was derived from Albert Einstein's theory of relativity expressed in the famous equation, $E = mc^2$, which states that matter and energy are equivalent with the proportional constant being the square of the velocity of light. In this equation the correct units are "E" energy, expressed in ergs; "m" mass, expressed in grams; and "c" velocity of light, expressed in centimeters per second. Einstein's theory was published in 1905, and from then until 1939, various experiments were performed all over the world which confirmed the existence of nuclear energy. In 1939, Otto Frisch and Lise Meitner interpreted the experimental results of Hahn and Strassmann to mean that when a nucleus of uranium absorbs a neutron it frequently splits into two parts of approximately equal mass with a simultaneous release of energy. Enrico Fermi, among others, pointed out that if this were true there would be an excess of neutrons present, over and above the neutrons required by the fission fragments. It was further reasoned that if a sufficient number of targets was available, the fission reaction could be self-sustaining. During the next few years it was shown that it would be theoretically possible to produce a continuous or chain reaction if sufficient uranium-235 were present to contribute the excess neutrons. As is well known, these theories were confirmed in December of 1942, when the first controlled chain reaction took place at the University of Chicago.

1.2 History

Nuclear energy in the United States was investigated during World War II by the military Manhattan Project as part of its overall purpose. When the war ended the Manhattan Project was terminated and its work was transferred to the civilian Atomic Energy Commission which was charged with the responsibility for developing peaceful uses of nuclear energy. To pursue a logical course, the Commission had to implement two programs simultaneously; first, to develop a feasible method of transforming nuclear energy into useful, controllable power and second, to investigate the effects of neutron bombardment on materials, and identify and/or develop the materials necessary to construct the power machines designed by the first program. The first power plant application was the commissioning of the submarine Nautilus in January, 1955, with a pressured-water nuclear power plant. The first commercial plants were started up in 1957 at Shippingport, Pennsylvania and at Calder Hall in Great Britain. Shippingport is a pressurized water plant, a direct outgrowth of the naval reactor program that producd the Nautilus. Calder Hall is a carbon-dioxide cooled, graphite moderated, reactor plant. Both reactors produced saturated steam in the 800 psig range. Present-day pressurized water reactors produce steam at approximately 1000 psig, while gas-cooled reactors produce steam at 2300 to 2500 psig.

The Atomic Energy Commission committed itself to a policy of reactor development in cooperation with private industry, offering a great deal of financial aid in the construction of these first generation reactors. All of the AEC unclassified technology was made available to industry without charge. As a result, in the last half of the 1950s many corporations attempted to enter the infant nuclear energy industry but the majority of these attempts were unsuccessful. Many withdrew until there were only six vendors left in the field: the General Electric Company which produces the boiling water reactor described in Chapter 3; the Westinghouse Electric Corp., Combustion Engineering, Inc., and the Babcock and Wilcox Company which produce the pressurized-water reactors described in Chapter 2; Gulf General Atomic, Inc., which produces a helium-cooled, graphite moderated reactor described in Chapter 4; and the Atomics International Division of North American Rockwell Corp. which designed and built a liquid sodium cooled prototype unit at Hallem, Nebraska; and which is now concentrating on the development of a sodium-cooled breeder reactor in which there will be more fuel produced than is consumed.

1.3 Reactor Concept

The reactor is a geometric array of fuel in an appropriate container, with a device for controlling the chain reaction and a method of moderating the energy level of the neutrons so as to be most efficient for the purpose intended. The reactor fuel is fissionable material, commonly, uranium-235, thorium-232, or plutonium-239, and the most economic array is a right circular cylinder. It should also be remembered that reactor fuel is made up of fissile atoms in any physical form—solid, liquid, or gas—so long as the proximity, shape, and quantity requirements for a chain reaction remain satisfied.

Commercial reactors employ fuels in the solid form, such as oxides or carbides, because it is the most convenient form to handle. Some work has been done on an experimental scale using liquid fuel either in the form of a uranium compound dissolved in an organic solvent or dissolved in a molten salt. However, the engineering requirements for these two concepts are still too expensive to justify commercial application.

The containers used are a steel pressure vessel for a water-cooled reactor and a prestressed concrete vessel for a gas-cooled reactor.

1.3.1 *Control Rods*—Calculations indicated that if there was no control on reaction, it would multiply almost instantaneously. Researchers, therefore, searched out and developed materials to absorb neutrons so that the reaction could be controlled. The original experiment in Chicago confirmed this part of the theory also, and the first of the absorbers used were cadmium and boron. These materials absorb neutrons and give off gamma rays and heat energy. They have to be cooled but they do not release neutrons. If too many absorber targets are available, however, the reaction rate will decrease because too many neutrons will be absorbed without fissioning an additional atom. If the number of absorber targets is decreased, the rate of fissioning will increase until an equilibrium situation is reached and the number of fissioning atoms per unit time is constant and all excess neutrons are absorbed. This, then, is the basic concept for the reactor control rod. To offer an excess of absorber targets the control rod is inserted; while to decrease the number of absorber targets, the control rod is withdrawn.

The term "control rod" infers a mechanical device with a fixed geometric shape, and the boiling water reactor uses neutron "poisons" in only that manner. But the designers of the pressurized water reactors have taken a different approach; a portion of their control targets are dissolved in their cooling fluid as boric acid; while the remainder are present as mechanical devices. The pressurized water reactors vary the position of the control rods as the hourly energy load on the reactor changes; the concentration of the dissolved boron is changed slowly as the uranium is consumed and neutron absorbing fission products are produced by the fuel.

1.3.2 *Moderators*—Original research and investigation revealed one more important characteristic that needed to be controlled. As the speed of a neutron is proportional to its energy, the relative speed of a neutron determines whether the neutron will enter a nucleus and create a different element or isotope, or split the nucleus and thereby release energy. A high-energy or "fast" neutron will enter a nucleus without causing fission; but a low energy or "thermal" neutron will split a nucleus. Therefore, fast neutrons have to be moderated or slowed down into thermal neutrons to participate in the fissioning process. The most common "moderators" are light water, heavy water, and graphite. The conventional, or light-water reactors, use demineralized light water for this purpose. It has the advantage that the fluid can be used for moderation and cooling so that only a single mechanical system is then required to cool the moderator and the fissioning uranium core. The helium cooled reactor described in Chapter 4 uses graphite for moderation, and the Canadian reactors use heavy water. Choice of moderator material is essentially based on economics and the state of the art. With the price of heavy water being upwards of $25 per pound there is a strong incentive to use either light water or graphite as the moderator.

In early power reactors, engineering problems such as determining the materials of construction and heat transfer were more easily solved by using light-water moderation than by using graphite; therefore initial development used light water. The graphite moderated power reactor is now catching up and has one American prototype (330 megawatts electric [MWe]) almost ready for commercial operation. This reactor is now being offered commercially in the 1100 MWe range.

1.3.3 *Fuel*—Nuclear fuel in fissioning reactors is some form of uranium and its natural sources are the ores of pitchblende and carnotite, which occur all over the world. Uranium occurs in a combined form, usually as an oxide. Natural uranium is a mixture of three isotopes: uranium-238, 99.274 percent; uranium-235 (U-235), 0.719 percent, and uranium-234, 0.0052 percent. U-235 is the easiest of three isotopes to fission and is the preferred starting fuel in the reactor. U-238 is converted to plutonium-239 which then fissions and develops energy. Natural uranium, either as a metal or an oxide, can be and has been used as a fuel. Canada, using heavy water for both moderation and cooling, uses a natural uranium fuel in the oxide form. The light-water reactors use a fuel enriched in U-235 content. Typical power reactor enrichments raise the U-235 content to 2½ to 3 percent. The maximum enrichment is 93.5 percent, which is commonly known as "fully enriched" uranium. The first enrichment method to reach engineering feasibility was gaseous diffusion. In this process enrichment is achieved through the use of a porous membrane and porous membrane technology is classified by the United States Government. The remainder of gaseous diffusion technology, however, was declassified in 1969. The gaseous diffusion process can be classed as an energy and capital intensive process. Currently in the United States, Japan, and Europe work is proceeding on a centrifugal process that shows much promise, as it will require less capital and energy than does the gaseous diffusion process.

The enrichment process is carried out in a series of thousands of minute stages each of which separates the incoming uranium compound into two portions; one of higher U-235 enrichment and one of lower U-235 enrichment, than the entering compound. The higher enrichment portion proceeds to the next stage for further enrichment, while the lower enrichment portion is recycled to the preceding stage. Feed can be added to any matching stage and product can be withdrawn at any stage. Thus, enrichment cost is directly related to the amount of enrichment required. At the present time enrichment cost is about ⅓ of the total fuel cost.

Fuel cost is calculated on: cost of the uranium to purchase and refine; conversion cost to gaseous uranium hexafluoride; enrichment; conversion to an appropriate solid form (usually UO_2); and then fabrication into a suitable fuel element. When the United States Government owns the uranium then conversion, enrichment, and fabrication charges must be paid in addition to "use charges" while the uranium is in the possession of the power utility. At the end of use, the fuel element is "reprocessed."

In reprocessing, the fuel element is completely dismantled and the known valuable elements: uranium, plutonium, neptunium, etc., are chemically separated and recovered. Plutonium is a trans-uranium element produced by the fast neutron absorption by U-238 and the subsequent transformation of the U-239. Recovered elements belong to the utility; the uranium is recycled back into the enrichment plant and is refabricated into a fuel element again. The plutonium can be sold to the government or saved to be used as a fuel.

The theoretical amount of energy available in a metric ton of U-235 is of the order of 200,000 megawatt days (MWD). Fuel in a light-water reactor is presently permitted to release 20,000 to 30,000 MWD per metric ton in the reactor before removal. This exposure is well within the safe limits determined by knowledge and experience with present-day fuel design and fabrication. Energy release will eventually increase as experience accumulates on the behavior of the various fuel designs under irradiation.

Fuel performance is restricted basically by two items: heat transfer and materials of construction. At present, heat transfer is handled very well and has yet to be a limiting factor for lack of knowledge. Mechanical properties of materials under high temperatures, stresses, and neutron fluxes are the controlling factors in design and anticipated performance of fuel elements.

In operation, uranium produces various fission products among which are gases. A typical water reactor fuel element is an array (usually square) of hollow capped zircalloy tubes. The tubes are filled with cylindrical pellets of uranium oxide held in place longitudinally by a spring. A void is provided in the tube for fission product gases. This void is filled with helium at assembly when the end caps are welded in place. Fission products are stored within the crystalline structure of the uranium oxide as well as in the gas void in the fuel tube. Fission products can only be released as a result of rupture of the fuel tube.

1.4 Atomic Energy Commission

The Atomic Energy Commission (AEC) is an agency of the Executive Branch of the United States Government. Organized in 1946 by Federal legislation, it was reorganized in 1954. A unique organization, it is responsible for the research and development of nuclear applications: nuclear energy, isotopic devices, weapons, high-energy physics, etc., and is also responsible for the control of fissionable material as well as for the licensing and safety of all nuclear installations in the United States. In the first role the Commission has established its National Laboratories; in exercising its control over fissionable materials, the Commission has its uranium enrichment plants and its plutonium programs. In its licensing and safety role, the Commission looks after the health and safety of the general public as they may be affected by nuclear installations.

Initially, licensing was carried on through an organization called the Advisory Committee on Reactor Safeguards (ACRS). At that time the first power reactor projects were just getting started. Technology advanced at a very rapid rate, however, and each installation was treated on an individual basis. In the middle 1960s possible conflicts of interest appeared when a single division within the AEC was responsible for both reactor development as well as for licensing and regulation. The responsibilities were then separated, a new Division of Reactor Licensing (DRL) was formed, and the procedures for licensing became more stylized.

Licensing is presently accomplished in two major steps:

Step 1. Shortly after selecting a reactor vendor (usually within three to six months) the owner prepares and submits a Preliminary Safety Analysis Report (PSAR), to DRL for review. The report is prepared cooperatively by the owner, the reactor vendor, and the consulting engineers responsible for the station design. DRL and ACRS will review the PSAR and ask any questions that may seem appropriate. A public hearing is part of the DRL review and any member of the public may ask to be heard at these hearings. After DRL is satisfied that the proposed plant will be adequately designed and that any questionable items have been specifically identified and design approaches and development programs indicated, they will then issue a construction permit. The average time from application to construction permit is approximately 18 months. However, during this period, the plant design is being accomplished.

Step 2. At the conclusion of design a Final Safety Analysis Report (FSAR) is filed and processed. The FSAR must now answer all matters outstanding in the PSAR, plus any questions that may have arisen during the design and construction period, which have not as yet been answered. An operating license is then issued which permits fuel to be loaded and the reactor to go critical. The operating license may be for an intermediate power level or for the final power level. During the period of design and construction, the Division of Compliance inspects both the design program and the construction program to insure that everything is in order. There is no specific time interval between inspections; visits are relatively frequent at the start of a job in order to build up a "level of confidence" for the particular project as to the quality of the work being done. Once this "level of confidence" is established, future inspection visits are tailored to the rating of the confidence level.

The Commission has published the requirements for a PSAR in Title 10, Part 50, of the Federal Code of Regulations. An important part of the requirements are the "General Design Criteria for Nuclear Power Plant Permits" which outline the minimum safety features required in a reactor installation. The features are those required to design and control the core, contain radioactive materials, and design protection systems and barriers. (There are fifty-five criteria and they are reproduced in this book as Appendix I.) From time to time the text will make reference to a specific criterion such as, "Criterion No. 50," in describing a design feature provided in answer to a particular criterion. These reports, both preliminary and final, are public documents available for reference in any AEC depository library.

1.5 Codes

The first reactors were designed in accordance with existing codes, principally the "Unfired Pressure Vessel Code," Section VIII, of the ASME Boiler and Pressure Vessel Code. Piping was in accordance with ASA B31.1, the Code for Pressure Piping. Both of these codes quickly proved to be unsuitable as written and attempts were made to rectify them by the nuclear code cases of the late 1950s and early 1960s. Meanwhile, the U.S. Navy produced its own code for nuclear vessels published as Dept. of Commerce PB-151987. This Navy code utilized completely new analytical techniques and also recognized the existence of different types of stresses and fatiguing due to operating cycles; requiring, in fact, a design report which went into minute detail in its stress analyses. Industry then recognized the inadequacy of Section VIII, and after some years' work produced its own new code, Section III, "Nuclear Vessels," which was issued in 1963. This new code used the techniques which were introduced in the Navy Code and classified its stresses into primary, secondary, and tertiary stresses.

The new code contained three classes of vessels. It is the responsibility of the owner or his engineers to classify the vessels. The highest quality vessel, Class A, is used for such things as a reactor pressure vessel or the primary side of a steam generator. As a guide line, those vessels used to contain a chain reaction, vessels whose removal from an operating system would endanger the chain reaction, or those vessels containing primary coolant that cannot be regularly inspected during a plant shutdown are classified as Class A.

Class B vessels are used to contain other vessels and equipment. A typical illustration is the containment structure in water reactor installations. Class C is used for vessels containing radioactive fluids that are not classified either Class A or Class B.

In November 1969, after a year of trial use and comment, ANSI B31.7, "Nuclear Power Piping," was officially issued. It contained three classes of piping: I, II, and III, that roughly parallel the three Classes in Section III of the ASME Code. Radioactive piping or potentially radioactive piping had to be built in accordance with American National Standard ANSI B31.7. Other piping, at the discretion of the designer, could be built under its rules. It became fairly common to design closed loop cooling systems, interspersed between a radioactive system and a final heat sink, to Class III of ANSI B31.7.

In the fall of 1968 a "Draft Code for Pumps and Valves for Nuclear Power" was issued by the American Society of Mechanical Engineers for trial use and comment. The Code was meant to be used in conjunction with ANSI B31.7. It had the same three classes as the piping code. The code, however, was incomplete as issued because design criteria for Class II and III pumps were not ready and could not be included in the original issue. The code concerns itself only with the design and inspection of the pressure boundary; no reference is made in the code to the operating characteristics of pumps or to the various types of valves. The engineer specifying a component and invoking this code must give just as much attention to the details of construction and operation as he ever did. As a specific illustration, Code Case N-2 of ANSI B31.1.0 requires that all valves in radioactive systems have two full sets of packing with a lantern ring between them and a packing box leakoff connection at the lantern ring. With the new Pump and Valve Code this requirement is lost; thus the engineer must make the feature a part of his procurement specification.

All three codes were complementary and, under the rules, it was possible for every component and piece of piping in a radioactive system to have a nuclear code stamp.

While an attempt was made to coordinate Section III of the ASME Code, ANSI B31.7 and the Pump and Valve Code, the effort was not completely successful. Since Class B, of Section III is specifically directed to containment vessels, it has no analogy in the other two codes. In like manner, the Class II piping has no analogy in Section III. Permissible materials are not always the same. In the winter of

1970, the addendum to Section III of the ASME Code, by specific reference, made the use of the other two codes mandatory and provided a system whereby a complete radioactive system could have code stamps on each of its individual parts and then also have a code stamp issued for the installed system as a complete entity.

However, despite this progress, there were still gaps in the system. The 1968 ASME Code was directed specifically to metal vessels; it did not recognize composite vessels, the most popular of which is concrete with an internal steel membrane or liner. The July, 1971 edition of Section III underwent a major change; its scope was changed to include all components formerly covered by Section III, ANSI B31.7 and the Pump and Valve Code. It now has three equipment classes: 1, 2, and 3—each of which includes vessels, piping, valves, and pumps. In addition, it includes subsections for reactor internals, metal containment structures, and equipment supports. A second division will be issued which will also cover concrete containment structures. Thus, at the present time, all nuclear mechanical systems can be built under a single set of rules.

The ASME Code has not yet recognized all the different manners in which a job is executed. It is normal practice that fabricators of vessels, valves, and pumps have their own engineering staffs so that when one of these companies is authorized to use a Code "N" stamp, the Engineering Department has also been checked. Piping is different. An independent engineering organization designs the piping system and specifies all the requirements such as material, dimensions, guides, supports, etc. A pipe fabricator, who has an "N" stamp, will fabricate the piping from the engineer's drawings. As can be seen, the missing link in the continuous chain is Code Authorization for the piping design agency. The piping design agency is not yet required to get a Code Authorization so that all aspects of the nuclear facility would have Code approval.

A fourth code, "Inservice Inspection of Nuclear Coolant Systems," was issued in the fall of 1968. This is a radical departure from previous practice in that this code is concerned only with inspection of nuclear plants in operation. Briefly, the Code requires that portions of the pressure boundary of the primary system be inspected at each shutdown and that an inspection cycle be completed in ten years. In addition, a "baseline" inspection is made of the new installation just prior to the start of operations, and it serves as the reference for the installation thereafter. The Code specifies the types, locations, and frequency of the inspections to be made. One of the purposes of this Code is to insure that some provision is made in the design to permit access to the primary coolant system so that inspections can be made during any scheduled plant shutdowns. The Code also stipulates the nondestructive test methods that can be used. It must be realized that these inspections will call for a combination of both remote and contact procedures. Thought was first given to the problems in 1969 and 1970, and nuclear power reactor proposals are now recognizing the "accessibility" aspect of the Code. The vendors are starting to offer research and development options with their reactors, to develop the tools and techniques to implement the Code.

It has been customary when ordering a reactor plant system, to specify that the codes to be used were those in effect on the date of the purchase order. In the succeeding period, until issuance of the construction permit code, improvements took place and were required by DRL in specific decisions. Licensing regulations were finally changed so that the code in effect in any nuclear plant, is the code in effect on the date of order of the *component,* not the system. As an illustration, a reactor system might be ordered in September, 1971, and its primary pump ordered by the reactor supplier in June, 1972. The code required for the pump is that in effect in June, 1972. (The text of this ruling is reproduced in this book as Appendix III.)

All the construction codes just reviewed are devoted basically to material qualifications, design requirements, and procedures and shop fabrication; little, if any, consideration has been given to field work. No codes exist for safeguarding the high quality of workmanship and cleanliness achieved in the shop in the interval between start of component shipment and plant start up. As may be expected, this uncontrolled gap has been a major source of difficulty as each engineer prepared specifications to the best of his knowledge and ability. Requirements vary from job to job and things are overlooked since there is usually neither sufficient time nor money available in a single project to do the thorough comprehensive job required. The Division of Compliance of the AEC has been known to threaten to halt work at a job site unless housekeeping and material storage were immediately upgraded.

This uncontrolled area has been recognized and the N-45 Committee of the American National Standards Institute has been charged with preparing the needed codes. Codes and standards are also in preparation for site storage and field quality control, and publication of the first of the codes can be expected in late 1971 or early 1972.

CHAPTER 2

Pressurized Water Reactors

2.1 General

A pressurized water nuclear power station contains a closed loop of pressurized water which removes the heat energy from the core, and transfers the energy to a second water system generating steam therein. The steam, in turn, drives a turbine generator set which produces electric power. This process is shown schematically in Fig. 2-1.

The reactor system consists of a pressure vessel containing the nuclear fuel which generates the heat energy, a steam generator in which the heat energy is used to generate steam, a circulating pump which circulates the coolant, and a pressurizer that maintains and controls system pressure. The coolant is demineralized water and typical operating conditions are 2300 psia and 600 degrees F—a greatly subcooled condition. Steam is generated in the range of 850 to 1000 psia and usually has 0.20 percent to 0.25 percent moisture. One manufacturer uses a different steam generator design and produces steam with 35 degrees F of superheat. A simplified diagram of the steam generating elements of the system is shown in Fig. 2-2. The reactor system is assembled with a series of coolant loops which radiate from the reactor vessel. Figure 2-3 shows a typical arrangement of a primary system.

A single reactor has only one pressurizer regardless of the number of loops. One loop is provided for each steam generator required and there are either one or two pumps per loop, depending on the manufacturer. In the 1969 designs, a 600 MWe unit has two loops, 800 MWe has three, and 1100 MWe has four loops. Figure 2-3 shows the relative position of the various components with respect to each other.

Pressurized water reactors are marketed in the United States by the Westinghouse Electric Corporation, Combustion Engineering, Inc., and the Babcock and Wilcox Company. Westinghouse also has licensees in Europe and Asia. Both Westinghouse and Combustion Engineering produce saturated steam using a U-tube steam generator as shown in Fig. 2-2,

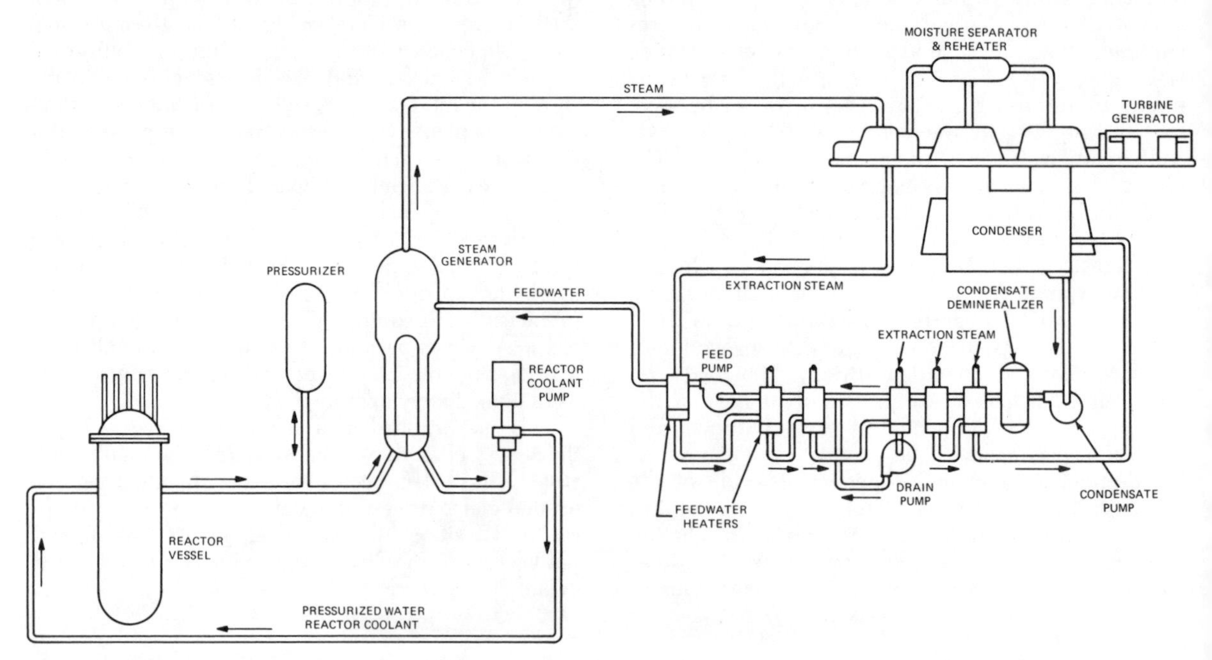

Fig. 2-1. Schematic arrangement of a PWR power plant showing the reactor primary loop and the turbine power train. The heaters and drain pumps are illustrative only and are not an actual cycle. The condensate demineralizer is an optional item in the cycle.

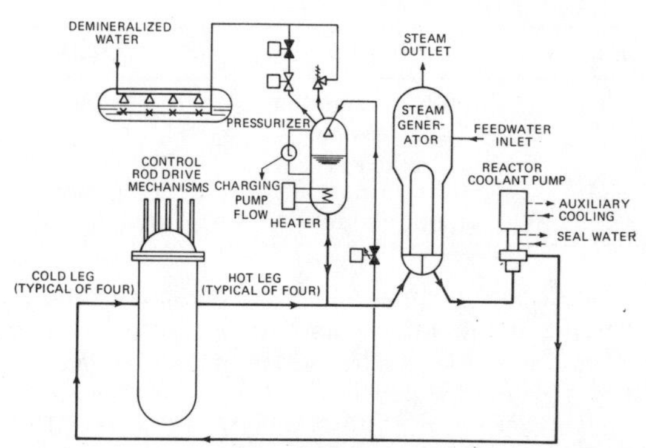

Courtesy, Westinghouse Electric Co.

Fig. 2-2. Schematic diagram of one loop of a primary coolant system. Reactors have from two to four loops, depending on size. Ratings range from 1785 to 3423 thermal megawatts.

while Babcock and Wilcox produce superheated steam using a once-through steam generator. Plants are available in three nominal ratings: 1750 Megawatts thermal (MWt); 2700 MWt; and 3100 to 3400 MWt which correspond to 550, 900, and 1150 Megawatts electric (MWe). These ratings will increase as experience accumulates.

Westinghouse manufactures its own circulating pumps and normally supplies one pump per loop. Combustion Engineering and Babcock and Wilcox purchase their pumps and supply two pumps per loop. However, all three suppliers manufacture their own steam generators and pressurizers.

2.2 Buildings

The nuclear island of a pressurized water reactor power station consists of three or four buildings, depending on the individual design. The three buildings that must always be present are the reactor containment building, the fuel handling building, and the auxiliary building. The fourth, optional, separate building, is the control building. The space provided in this last building is always required, but some designs add it on as an additional wing to another, already existant building.

2.2.1 *Reactor Building*—The reactor building is a cylindrical structure with a hemispherical or elliptical top head. The bottom head may be either of the two shapes just mentioned or a flat head, depending on the structural design. The entire structure is Seismic Class I (see Chapter 6). The building contains all of the primary system equipment shown in Fig. 2-3, plus the accumulator tanks shown in Fig. 2-11, and it also has a polar crane, usually of 100-ton capacity, just beneath the spring line of the top head. The crane is used during normal refueling, in-service inspection, and for general maintenance. Additionally, the air-handling equipment for normal and emergency cooling, and any other equipment needed in fuel manipulation, is housed in the reactor building.

2.2.2 *Fuel Handling Building*—The fuel handling building contains the spent fuel storage pool, a bridge crane over the pool for handling the fuel shipping cask (100 to 150 tons), and all of the spent fuel cooling system equipment shown in Fig. 2-9. This building is connected to the reactor building by the fuel transfer tube shown in Figs. 2-9 and 2-15. The fuel storage pool must be Seismic Class I, and is usually designed to have about twenty-five feet of water shielding above the stored spent fuel. No other portion of the structure or the bridge crane need be Seismic Class I; however if the design is Seismic Class II, the design must be such as to guarantee that

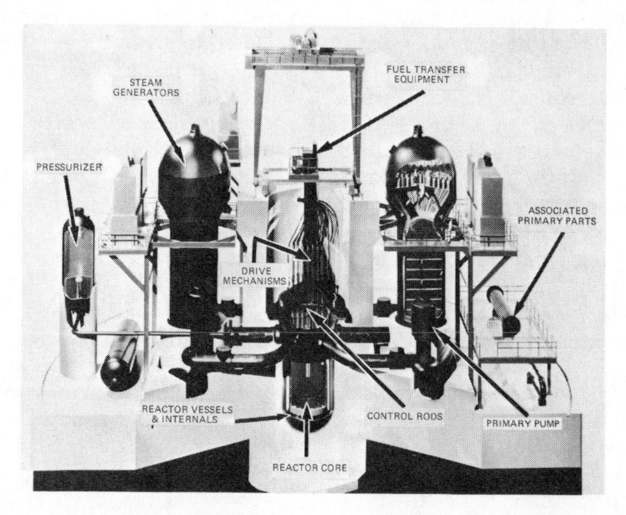

Courtesy, Combustion Engineering Corp.

Fig. 2-3. Typical arrangement of a PWR system. Note the relative elevations of the steam generators and the reactor vessel. A pipeline failure cannot drain coolant from the vessel.

failure of lesser class equipment in a seismic occurrence will not endanger the integrity of the storage pool nor the fuel stored in it. It is usually simpler to build the entire structure and crane as Seismic Class I than to provide the guarantees for lower class designs.

2.2.3 *Auxiliary Building*—The auxiliary building is a multiclass building of reinforced concrete. It houses the residual heat removal system, Fig. 2-8; the chemical and volume control system, Fig. 2-6; the safety injection systems, Fig. 2-7; the radioactive waste system, Fig. 12-2; the closed loop cooling system, Fig. 2-11; plus any necessary air-handling and cooling equipment for the building. The process systems listed above are a combination of Seismic Classes I and II. The building and equipment arrangement must reflect an economical structural design, considering the requirements for two seismic classes in a single structure. In addition, the building must contain "cells" for certain equipment to provide biological shielding for operating personnel. In arranging this building, consideration must be given at all times to the separation of equipment, piping, and wiring of safety systems so that no single failure of a structure, equipment, piping, or wiring can prevent fulfillment of the safety function. In conformance with this separation criterion, the addition of equipment cells or protection walls, and cooling devices for rooms containing motors that must run in an emergency, will be required.

2.2.4 *Control building*—The control building, or wing, contains the central control room with its console and control panels, as well as the relay room with its relays and controllers. The entire plant is controlled from this spot; it is, therefore, an emergency area and must be designated Seismic Class I. The control room and relay rack room share an air-conditioning system to cool the electronic equipment and to protect all personnel from any outdoor radiation that might exist after a nuclear incident. The building also contains the instrument air compression system, and the duplicate DC power sources—two battery sets and a motor generator set. The building also has a "cable room" beneath the relay rack room. The cable room is the termination point for all cables coming from the plant to the control room. All cables from the plant are fanned out from the cable room to their individual destinations in either the control room or the relay rack room.

2.3 Safety Classes

The N-18 Committee of the American Nuclear Society (ANS) has developed four categories that describe the complete range of plant operating conditions. The words used by the American Nuclear Society are slightly different than those used in Section III(ASME) of the Code, but they mean the same thing. The four conditions with both ANS and ASME terminology are as follows:

Condition	ANS	ASME
I	Normal operation	Normal operation
II	Faults of moderate frequency	Upset
III	Infrequent faults	Emergency
IV	Limiting faults	Faulted

Condition I includes all normal operating situations including temperature and/or pressure transients within the scope of the design power range. Condition II includes deviations from normal conditions anticipated to occur often enough so that the design should be capable of withstanding the condition without operational impairment. (A Condition II situation either does not require a forced outage, or if a forced outage occurs, the correction does not require repair of mechanical damage.) Condition III includes forced outages of low probability that require minor repair to mechanical damage but has assurance of the absence of gross failure of structural integrity. Condition IV is an extremely low probability event; it can include gross structural failure and may involve consideration of the health and safety of the public. These conditions are associated with Safety Classes as follows:

Safety Class 1. SC-1 applies to reactor coolant system components whose failure could cause a Condition III or IV loss of reactor coolant. "Loss of reactor coolant" in this context shall mean rate-of-loss of coolant beyond the capacity of normal reactor coolant makeup systems.

Safety Class 2. SC-2 applies to those structures and components of safety systems required to fulfill a safety function. In this context a safety system is any system that functions to shut down the reactor, cool the core, cool another safety system, or cool containment; and contains, reduces, or controls radioactivity released in an accident. This class is divided into two subgroups, which are termed Safety Class 2a and Safety Class 2b.

Safety Class 2a. Safety Class 2a applies to containment and to components of those safety systems, or portions thereof, through which reactor coolant water flows directly from the reactor coolant system or from the containment sump.

Safety Class 2b. Safety Class 2b applies to all other components of Safety Class 2. Note that these two subgroups permit a multichamber vessel such as a heat exchanger to be in two classes at the same time so that the designer may specify the complete vessel for the single highest code classification or assign different code classifications to discrete portions of the vessel.

Safety Class 3. SC-3 applies to components not in Safety Class 1 or 2, the failure of which would result in release to the environment of gaseous radioactivity normally held within the facility for decay.

From these descriptions the following table illustrates the logical correlation between N-18 Safety Class and Section III Code Class:

N-18 Safety Class	Section III Code Class
1	1
2a	2 and MC
2b	3
3	3

Due to the restrictions of the ASME Code, the code classes do not cover vessels whose design pressure falls within the range from atmospheric pressure up to just below 15 psig. For large vessels in this category it is suggested that the designer use ANSI B96.1, American Petroleum Institute API-620 or 650, or American Water Works Association AWWA-D100, whichever is most appropriate for the vessel intended. Additional requirements in the form of seismic criteria and weld inspection must be added to any of these standards.

2.4 Reactor Assembly

The reactor assembly consists of a pressure vessel with a thermal shield, core support plates, core barrel, control rods, etc., as shown in Fig. 2-4. Coolant enters the vessel and flows down the annulus between the core barrel and the vessel wall; in the process it cools the thermal shield. The coolant turns, flows up through the fuel elements and out of the vessel to the steam generator. Control rods enter through the top head and the control rod drive packages are mounted on the top head. The drive packages and top head are handled as one package during plant refueling.

The neutron detectors entering the bottom head are combined with thermocouples which measure coolant temperature rise through an individual fuel assembly. Together they are known as "in-core instrumentation." They are information instrumentation only and provide data with which to calculate the burnup of the various fuel assemblies and the power produced in the different regions of the core. The in-core instrumentation is classed as a "consumable" because neutron damage limits their life to three or four years.

In the refueling operation, all internals inside the core barrel above the upper core plate are removed. This permits direct access to the fuel assemblies for manipulation.

The thermal shield is provided to shield the vessel wall from the core and minimize neutron impingement and thermal stress in the vessel. Combustion Engineering, Inc., however, offers some plant designs without a thermal shield since their calculations indicate that only limited transformation will be incurred by the reactor vessel material in the absence of the shield.

All reactor internals are made of austenitic stainless steel, Inconel, or the zirconium alloys. The vessel is a low-alloy steel clad with either austenitic stainless steel or Inconel. Vessel design is Class 1, of Section III of the ASME Code, plus Seismic Class I. (See Chapter 6 for seismic design considerations.)

2.4.1 *Core*—The core is made up of sintered cylindrical uranium dioxide pellets in a Zircalloy tube. Actual pellet diameter varies between manufacturers, but it is of the order of 0.5 inch. Generally, an individual pellet is about as long as its diameter. The pellets are stacked inside the Zircalloy tube so that contact actually occurs between the fuel pellet and the tube at operating temperatures to enhance heat transfer. A plenum is provided at the top of the fuel rod for collection of fission product gases; but during manufacture, the plenum is usually filled with helium. The plenum volume is selected to prevent excessive pressure buildup during the active life of the fuel element (3 to 4 years). The ends of the tubes are sealed with welded plugs. Active core height is from 10 ft to 12 ft depending on the manufacturer. The fuel rod is generally the same as that shown in Fig. 3-3.

These fuel rods are assembled into grids or clips in a 14 x 14- or 15 x 15-square array.

The rods in the grid are assembled on a pitch that is slightly larger—perhaps 0.10 inch to 0.20 inch—than the diameter of the rod. One assemblage of fuel rods into grids is called a "fuel element assembly" and this is the smallest fuel unit handled in a power station. A core is built up by grouping fuel assemblies side-by-side in a cylindrical shape. The higher the power level desired, the more fuel element assemblies are used. Core diameters run from 9 ft 0 inch to 11 ft 0 inch, depending on the power level. Core height is always the same.

2.4.2 *Control Rods and Drives*—Control rods are cylindrical rods located inside control-rod channels in various locations in a fuel element assembly, in place of fuel rods. The manufacturers do not use the same neutron adsorber nor do they have the same number of rods. Boron carbide is used by one manufacturer as the adsorber while another uses a combination of silver, indium, and cadmium. The manufacturers use a cluster of control rods fastened to a spider at the top, as shown in Fig. 2-5. A cluster may have as few as five or as many as twenty rods. One control rod drive package is attached to each cluster of control rods.

Control rod drives are mechanical, and depending on the manufacturer, are either infinitely positionable or move in a series of discrete steps. All rod drives, regardless of manufacturer, incorporate a magnetic device which is de-energized for scram. (See 2.8 for definition of "scram.") The control rods withdraw upward and drop into the core by gravity during scram. The control rod drives enter the reactor pressure vessel through the top head.

The infinitely positionable rod drive has a con-

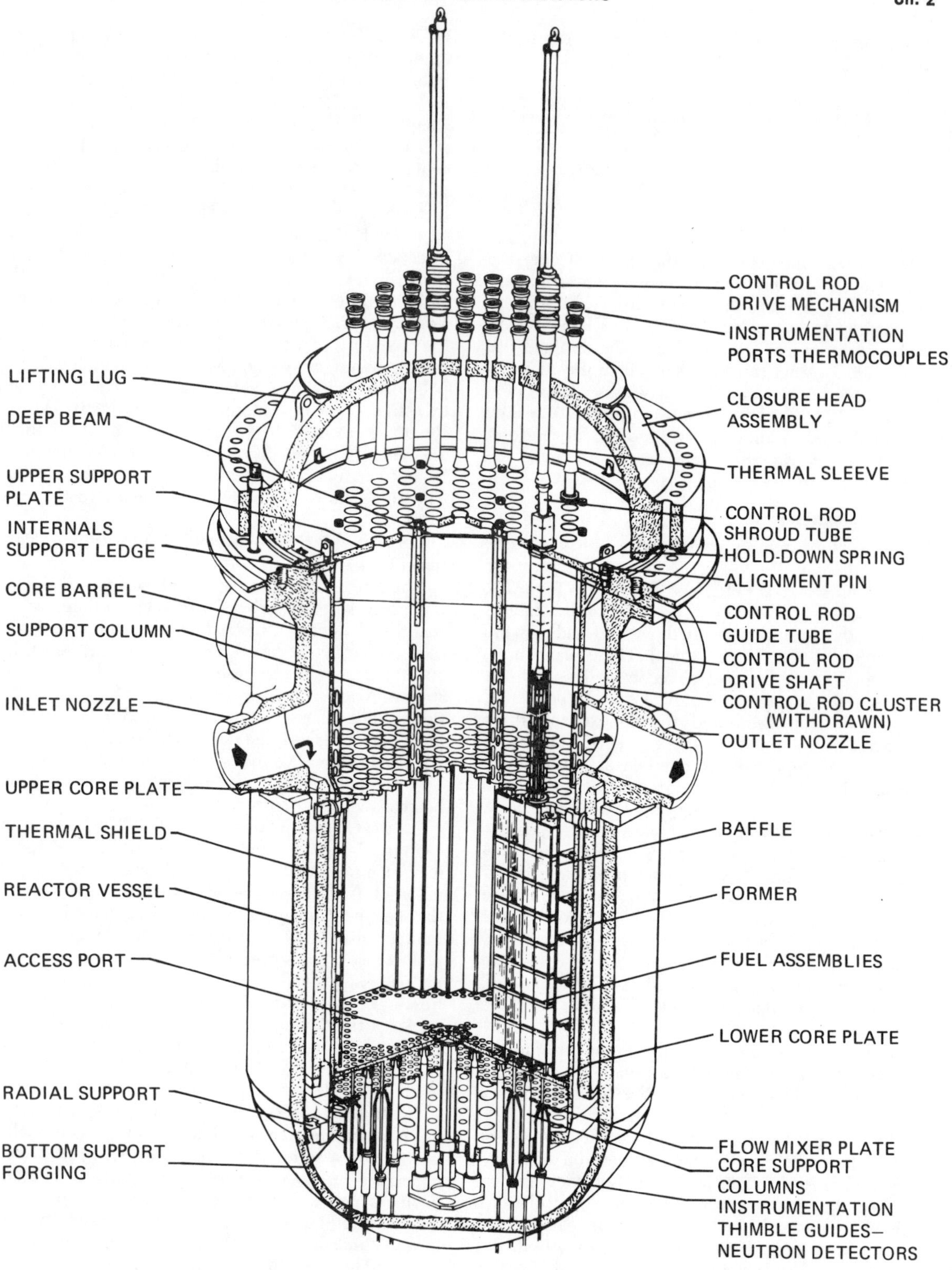

Courtesy, Westinghouse Electric Corp.

Fig. 2-4. Assembly of a pressurized water reactor.

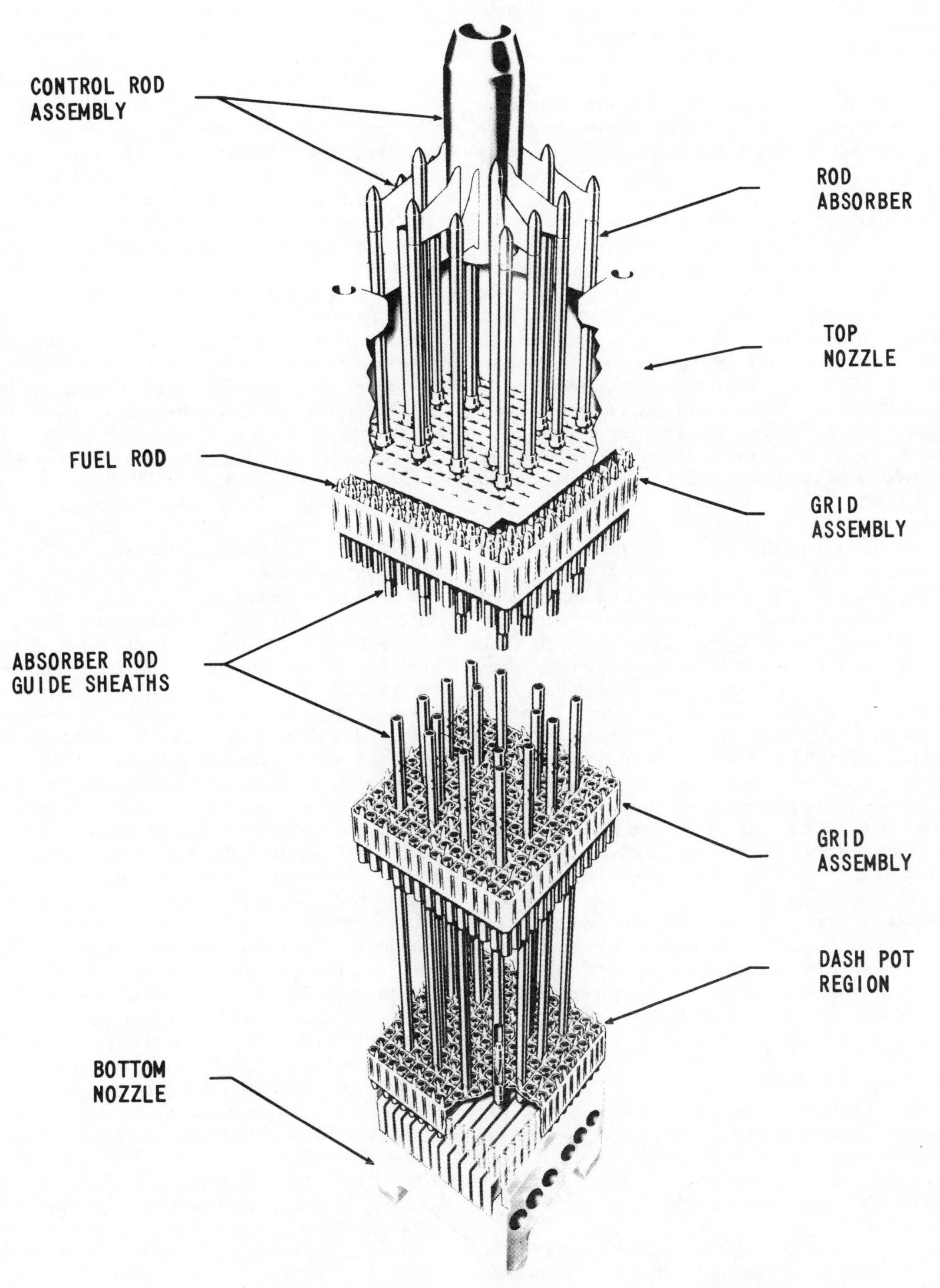

Courtesy, Westinghouse Electric Corp.

Fig. 2-5. Diagrammatic assembly of a control rod cluster and its guide sheaths. The guide sheaths are actually part of the fuel element.

tinuous position indicator that follows the rod position at all times. The stepping rod drive has a series of switches that indicate position every inch or one-and-a-half inches.

The number of control rod drives may vary from forty to seventy drive packages depending on the reactor power rating and the manufacturer.

2.5 Primary Loop

The typical loop of a pressurized water reactor is shown in Fig. 2-2. The coolant pump supplies the dynamic head required for circulation. Pump power runs from 4000 to 6000 hp each, depending on the size. The pressurizer controls system pressure and, within limits, provides system expansion volume.

Pressure is created by boiling water in the pressurizer with electric heaters. It is controlled by alternate use of the heaters to increase pressure, and a "cold" water spray, taken from the circulating pump discharge, to lower pressure. Water is added to or removed from the system by the pressurizer liquid level control when the liquid level exceeds prescribed limits. Thus, the water in the pressurizer is at saturation temperature while the rest of the system is subcooled.

The pressurizer is equipped with two sets of relief valves. One set of valves is the code required mechanical safety valve set at design pressure. The second set is composed of motorized relief valves set at a slightly lower pressure and responding to a pressure signal instrumentation system. The relief valves are designed to remain tight under repeated cycling. Relief-valve effluent, which is radioactive, is caught in a relief tank filled with quench water and cooled. The relief valves prevent the safety valves from lifting during a step load reduction such as when the turbine sheds its load.

The primary loop has stainless steel in contact with the primary coolant. Whether the material is stainless steel clad or solid stainless steel is decided solely on the basis of procurement economics. The steam generator tubes are of Inconel. All equipment and piping are Class I Seismic design. Vessels, pumps, valves, and piping are in accordance with Class 1, of Section III of the ASME Code.

2.6 Auxiliary Systems

2.6.1 *Chemical and Volume Control System*—In order for the reactor and primary coolant system to operate for any appreciable length of time the Chemical and Volume Control System (CVCS) must also operate. The CVCS performs three major duties:

a. It maintains primary system purity and prevents build-up of system radioactivity by means of a continuous primary coolant stream passed through a mixed bed demineralizer.
b. It provides a storage location for excess primary system liquid as it expands during reactor warm-up. Conversely, it also acts as a water supply for the primary system during reactor cooldown.
c. The system injects the long term reactivity control compound (boric acid) as required, and under manual authorization, removes and recovers the poison for reuse. The system also injects chemicals for pH control and oxygen scavenging when necessary. Radiolytic dissociation of coolant is controlled by dissolving hydrogen in the make-up coolant.

In addition, the system supplies seal water if and when required by the primary pump seals. The last function of the system is to pressurize the primary system, when it is cold, to a pressure high enough to supply the net positive suction head required to permit primary coolant pump operation. A typical flow diagram for a chemical and volume control system is shown in Fig. 2-6. The system is a bleed-and-feed system. Water is bled off at the primary pump discharge of one loop and taken through a throttling valve controlled by pressurizer liquid level. The liquid is cooled in a regenerative heat exchanger to conserve heat; it next passes through a let-down heat exchanger for additional cooling (to 140 degrees F, or less) and is reduced in pressure. The fluid is then purified in a mixed-bed demineralizer plant and sprayed into the volume control tank. The volume control tank is pressurized with about one atmosphere of hydrogen pressure so the water becomes saturated with hydrogen to retard radiolytic dissociation of water. As the fuel is consumed, the reaction forms neutron poisons and boron must be removed to compensate. Some of the bleed fluid is bypassed to the boron recovery system and boron-free demineralized water replaces the extracted solution. Clean water, injected by the charging pumps, splits into two paths. A small percentage is injected into the primary pump seals to prevent leakage of contaminated water; the remainder enters the primary system via the regenerative heat exchanger. In the event that the main bleed path is unavailable, an alternate path, using the excess let-down heat exchanger can be used. When this path is used the system circulation rate is reduced to that of the seal injection circuit.

Materials of construction are austenitic stainless steels. Vessels are designed in accordance with Class 3, Section III of the ASME Code for vessels with radioactive fluids and Section VIII for non-radioactive fluids. Piping is Section III, Class 2 or Class 3 or ANSI B31.1.0, depending on the fluid and its temperature and pressure. Piping, pumps, and valves for radioactive fluids are in accordance with Classes 2 and 3 of Section III of the ASME Code or fossil plant standards for non-radioactive fluids.

2.6.2 *Safety Injection Systems*—The safety injection systems are provided to guarantee core cooling and prevent fuel-cladding melting in the unlikely

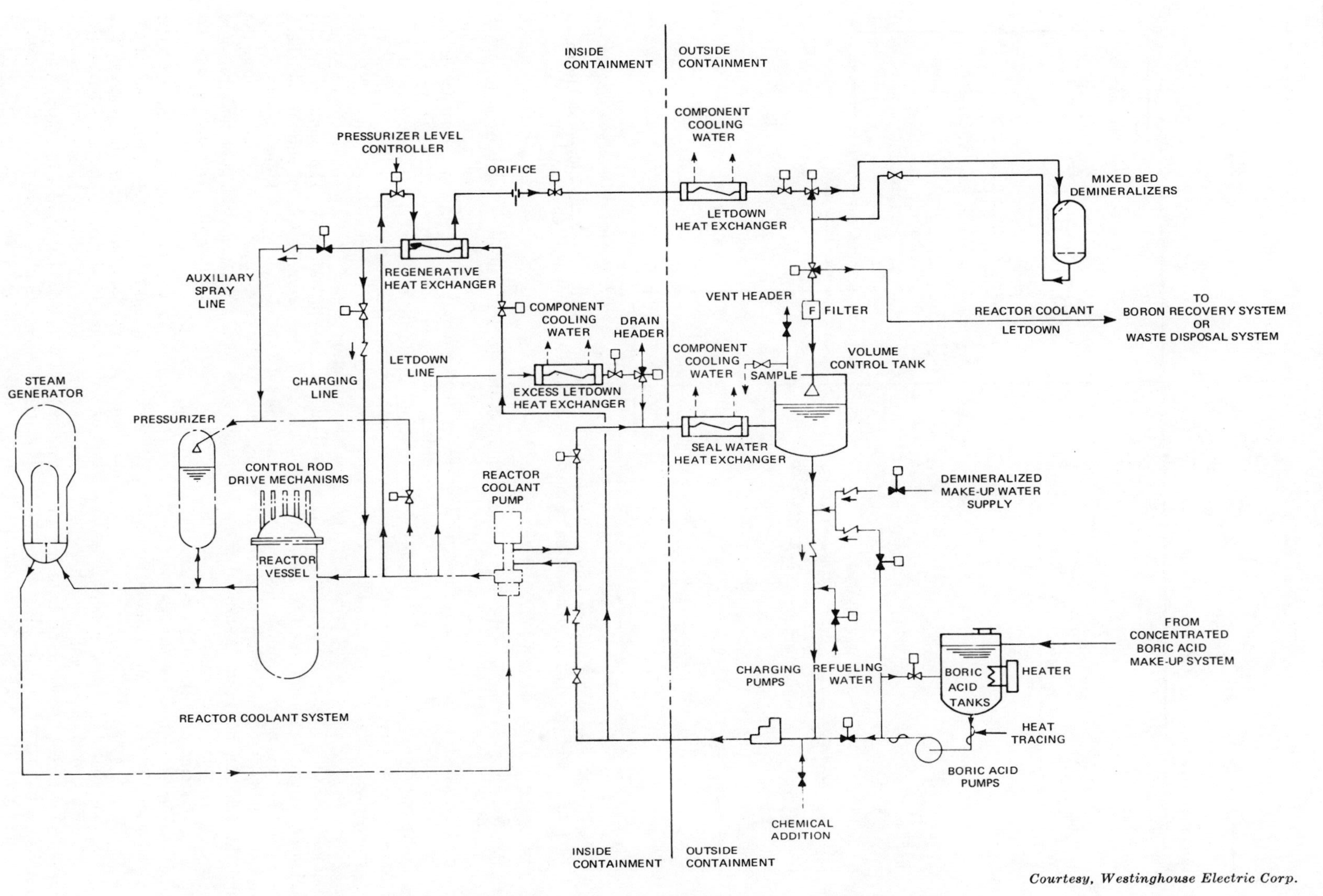

Courtesy, Westinghouse Electric Corp.

Fig. 2-6. Process diagram of a chemical and volume-control system. One system is required per plant. Present trend is toward one system size regardless of plant thermal rating.

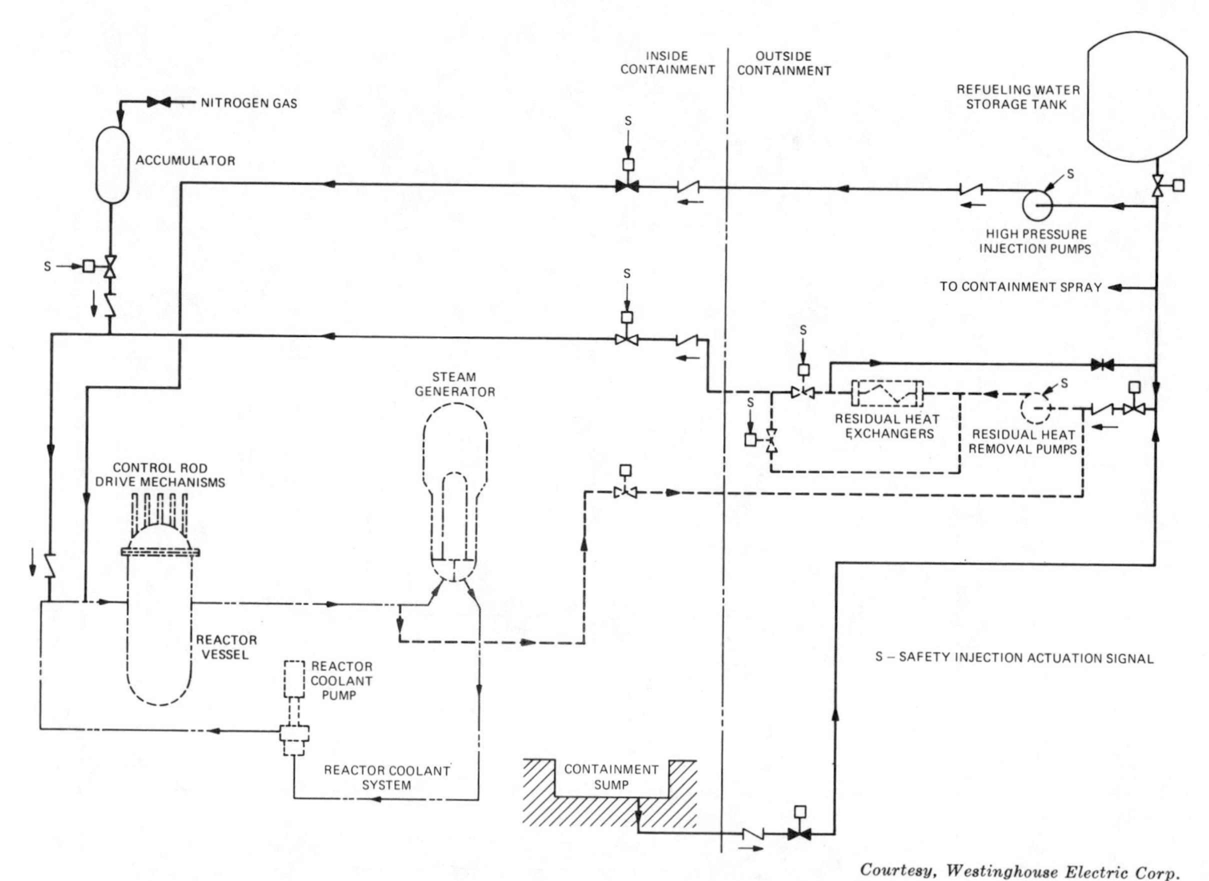

Courtesy, Westinghouse Electric Corp.

Fig. 2-7. Safety injection system process diagram. The containment sump collects all water leaking out of broken pipelines and closes the loop to permit continuous core cooling with a ruptured primary system.

event of a double-ended pipe break of the largest line in one coolant loop. The definition of a "double-ended pipe break" says that the pipe has sheared off and an unrestricted stream of water at full system pressure head is jetting out of each side of the break. The safety injection systems must cool the core and limit core damage to negligible amounts under these conditions. They must cover the full pressure range from operating pressure down to atmospheric. A combination of systems is usually used to meet the full pressure range requirement. Two parallel systems are provided so that if one fails the second system will fulfill the function by itself. A typical single system (one of the pair) is shown in Fig. 2-7. The complete set of systems uses the RHR system, the water supply in the refueling water storage tank, and a pressurized water supply in an accumulator. All water supplies store borated water.

The description that follows is generic; actual pressures at which events occur will vary between suppliers, and an engineer reviewing or designing an actual installation must get certified values from the vendor in question. The high-pressure injection pumps are sized for small or medium leaks. Their basic purpose is to maintain pressure at some value, lower than normal, while the plant comes to an orderly shutdown and depressurizes. These pumps start automatically at about 1800 psi on receipt of the safety injection signal after the reactor has tripped (See 2.8 for definition of "tripped"). The safety injection signal also opens the block valve sealing the accumulator tank. The contents of the accumulator tank are now isolated only by the check valve being held closed by the higher reactor-vessel pressure. As the vessel pressure continues to drop, the check valve opens when accumulator tank pressure exceeds reactor-vessel pressure. This occurs between 600 and 800 psig.

The safety injection signal also lines up the RHR system so that it can take its suction from the refueling water storage tank and starts the pumps. The system will begin to deliver water as soon as the reactor-vessel pressure drops below RHR pump shutoff head which is of the order of 300 to 350 psi. The RHR system supplies water from the refueling water storage tank until the supply is exhausted; at that time the suction is switched to the containment building sump and recirculation is continued for as long as is necessary, using the RHR heat exchangers to dissipate the heat energy. Note that in Fig. 2-7 a connection is provided so that the water supply from the refueling water storage tank can be made available to the spray system if the reactor operator desires.

In summary then, on receipt of manual or automatic safety injection signal, all stages of emergency injection are started. If the leak is small to moderate, the high pressure injection will maintain the pressure and neither of the other two stages will deliver water. If the pressure holds, the reactor operator can turn off the RHR pumps after 10 or 15 minutes, if he wishes. If the leak is too large for the high-pressure pumps, pressure will continue to fall and then, in order, the accumulator tanks will deliver their water supply and, in turn, the RHR system will pick up the load.

2.6.3 *Residual Heat Removal System (RHR)*—The main function of the RHR is to remove decay heat generated in the reactor core after shutdown. The decay heat is caused by fission product decay in the fuel elements and gamma activity in the reactor internals due to neutron irradiation. An equation for a decay heat curve is given in Chapter 12. Residual heat removal is accomplished by a combination of steam condensation in the main condenser and then, when coolant temperature and pressure have dropped low enough, the RHR is put into operation. A typical RHR is shown in Fig. 2-8. The manufacturers vary the point at which they start RHR operation, but a typical value has the primary system at 350 degrees F and 350 psig. The system is designed to minimize the time necessary to drop to refueling temperature (140 degrees F) and pressure (atmospheric). The equipment is split into two subsystems—heat exchangers and circulating pumps. Each heat exchanger-pump combination is checked to insure that it alone can guarantee fuel integrity if the other half system should fail. This provides a redundant RHR system. Westinghouse Electric Corp. designs its total system to reduce temperature from 350 degrees F to 140 degrees F in twenty hours, with both pumps and heat exchangers in operation.

During reactor cooldown, all primary pumps operate. When the RHR system is turned on, all primary pumps except one are turned off. The one pump runs to insure even cooling of the primary system.

All material in contact with primary coolant is austenitic stainless steel. Vessels, pumps, valves, and piping conform to Class 2 of Section III of the ASME Code. Since the system is essential to the safe shutdown of the reactor it conforms to Class I seismic design.

2.6.4 *Spent Fuel Storage and Cooling*—Spent fuel (fuel elements removed from the core) is too hot, both thermally and radioactively, to be shipped for reprocessing immediately after removal from the reactor. These spent-fuel elements must be stored in the reactor facility for three to six months in a storage pool to cool off. The storage pool is located in a separate fuel storage building which is provided with special facilities for storage and handling. The ele-

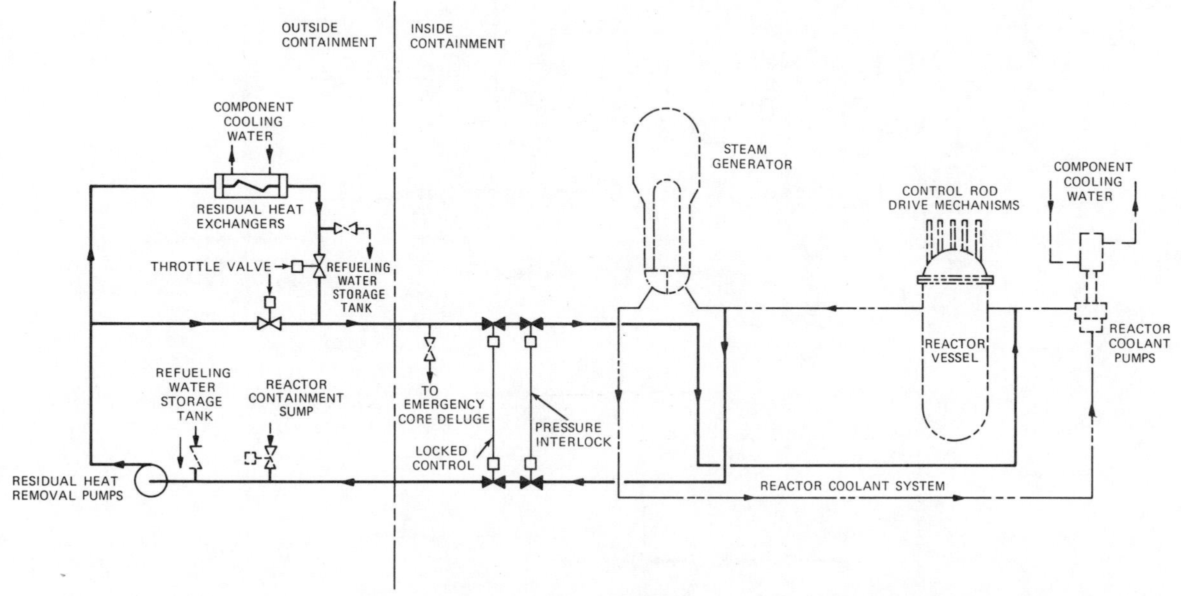

Courtesy, Westinghouse Electric Corp.

Fig. 2-8. Process diagram for a residual heat removal system. The pumps and heat exchangers of this system are also connected to the containment spray system.

ments are stored in racks at the bottom of a pool which provides cooling and radiation shielding. Fuel element locations in a rack are so arranged that the elements cannot interact with each other to cause criticality. The racks themselves are designed in such a manner that they cannot interact with each other.

The storage pool water circulating system has two capabilities; it cools the pool (rejecting fuel element decay heat to a closed loop cooling circuit) and it maintains water purity. A typical spent-fuel cooling system is shown in Fig. 2-9. Note in this illustration that the system includes connections into both the fuel transfer canal and the RHR system. With these connections the heat load of the spent fuel elements can be followed as the fuel moves out of the reactor through the canals and transfer devices into storage racks in the fuel storage building.

Equipment in the system is duplicated so that failure of a component will not jeopardize system function. All equipment and structures required by the system are built in accordance with Class I Seismic criteria. Vessels, pumps, valves, and piping are all in conformance with Class 3 of Section III, of the ASME Code.

All materials of construction are corrosion resistant. Stainless steel is used although fuel storage racks are frequently of aluminum. The storage pool is lined with stainless steel to complete the system corrosion resistance.

2.6.5 *Containment Cooling*—The containment structure is provided with two separate 100-percent capacity cooling systems for use after an accident, in conformance with Criterion Number 38. The two systems are based on different principles. One system is an air recirculation system which has at least three 50-percent capacity fan-coil units, each of which normally has two cooling coils. One coil is suitable for heating or cooling using the normal water supply and the second coil is served by the emergency systems. If the coil is served by an emergency closed loop system, it is also backed up by a direct connection from the emergency service water system.

The second containment cooling system is a direct spray into the structure. For this purpose connections are provided from the refueling-water storage tank and/or the condensate storage tank.

Piping connections are provided so that the spray pumps can take their suction from the reactor building sump if the water supply from the storage tanks runs out. The connections also may be arranged to use the RHR heat exchangers to cool the recycling spray fluid. In addition, a concentrated sodium solution is mixed with the spray water so that the actual spray is a sodium solution. The purpose of the sodium solution in the spray is to remove any iodine-131 (I_2^{131}) in the air by forming sodium iodide which then dissolves in water. Initially the compound used was sodium thiosulfate. Some objections have been made to the elemental sulfur formed in the reaction and later installations have sometimes substituted caustic soda. When a sodium spray solution is used, however, aluminum cannot be

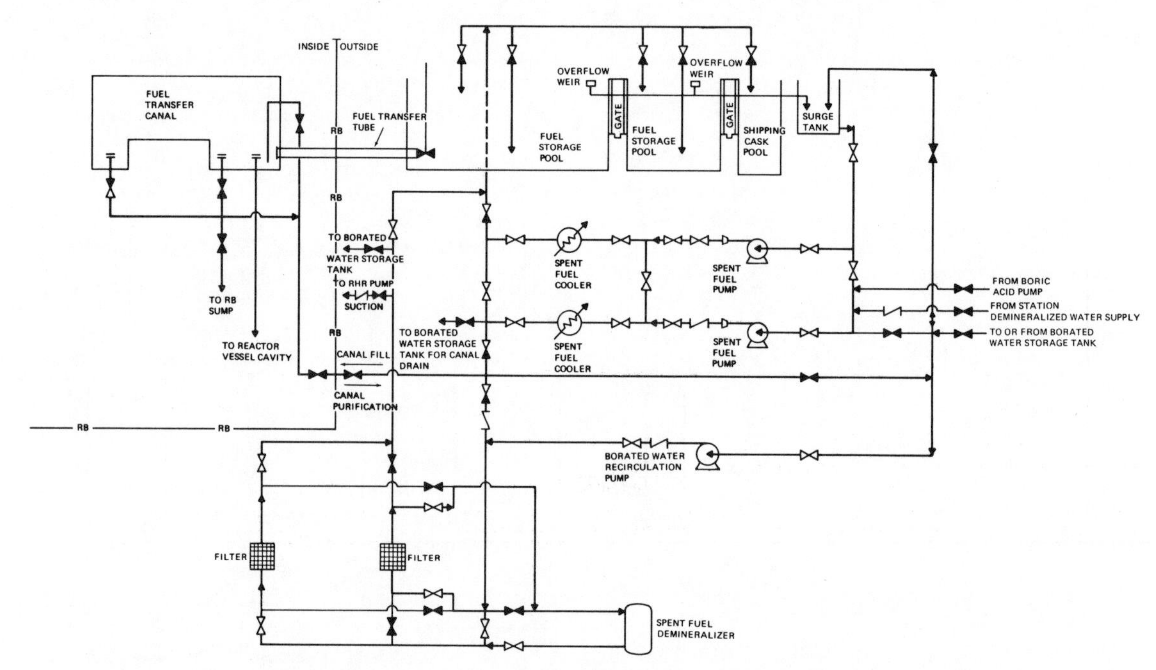

Fig. 2-9. Spent fuel cooling system process diagram. The connections permit following the heat load of a fuel element as it moves from the reactor vessel into storage.

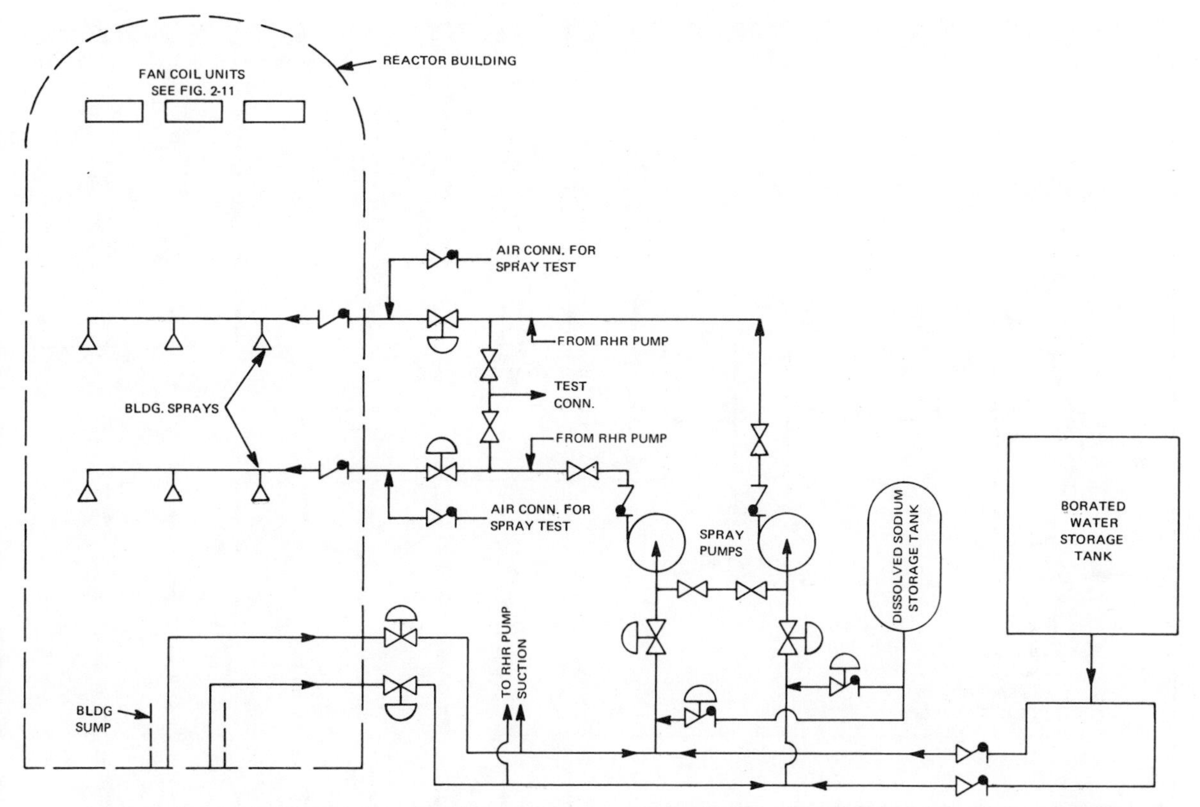

Fig. 2-10. Process diagram for a containment cooling system. The thermal cooling capacity of this system is the same as the capacity of the fan coil units at the top of the reactor building.

used for any purpose where it will be subject to alkali spray solution after an accident.

Equipment and piping design must conform to the requirements of Class I seismic design. In addition, the fan coil units and ductwork must accept any transient pressure differentials that may exist during the building pressure rise after an accident. Containment cooling system piping conforms to Class 2 of Section III of the ASME Code.

A typical set of systems is shown in Fig. 2-10.

2.6.6 *Closed-loop Cooling Systems*—All heat rejected from the reactor auxiliary systems is rejected via closed loop heat transport systems using inhibited demineralized water. The closed loops are provided so that if a radioactive heat exchanger should develop a leak, there is a second barrier, the closed loop, between the leak and the environment. A typical closed loop is shown in Fig. 2-11 which also shows the spent fuel pumps and coolers as part of the essential services of the facility. The earliest plants were not designed in this way; but current thinking tends toward including the spent fuel cooling system as an essential service.

Also as seen in Fig. 2-11, a closed loop is composed of a bank of energy-absorbing heat exchangers, an energy-rejecting heat exchanger, and circulating pumps. The complete plant must be analyzed for the individual heat loads to be serviced under operation, shutdown, and emergency conditions before choosing the system or systems arrangement to be used for any given plant. It is more and more common to provide a separate Residual Heat Removal System closed loop and service water loop in place of the arrangements shown in Figs. 2-11 and 2-12. In evaluating the systems arrangements, consideration must be given to the cost of the emergency power supply as well as to the complexity of the control system. Any system meeting the requirements of Class I seismic duty must be designed in accordance with Seismic Class I criteria.

Energy absorbing exchangers carry the name of the energy producing system (RHR exchanger, seal water exchanger, etc.) and the energy rejecting exchanger is called the closed loop heat exchanger. The closed loop coolant fluid is normally on the shell side of the heat exchangers. All piping, pumps, valves, and heat exchanger shells are made of carbon steel, but the heat-exchanger tube material is governed by the type of fluid inside the tube. Section III Class 3 of the ASME Code is followed in the design of these systems.

The system is continuously monitored for radio-

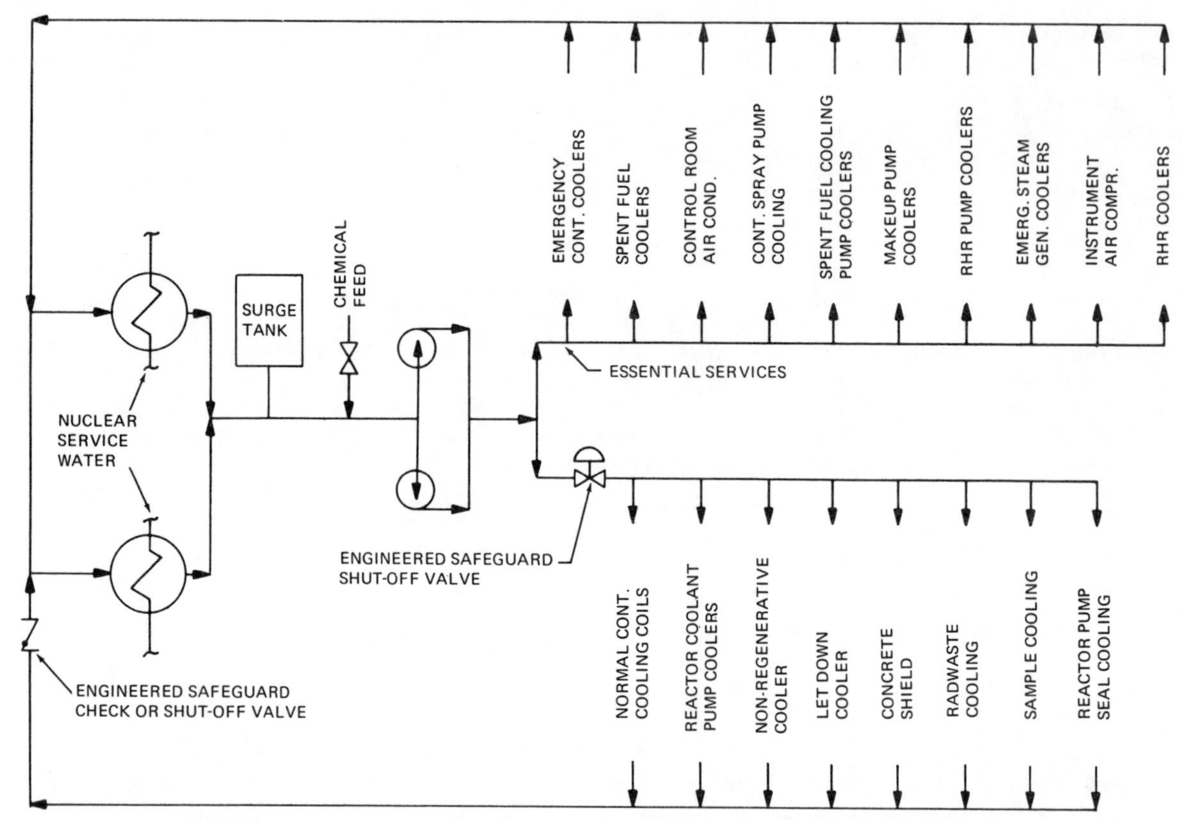

Fig. 2-11. Closed loop cooling system process diagram. This system is an additional barrier between a radioactive fluid system and the final heat sink. This diagram is one of the many satisfactory system arrangements.

activity. Any trace of activity levels higher than background is evidence of a leaking heat exchanger.

2.6.7 *Service Water Systems*—The service water systems are the final heat sink for all heat loads other than the primary loop. The fluid may be ocean water, river water, lake water, or cooling tower water, depending on the plant location. The systems absorb all heat energy from the closed loops, plus containment cooling, in an emergency. The systems are usually split into a "Nuclear Service Water System" and an "Emergency Service Water System." The "Emergency Service Water System" cools all loads that must be guaranteed under accident conditions, such as the RHR coolers, spent fuel coolers, emergency containment cooling coil, and any closed loop load required to operate. Design criteria require two complete 100-percent capacity systems so that if one system fails, the duty can still be fulfilled by the other.

The systems are designed in conformance with Section III Class 3 of the ASME Code. Additionally, the emergency system conforms to Class I seismic design. Materials of construction must satisfy the requirements of the local water supply. A typical set of service water systems is shown in Fig. 2-12. Note that in the figure, the emergency service water system has two completely separate and parallel systems from the intake structure as far out as the outfall line. The pumping loads are split so that loss of a unit will not mean total loss of the system. The separation carries completely back through the power cables, bus bars, and power supplies.

2.6.8 *Ventilation*—Ventilation and cooling in containment are maintained by water-cooled fan coil recirculating units and the conditions are dictated by the needs of the electrical gear. Access during operation is limited to an instrument room for two or three hours per week and special consideration is given to this area for temperature control and air flow patterns. The total number and distribution of fan coil units needed for normal cooling and ventilation includes those previously mentioned for emergency service after an accident. Fan coil units not needed for emergency service are conventional industrial units with a single-source power supply and no backup.

Additional fan units are placed inside containment at strategic locations to insure adequate flow of air for cooling purposes over the control rod drive units and down into the reactor vessel cavity to cool the concrete. If the primary coolant circulation pumps are air cooled (they may be water cooled if desired), special consideration must be given to them to insure adequate cooling.

2.7 Control

The saturated steam PWR is operated as a constant pressure-constant flow device with a programmed average temperature that rises with increasing load. Reactor control is maintained by a combination of mechanical control rods responding automatically to load changes, and the soluble neutron poison, boric acid, responding to fuel burnup changes.

The plant is basically a "boiler follows turbine" arrangement. Primary plant load changes are directed to the turbine and the reactor power controller. The reactor power controller compares the actual average coolant temperature with the desired average coolant temperature and develops an error signal. Control rod movement causes the reactor power level to change as directed by the temperature comparison error signal.

At the same time, the turbine responds to the load change by a change in throttle valve position which, in turn, affects steam pressure at the outlet of the first stage. The error signal generated by the pressure change corrects the feedwater flow rate.

The superheated steam PWR, is controlled in the same way, but here the average temperature is constant. In both cases, because circulation is constant, the temperature differential across the reactor core must increase to transport more energy, as shown in Fig. 2-13A and 2-13B. The difference in the temperature programming of the two types of

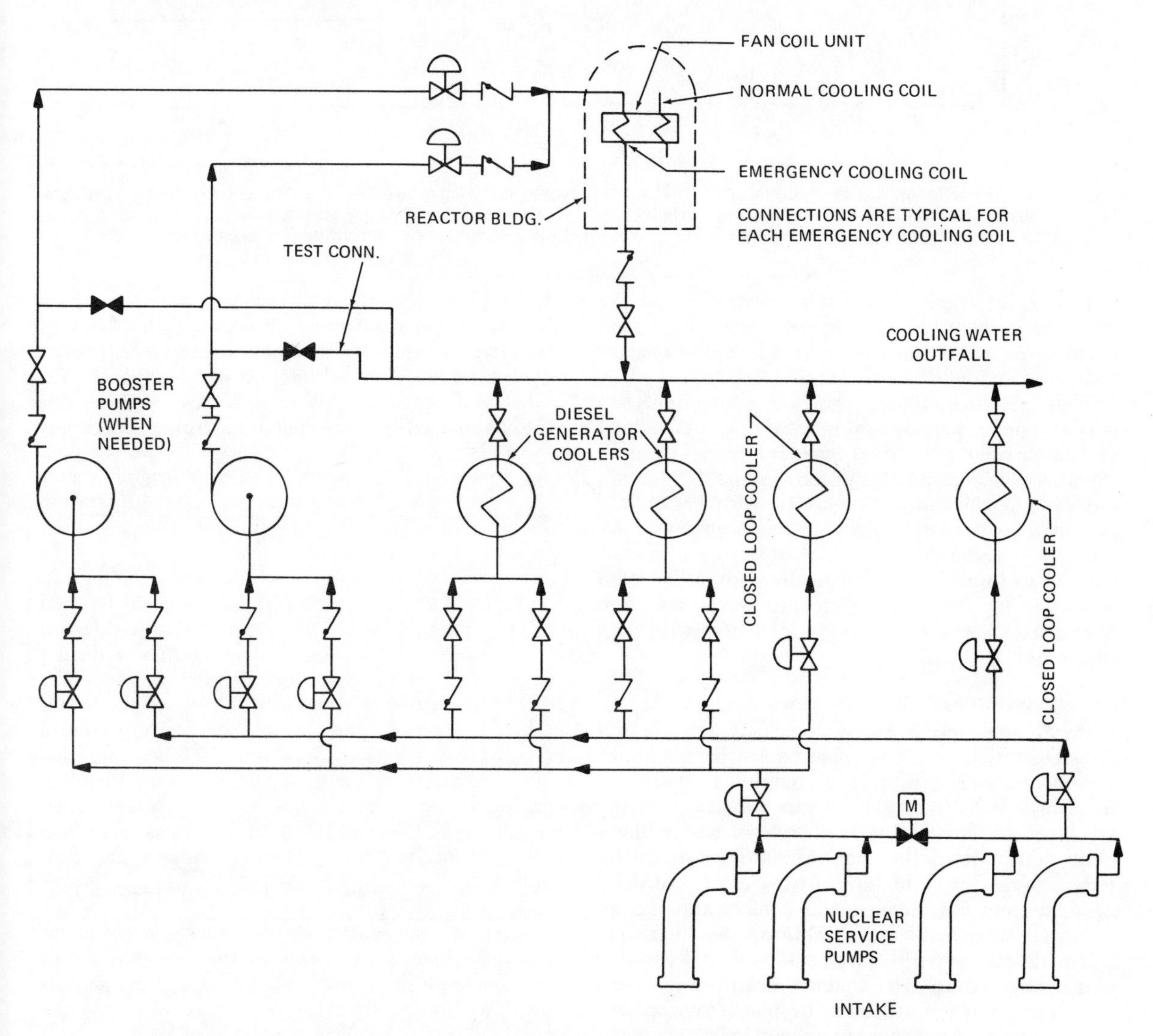

Fig. 2-12. Process diagram of a nuclear service water system. This system is the final heat sink for all heat loads in the nuclear island except the primary cooling system.

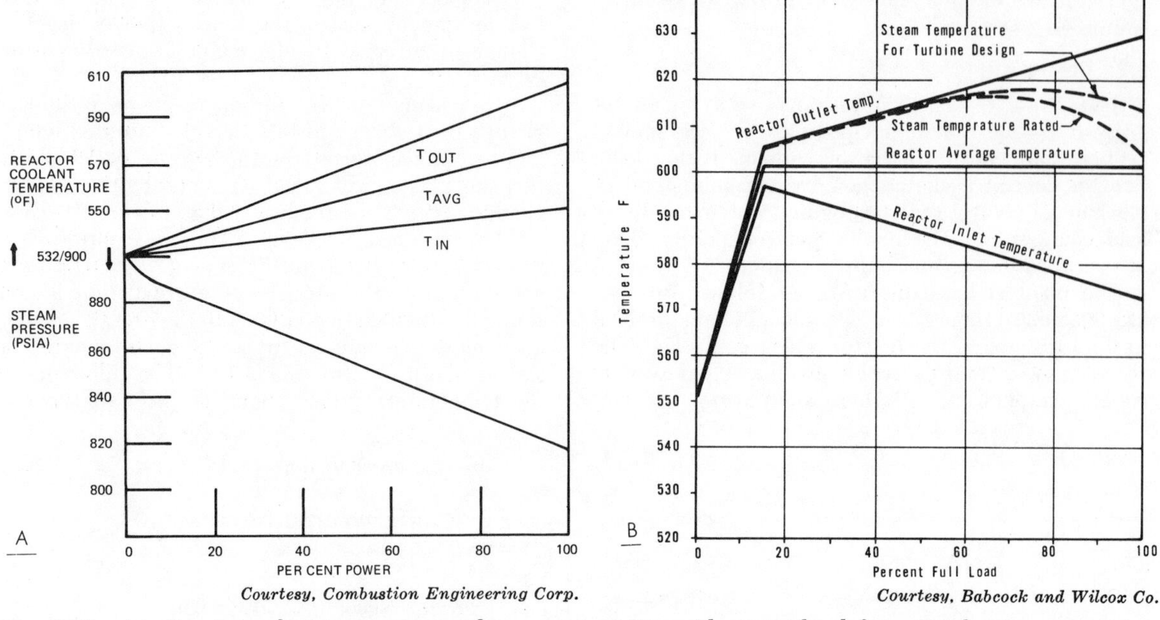

Courtesy, Combustion Engineering Corp. *Courtesy, Babcock and Wilcox Co.*

Fig. 2-13. A. Reactor coolant temperature and pressure variation with power level for a U-tube steam generator. B. Reactor coolant and steam temperature variation with power level for a "once-through" steam generator. Steam temperature decreases with rising power level but remains above saturation temperature.

PWRs is due to heat transfer considerations in the steam generator. The U-tube steam generator used by the saturated steam PWR has essentially constant heat-transfer conditions at all times.

The superheat PWR, using a counter-current, once-through steam generator, varies the percentage of heat-transfer surface assigned to boiling and to superheating as required by load. Figure 2-14 shows the variation in surface alignment with load. Note that at loads above 75 percent, more surface is required for boiling than for superheating. Surface variation is accomplished automatically within the steam generator by the heat-transfer requirements. The parameter allowed to vary with load is superheat as shown in Fig. 2-13B.

2.8 Protection System

Plant protection is based on a logic system designed to achieve two objectives: to minimize spurious reactor trips and to guarantee shutdown from unsafe conditions. The terms "trip" and "scram" are used interchangeably. Either term designates a rapid, unscheduled shutdown of the reactor. In the 1950s, "scram" was the term used because it usually denoted rapid evacuation of the facility also. Today "trip" is more generally used. Multiple measurements are made of each trip parameter and at least two measurements must agree that the parameter is unsafe in order to trip the reactor. The manufacturers use two-out-of-three and two-out-of-four logic. The two-out-of-four logic is designed so that on failure of one of the channels it can be converted to a two-out-of-three system. Note that this two-out-of-four logic system is a true two-out-of-four system rather than the selective system described in Chapter 3. All instrument channels and circuits are designed on a "fail-safe" philosophy. A loss of signal or current places the channel in the equivalent of the tripped condition.

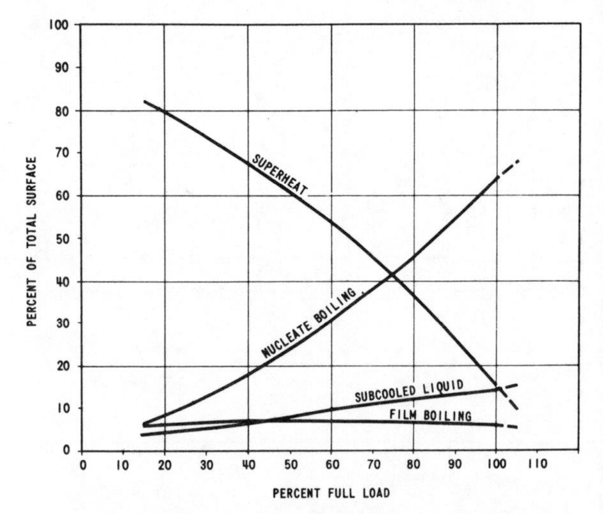

Courtesy, Babcock and Wilcox Co.

Fig. 2-14. Distribution of heat transfer surface with load in a "once-through" steam generator producing superheated steam. Surface is split evenly between boiling and superheating at 75 percent load.

The manufacturers do not all use the same trip parameters, but normally select them from the following list:

Manual trip
High neutron flux (start-up channels and power channels)
High rate-of-power increase (start-up channels and power channels)
Short period
High power
High coolant temperature
High system pressure
Low system pressure
Low or loss-of-coolant flow
Low steam generator water level
Turbine generator trip
Low steam generator pressure
Safety injection initiation

2.9 Plant Loading

The plants designed today, as the newest on the system, produce the cheapest power; hence, they run base-loaded. This will continue for fifteen to twenty years until newer plants produce cheaper power. At that time, today's plants will be operated in a load-following mode. These plants can accept the following load changes without tripping:

(1) A step change of ±10 percent
(2) A ramp change of ±5 percent per minute between 15 percent and 100 percent of full power.

2.10 Refueling

The reactors are refueled once each year. At that time 20 percent to 33 percent of the fuel is changed. Exact quantities of fuel replaced depend on the reactor vendor's design and the owner's economics. Refueling is carried out with the reactor shut down and depressurized.

The reactor pressure vessel is at the bottom of a refueling cavity similar to that shown in Fig. 3-26. Space is provided in the containment building for laydown space of the reactor internals while spent fuel is stored in a Spent Fuel Storage Building. A transfer device, shown schematically in Fig. 2-15, connects the two buildings. A transfer canal in the containment structure provides the necessary manipulating at the reactor.

The steps in a normal refueling are:

1. Cool down the reactor and depressurize. Enter containment as soon as radiation levels permit.
2. Check out all refueling gear. Remove top missile shield.
3. Remove all reactor-vessel head insulation and control-rod drive connections. Drain reactor vessel to below parting line of the top closure head.
4. Remove reactor vessel studs and nuts. Install reactor vessel-concrete well seal.
5. Remove and store top closure head. Inspect as necessary.
6. Fill the refueling canal. Steps 5 and 6 are executed simultaneously to maintain shielding value of closure head steel.
7. Disconnect control rod drive shafts.
8. Retract in-core instrumentation.
9. Remove and store upper core support assembly.
10. Remove spent fuel assemblies. Reposition the remainder of the assemblies as planned. Install new fuel elements. Transfer new and spent fuel assemblies between reactor building and fuel handling building.
11. Reinstall upper core support assembly. Reposition in-core instrumentation. In-core instrumentation is replaced as required.
12. Connect control rod drive shafts.
13. Replace reactor vessel head.
14. Drain refueling canal. (Steps 13 and 14 are executed simultaneously.)
15. Inspect and clean reactor-vessel sealing surface. Clean and decontaminate refueling canal.
16. Remove reactor vessel to concrete seal.
17. Replace and tighten reactor-vessel nuts and studs.
18. Refill and vent the primary system and pressure test.
19. Remake all connections to the top closure head.
20. Replace top missile shield.
21. Test control system.
22. Heat-up and pressurize.
23. Check dissolved neutron poison.
24. Recheck control system.
25. Come up to power and assume the load.

Refueling takes between 14 and 21 days depending on the reactor size.

2.11 Water Chemistry

Water chemistry is controlled for the following reasons:

1. To minimize corrosion on equipment and fuel
2. To control radiolytic dissociation of water
3. To minimize activity levels in the primary system.

Specific conditions vary from vendor to vendor, but typical conditions are:

pH	4.5 to 10.5
Conductivity	1 to 30 micromhos/ cm. at 25°C
Oxygen, ppm, max.	0.1
Chloride ion, ppm, max.	0.15
Fluoride ion, ppm, max.	0.1
Hydrogen cc (STP)/kg H_2O	25-35
Total suspended solids, ppm, max.	1.0
Boric acid, as ppm B	0 to 4000-4500

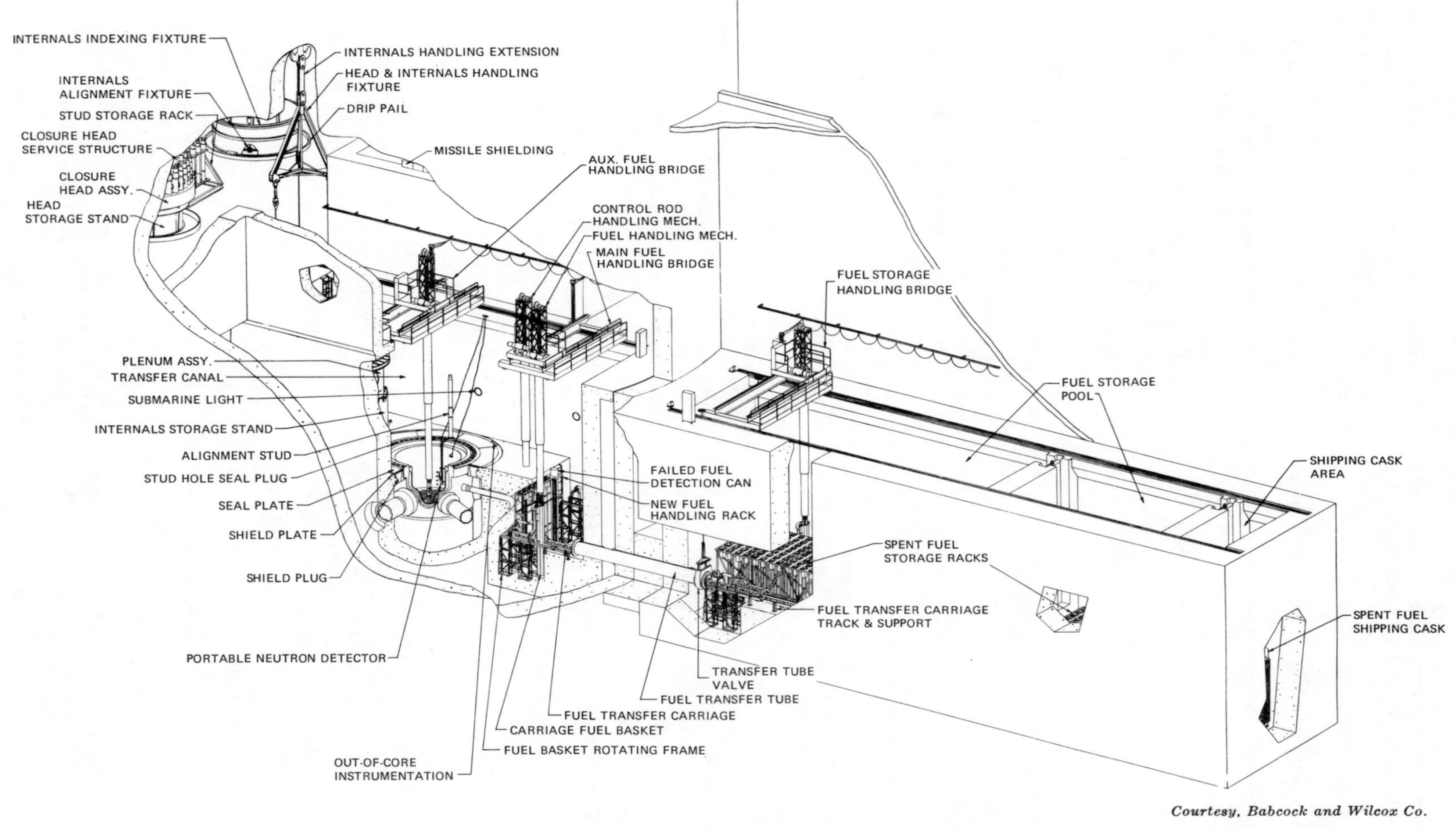

Fig. 2-15. Typical arrangement of fuel handling and storage facilities.

The boric acid (neutron poison) varies with core life from the maximum immediately after refueling down to essentially zero just before the next refueling.

Make-up water to the primary system must meet the following specifications:

pH	6.0 to 8.0
Conductivity	< 2.0 micromhos at 25°C
Oxygen, ppm	< 0.1
Chloride ion, ppm	< 0.15
Fluoride ion, ppm	< 0.1
Total dissolved solids, ppm	< 0.5
CO_2, ppm	< 2.0
Solids filtration	< 25 microns

The make-up water demineralizer plant is custom designed to the water supply available, but must always produce water of the quality just given, or better.

2.12 Containment

Two types of containment for PWRs are described in Chapter 5, a dry containment and an ice-condenser containment. In the ice-condenser containment the cooling duties of the emergency containment air-recirculation system are taken over by the ice bank. The ice is a borated ice to which caustic has been added so that both a neutron poison and an iodine scavenger will be available in the melted liquid. It has been demonstrated to the satisfaction of the licensing authorities that the cooling capability of the ice bank operates as predicted and the only remaining requirements for the air-recirculation system are the normal cooling and ventilating duties. The second cooling system is the containment spray system. Because the ice containment limits the peak accident pressures in the structure, emergency power requirements after an accident are reduced in the order of 70 to 80 percent.

2.13 Radioactive Waste

Radioactive waste systems were furnished by the reactor suppliers in the early 1960s. With the advent of the "nuclear steam supply system" concept and the increased experience of consulting engineers, the reactor suppliers were quite willing to restrict their scope of work to specifying the sizing criteria only, for the system. At the present time the consulting engineers designing the overall facility include the radioactive waste system in their scope. However, a reactor vendor will still furnish a radioactive waste system if requested. A full discussion of the subject appears in Chapter 11.

CHAPTER 3

Boiling Water Reactors

3.1 General

The boiling water reactor (BWR) power system marketed today is a direct-cycle steam generating system; it is shown in its simplest form in Fig. 3-1. Water is boiled by the fissioning core in the reactor vessel and the wet steam passes up through moisture separators and steam dryers, leaving the reactor vessel with 0.30 percent moisture at 970 to 1000 psia and goes on to the turbine. After passing through a suitable heat cycle, the steam is condensed. From the condenser, the condensate is pumped through full-flow polishing demineralizers, feedwater heaters, and then back to the reactor.

The industrial boiling water reactor was developed and marketed by the General Electric Company, and is available in sizes that range from 1500 to 3300 MWt. Heat energy is removed from the core by the evaporation of water; however, this mass flow reflects only a portion of the total flow through the core. The remainder of the flow represents a recirculation loop whose primary function is reactor control rather than cooling.

3.2 Buildings

The nuclear island of a boiling-water reactor facility consists of a "compound" building plus two other "simple" buildings. The compound building is the reactor containment structure centered inside a limited leakage building. Figure 3-26 shows an arrangement of the compound building. Common terminology calls the containment structure "containment" and the enveloping structure, the "reactor building." Note the difference in the meaning of the term

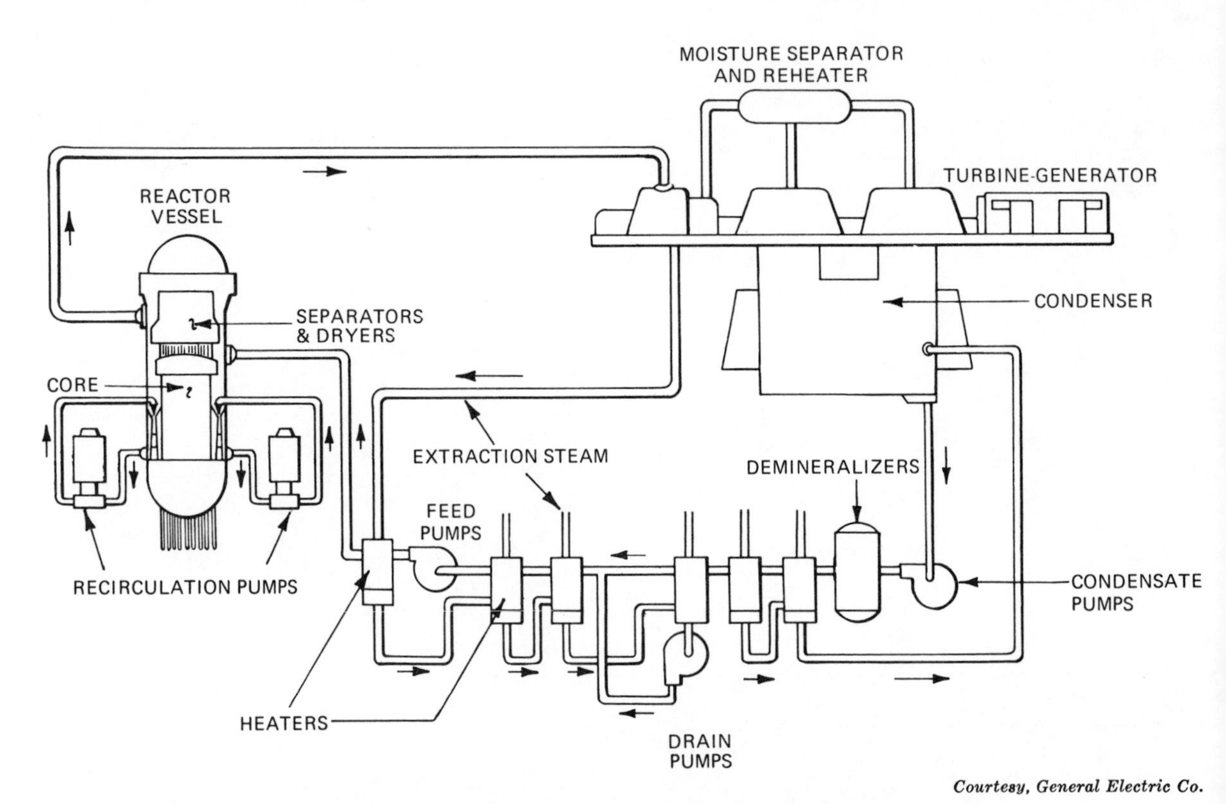

Courtesy, General Electric Co.

Fig. 3-1. Direct cycle reactor power system. Steam generated in the reactor vessel is used directly to drive the turbine-generator.

"reactor building" as used in a BWR facility and as used in a PWR facility. The remaining two buildings in this type of facility are the "radwaste" building and the "control" building.

3.2.1 *Containment*—Containment is a frustum of a cone on top of a cylinder. The reactor is high up in the conical portion of the structure while the pressure suppression pool is down below in the bottom cylindrical part of the structure. A general view of the structure is shown in Fig. 6-1. This structure contains all the equipment shown later, in Fig. 3-7, plus the inner containment isolation valves of the steam and feedwater lines and other auxiliary systems and air-handling equipment. The top of containment is a removable head to permit access to the reactor top head for refueling. The parting line of the containment head and the reactor top head are at the same elevation to accommodate sealing, from containment across to the reactor vessel, to permit flooding for refueling operations.

3.2.2 *Reactor Building*—The reactor building is either a square or circular limited leakage building that entirely surrounds containment and has four or five stories. The top floor, called the refueling floor, has a spent fuel storage pool, refueling equipment, and a reactor service crane of at least 100 tons capacity. The remainder of the building houses the following systems:

Control rod drive hydraulic system*	Fig. 3-8
Reactor water cleanup system	Fig. 3-11
Standby liquid control system*	Fig. 3-12
High-pressure core spray system*	Fig. 3-13
Low-pressure core spray system*	Fig. 3-14
Automatic blowdown system*	
Residual heat removal system*	Fig. 3-15
Reactor core isolation system	Fig. 3-20
Spent fuel cooling system	Fig. 3-21
Closed loop cooling system*	Fig. 3-22
Standby gas treatment system*	Fig. 3-24
Ventilation system	Figs. 5-9 and 5-10

The systems marked with an asterisk have safety functions, thus the structure; equipment; piping; and wiring arrangement must be so designed that no single failure of any element can prevent the system from fulfilling its safety function. As in the PWR this requires equipment cells, separation walls, separate cable trays, etc.

3.2.3 *Radwaste Building*—This is a two- or three-story structure housing the radioactive waste system (Fig. 12-3), and the necessary ventilation equipment. In the new "zero release" concept it is very probable that the equipment in the off-gas system (Fig. 12-4), will also be housed in this building.

3.2.4 *Control Building*—The control building contains the control consoles, instrument panels, and relay racks. Additionally, the building houses the two battery sets, D.C. motor generator set, instrument air compression and storage system, plus the normal air-conditioning and ventilation equipment needed for the building. Since the plant is controlled from this control room, and all the other equipment listed has safety functions; the building, therefore, is Seismic Class I. The air-conditioning and ventilation systems are designed to protect personnel and equipment from any outdoor radiation that may exist after a nuclear incident. It is common practice for the relay rack room to have a cable room beneath it as the terminal location for cabling coming in from the plant. The individual conductors in each cable fan out from here to their individual destinations in the control room or relay rack room.

3.3 Safety Classes

The American Nuclear Society has formed the N-22 Committee to prepare safety design criteria that will parallel the scope of the N-18 Committee referred to in Chapter 2. Their work is still (Fall, 1971) preliminary and nothing, as yet, has been published. Reviews of recent Safety Analysis Reports permit the following conjectures to be made:

1. The criteria will align itself in some fashion with the four plant conditions previously described in Chapter 2.
2. The criteria will contain three classes with no subdivisions within a class.

Safety Class 1 will apply to systems or equipment in which a single failure can cause major fuel damage. This will include fuel assemblies, reactor coolant pressure boundary and various supporting structures. Safety Class 2 will include systems, components, and structures of engineered safety features that perform a direct safety function. This should be interpreted to mean reactor containment or emergency cooling of the core, or containment. Safety Class 3 will include systems, components, or structures necessary to accomplish safety functions. This would include service water systems, fuel storage facilities, and fuel handling facilities. It is also probable that portions of the radioactive waste systems could be included here, similar to the PWR.

3.4 Reactor Assembly

The reactor assembly consists of the reactor vessel, the internals, such as the steam separator and dryer, core barrel, etc., and the control rod drives and housings. In the operating configuration the fuel elements and control rods are added to the array. A cutaway view of a typical boiling water reactor is shown in Fig. 3-2. The core volume can be seen in the lower portion of the vessel, while the steam separators and dryers can be seen in the upper portion. The temporary control curtains are used to supply additional neutron poison until the first refueling. After the refueling, the irradiated fuel has developed enough poisons, and the temporary curtains are removed. (Location of the control curtains is shown later in Fig. 3-5.)

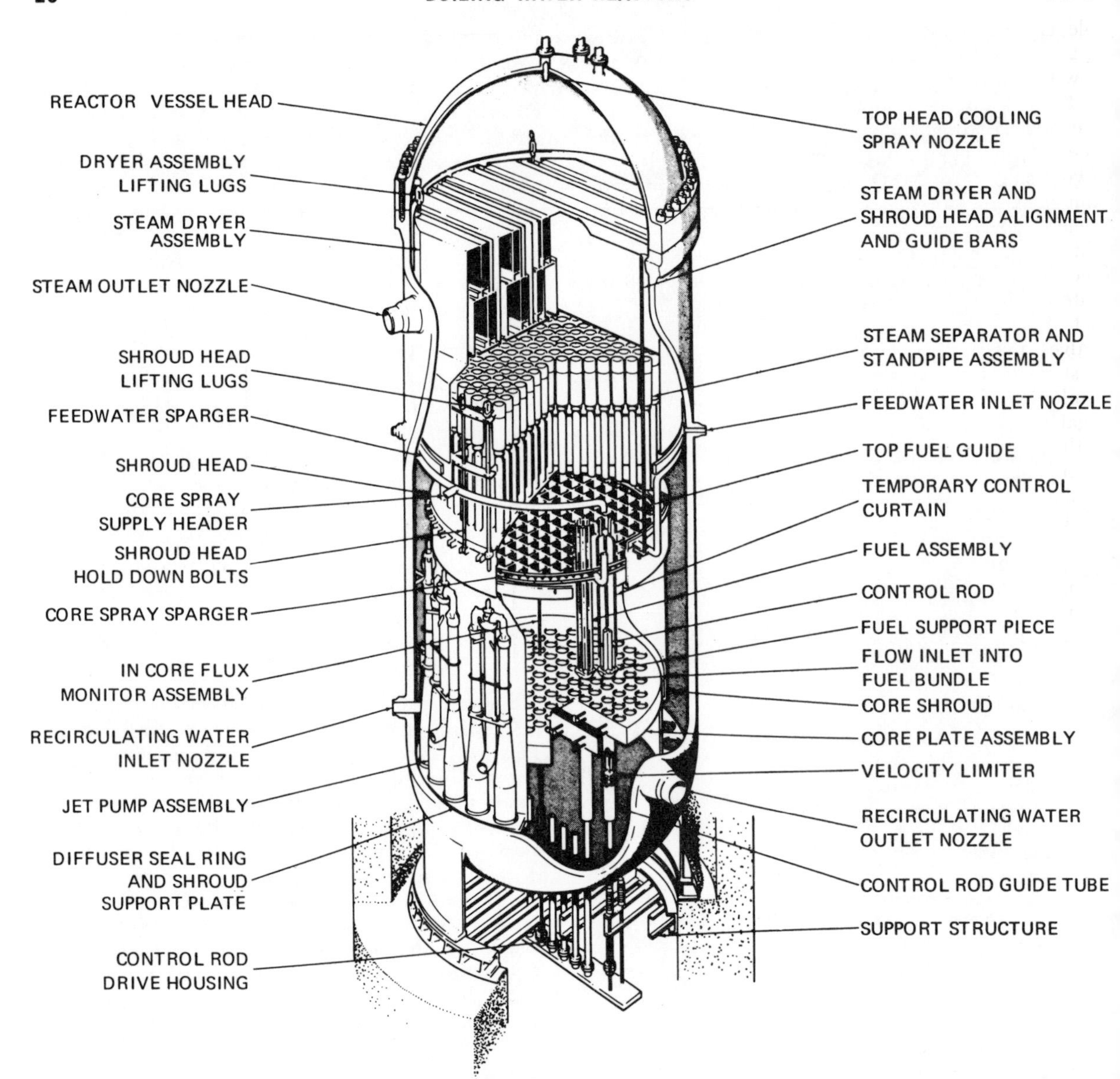

Courtesy, General Electric Co.

Fig. 3-2. Cutaway drawing of an assembled boiling water reactor. Control rods enter through the bottom to avoid the complexities of passing through the steam separating and drying equipment.

There are three flow paths in the vessel: *a*. Feedwater enters the vessel through the feedwater nozzle, is mixed with internal recirculating water and flows into the annulus between the vessel wall and core shroud; *b*. From the annulus, a portion leaves the vessel via the recirculating water outlet nozzle; and *c*. The remainder enters the suction inlets of the jet pump assemblies. The external recirculation water is pumped through the recirculation loops and returns to the nozzle inlets of the jet pumps shown in Fig. 3-6. All three streams join together in the pump diffuser and discharge into the plenum below the core. The fluid then flows up through the core where a portion of the liquid is evaporated to steam. The two-phase mixture passes up through the steam separators and dryers. The separated liquid flows back, down to the annulus, to recycle, and the steam passes out to the turbine.

The core spray sparger is part of an engineered consequence-limiting system for emergency cooling. A nozzle for injection of a back-up shutdown solution—sodium pentaborate—is located in the liquid plenum below the core, but it is not shown in Fig. 3-2. Water level in the assembly is held slightly below the top of the steam separators.

Saturated steam leaves the reactor via a series of steam headers which penetrate containment before they are manifolded together. Each steam line has two isolation valves—one inside and one outside containment. In addition, each line has two groups of relief valves—the first group is the self-actuated safety valves required by the ASME Code and the second group is remotely controlled power-actuated safety relief valves. This second group of valves is set at a pressure lower than the self-actuated safety valves and is interlocked with the standby cooling systems. The power-operated valves are designed for multiple-cycle operation and protect the safety valves against excessive operation. Steam, vented through either group, is discharged into the pressure suppression pool.

3.4.1 *Core*—The core is comprised of a group of rectangular fuel assemblies arranged to form a right circular cylinder. An individual fuel assembly has 49 fuel rods arranged in a 7 by 7 square array. The fuel rods are surrounded by a Zircalloy-4 channel. A fuel rod is a Zircalloy tube sealed at each end, containing uranium oxide cylindrical pellets. The top end of the rod cavity has a gas plenum and a spring; the plenum

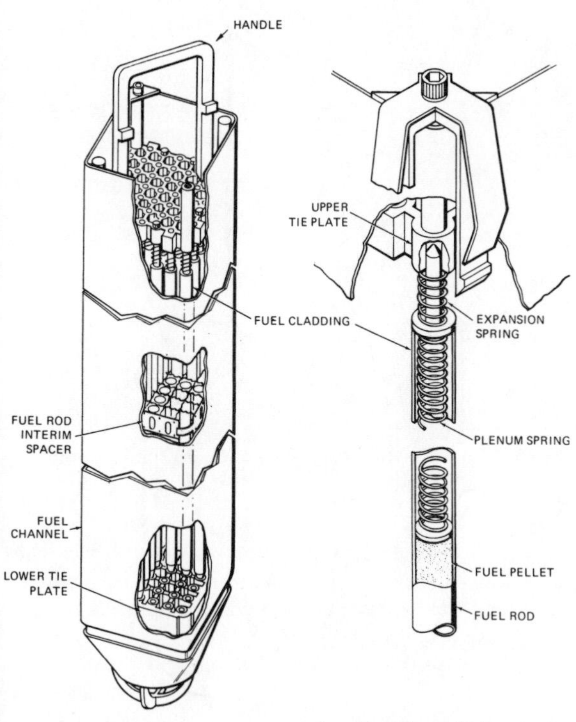

Courtesy, General Electric Co.

Fig. 3-3. Fuel assembly. The fuel element is assembled into a fuel channel at the reactor site; the fuel channel is recycled.

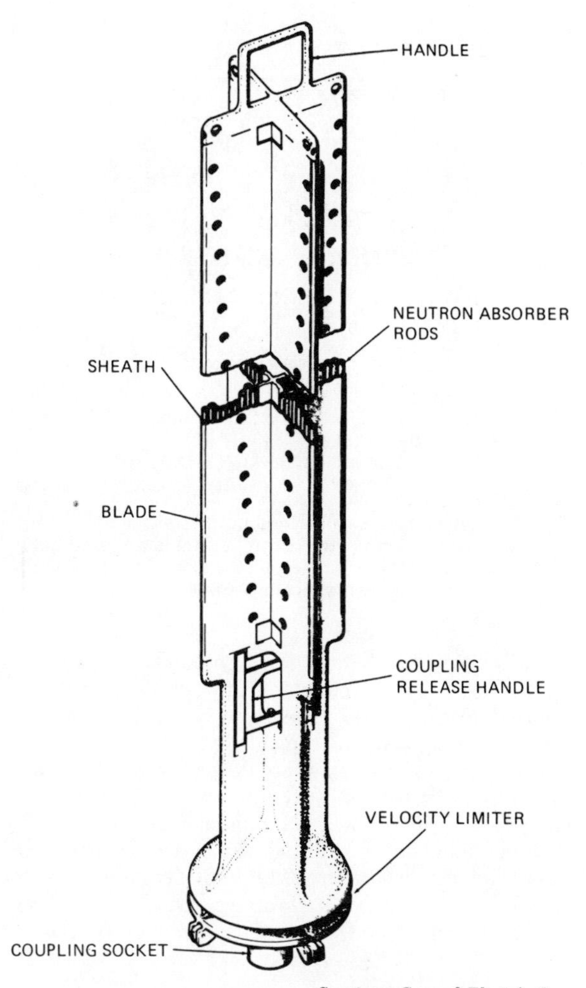

Courtesy, General Electric Co.

Fig. 3-4. Control rod. Neutron absorber rods are stainless steel tubes filled with boron carbide.

serving as a volume to collect any gases produced by fissioning of the core. The spring holds the cylindrical pellets together, while at the same time, permitting longitudinal expansion of the pellets. The plenum and spring arrangement and a complete fuel assembly are all shown in Fig. 3-3. The fuel assembly is the unit handled at the plant site.

The control rod is cruciform in shape, as seen in Fig. 3-4. It is comprised of stainless steel tubes, containing compacted boron carbide powder (neutron poison), arranged in a stainless steel sheath. The holes shown in Fig. 3-4 are to permit water to flow through the assembly and cool the rod. The control rods are arranged on the core on 12-in. pitch between the fuel assemblies, as shown in Fig. 3-5.

3.4.2 *Recirculation*—Total core recirculation is approximately 180,000 gpm for a 2460 MWt (800 MWe) plant, about half of which is circulated ex-

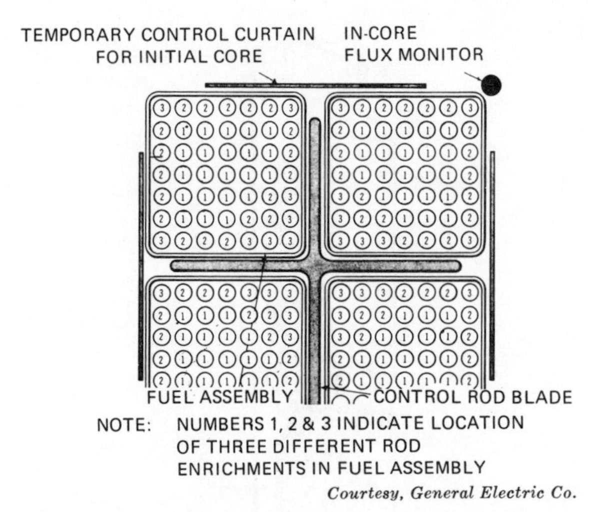

Courtesy, General Electric Co.

Fig. 3-5. Arrangement of a cruciform control rod in the reactor core. The temporary control curtains are used only during the first fuel cycle before the core has generated its own neutron poisons.

ternally through the mechanical recirculation pumps, while the second half circulates through the jet pumps internal to the vessel. The dynamic head required for the driving water is typically about 200 psi.

Initially, all recirculation loops were external, driven by electric pumps; the number of loops varied from two to five. Then jet pumps were developed and the number of external loops was reduced to two. The next step, in 1969, replaced a variable-speed flow control with throttle valve flow control, plus a constant-speed pump. The same control range is available with either method. The recirculation loops are shown diagrammatically in Fig. 3-6. Notice, in Fig. 3-6, that if a circulation pipeline ruptures, the core shroud and the jet pumps, together, form a "vessel" and keep the core covered.

The flow is throttled by a valve on the discharge side of the pump. Tests have indicated that one valve could not be used for both throttling and shutoff, therefore a recirculation loop has two shutoff valves and also one throttling valve. These valves are electric motor operated. The two loops feed into a common manifold, which in turn supplies the jet pumps, as shown in Fig. 3-7. The recirculation loop is constructed of austenitic stainless steel.

3.4.3 *Control Rod Drive System*—The control rod drives enter the reactor vessel through the bottom head. The rods withdraw in the downward direction and so have a completely powered scram stroke against the force of gravity. Drives are hydraulic, and bascially, they are hydraulic cylinders controlled by a "four-way valve with a closed center." The hydraulic fluid is demineralized water. The

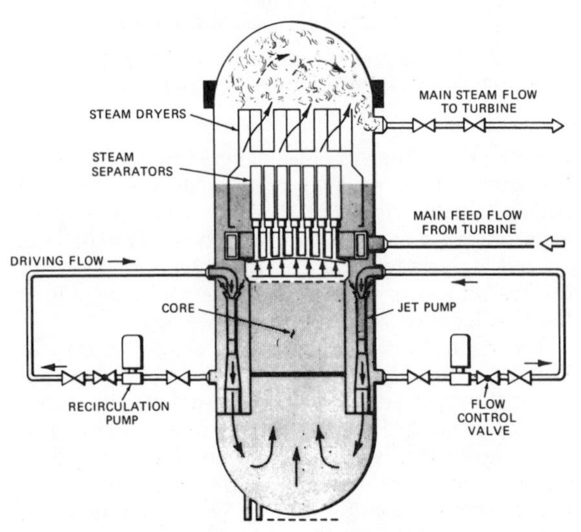

Courtesy, General Electric Co.

Fig. 3-6. Recirculation system. The "tank" defined by the core shroud, support plate, and jet pump keeps the core flooded if an external recirculation pipe ruptures.

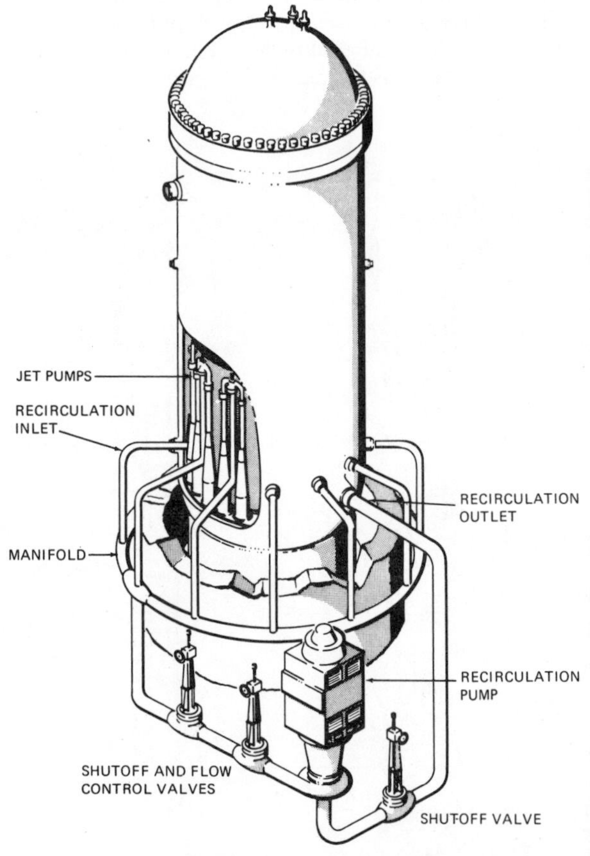

Courtesy, General Electric Co.

Fig. 3-7. Arrangement of the jet-pump recirculation system. The shutoff and flow-control valves are separate components, as shown.

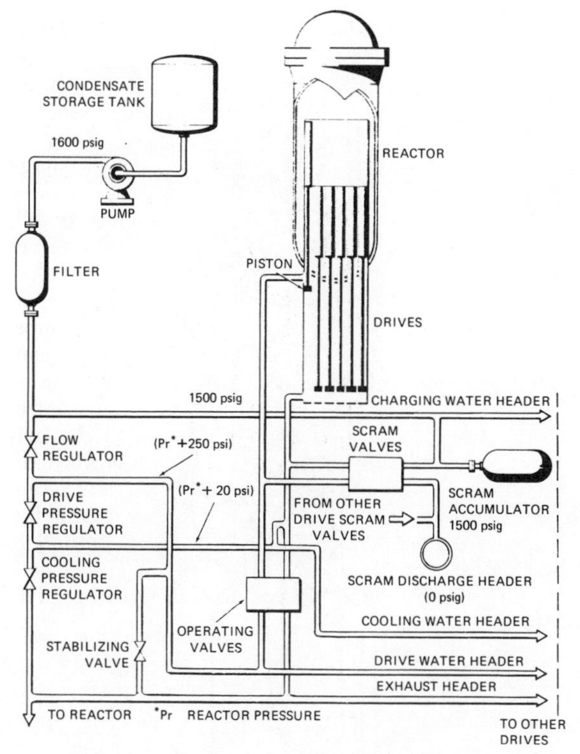

Courtesy, General Electric Co.

Fig. 3-8. Conceptual diagram of the control rod hydraulic drive system. The operating and scram valves are actually multiple component items, as shown in Fig. 3-9.

"closed center" permits the control rod to be locked in any position. The complete control system is shown in Fig. 3-8, while the actual control-valve arrangement is shown in Fig. 3-9.

In Fig. 3-9, the four "withdraw" and "insert" valves together make up the "four-way valve" referred to above. Drive water is 250 psi higher than reactor pressure. The cooling water is 20 psi higher than the reactor pressure and is adjusted to leak about 0.29 gpm, through each drive seal into the reactor. To insert a rod, the two insert valves are opened and high-pressure drive-water is then admitted beneath the drive piston. Speed is controlled by the insert speed-control valve. When the control rod reaches the desired location, the insert valves are closed again, and the latch is driven into place by its spring. Coolant flow continues to leak into the reactor vessel. The control rod is now locked into place both hydraulically and mechanically.

Rod withdrawal, however, is more complicated and is, therefore, programmed automatically. The insert valves are opened momentarily, the rod lifts, clearing the latch, then the high drive-pressure, acting on the collet piston, drives the latch back out of the way. The insert valves close, the withdraw valves open and the rod now withdraws to the new position.

The other function required of the drive packages is to scram the reactor. For this action the drives are independent of any outside power source; all the necessary power is contained in the accumulator and

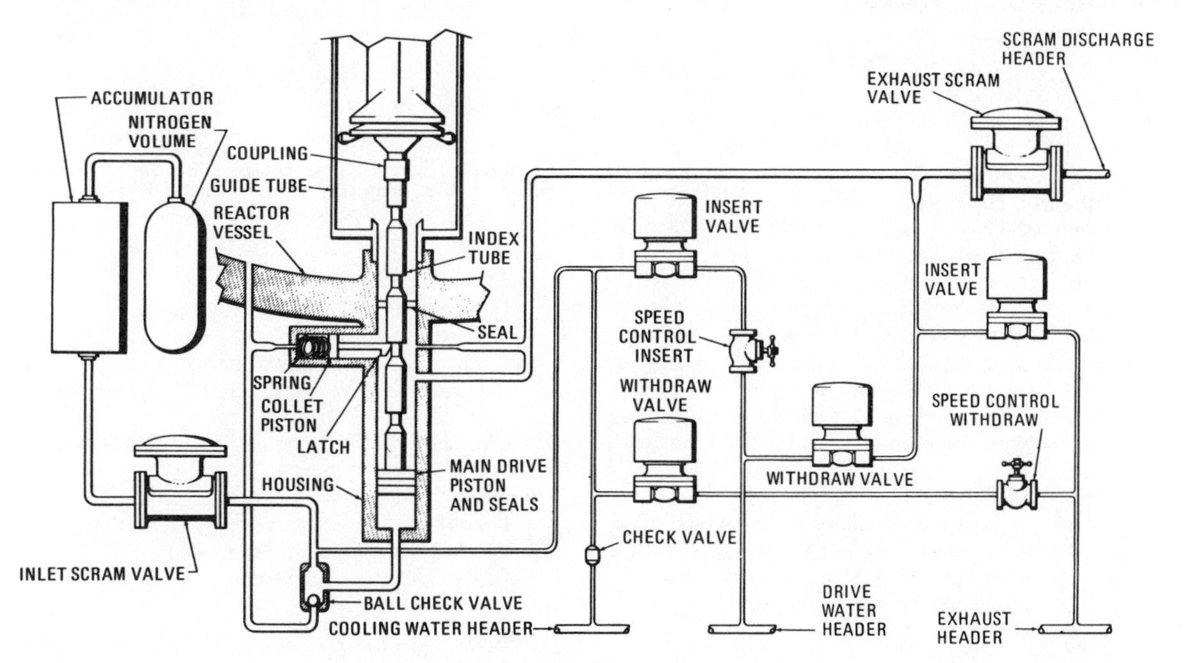

Courtesy, General Electric Co.

Fig. 3-9. Valving details of the control rod hydraulic drive system.

the reactor vessel. On a scram signal the top of the piston is vented to an atmospheric header and the liquid in the accumulator, at 1500 psig (about 500 psi higher than the reactor vessel pressure), is admitted beneath the piston. The piston accelerates in the insert direction dropping the pressure in the accumulator. When the accumulator pressure drops below reactor pressure, the ball in the check valve shifts, closing off the accumulator, admitting reactor vessel pressure to complete the scram stroke. Accumulators are charged with nitrogen from gas cylinders and experience has shown that they need topping off about every six months; the actual recharging requires only ten minutes and does not interfere with plant operation. Nitrogen pressure is monitored and a pressure switch actuates an alarm whenever the accumulator needs recharging.

The accumulator, operating valves, and interconnecting piping for each drive package are preassembled, as shown in Fig. 3-10, and are tested at the factory. Each assembly shipped to the field then needs only pipe and electrical connections at the site.

Control rod position indication is not continuous. Position is indicated every three inches by a series of glass-reed switches actuated by a magnet on the main drive piston.

Hydraulic fluid is supplied by one of two 100-percent capacity pumps. Excess fluid, not required for drive motion or cooling, is discharged into the reactor vessel. The fluid is supplied from the condensate storage tank.

3.5 Main Steam and Feedwater

A BWR does not have a primary system in the same manner as a PWR. The limits of a BWR primary system have been defined (see Appendix III) as the outermost of the two containment isolation valves on the main steam lines to the turbine and on the reactor feedwater lines from the reactor feedwater pumps. The piping, valves, supports, etc., are all considered Class 1 of Section III of the ASME Code. A schematic arrangement of these lines is shown in Fig. 3-6. In actual practice, these lines each connect to the reactor through four or more nozzles and each steam connection is brought out through containment, then manifolded in the reactor building. The feedwater manifold may be inside or outside containment depending on the space available and on the stresses in the piping.

3.6 Auxiliary Systems

The reactor requires auxiliary water systems of various types, some of which are required during reactor operation and others that are on standby for shutdown, backup, or emergency duty. The systems needed during normal operation are:

Reactor water cleanup system
Fuel storage pool cooling system

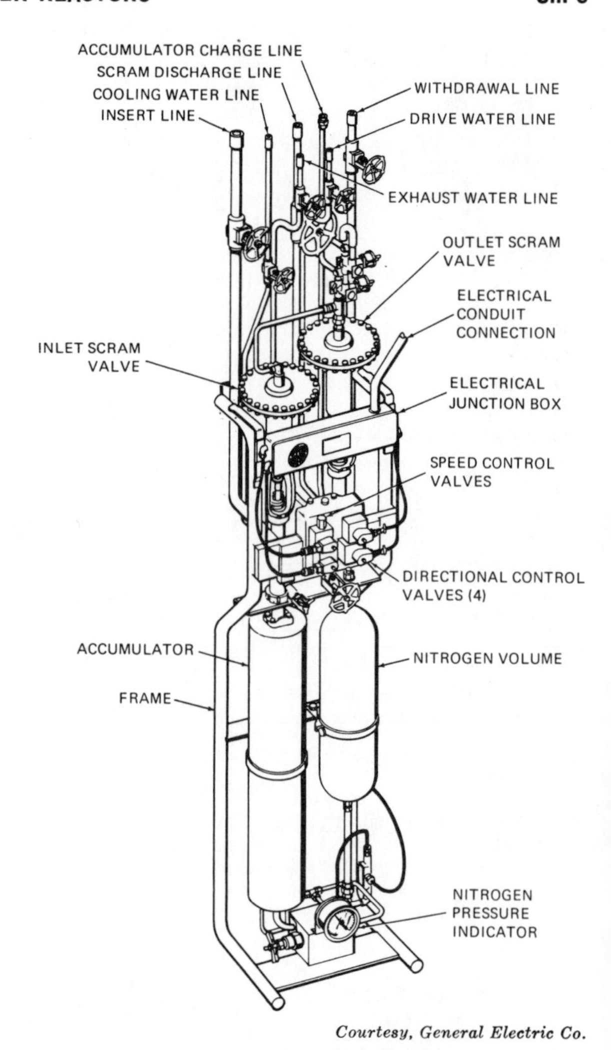

Courtesy, General Electric Co.

Fig. 3-10. Preassembled control rod drive, hydraulic control unit as furnished by the General Electric Co., to a power station.

Closed loop cooling system
Radioactive waste disposal system
Service water system.

Systems used for shutdown service are:

Residual heat removal (RHR) system in the shutdown mode
Closed loop cooling system
Service water system.

Standby systems for core backup cooling are:

Reactor core isolation cooling (RCIC) system
RHR system (hot standby mode)
Closed loop cooling system
Service water system.

Systems required for emergency core or containment cooling are:

Standby liquid control system

RHR system (containment cooling mode)
RHR system (low-pressure coolant injection mode)
High-pressure core spray system
Low-pressure core spray system
Blowdown system
Service water system
Closed loop cooling system.

The closed loop cooling system and the service water system appear in each of the four categories because the closed loop serves all the pump bearing cooling heat exchangers and the service water is the final heat sink of the plant for auxiliary and emergency systems. It should be noted that if the service water quality is good enough for RHR pump and bearing cooling, then the closed loop cooling system is not needed for emergency duty.

3.6.1 *Reactor Water Cleanup System*—Although the BWR is a direct steam cycle reactor plant with a full flow condensate demineralizer, about 80 percent of the primary water recirculates within the reactor vessel and the recirculation loops, and never sees the condensate demineralizers. A by-pass demineralization plant, shown in Fig. 3-11, is required to maintain proper reactor water quality. Its purpose is to reduce fission product concentration and clean up reactor water before and during reactor refueling. Additionally, it removes any products of corrosion and trace impurities introduced with make-up water. The system is sized to circulate one complete reactor system volume in 4½ hours and operates at reactor pressure. Constructed 100 percent of stainless steel, it is comprised of heat exchangers, pumps, and a filter demineralizer. Supply is taken from the suction side of a recirculation loop and returned through the reactor feed water line. The process liquid is cooled enough (to 120 degrees F) to avoid thermal damage to the demineralizer resin. Heat is recovered by the use of the regenerative heat exchanger before the nonregenerative unit. The system is provided with remotely controlled isolation valves so that it can be removed from service during operation. Since the entire system can be isolated from the reactor and is not essential to reactor safety, all vessels; valves; pumps; and piping are built to Class 3, Section III of the ASME Code even though they contain reactor coolant. The system has two 50-percent capacity pumps, two heat exchangers of 100-percent capacity each, and two 50-percent capacity filter demineralizers. Water quality is maintained at a conductivity less than 1.0 micromho per cm^3, 5.6 to 8 6 pH, and less than 0.1 ppm chloride ion. The system can be operated independently of the reactor and can act as a "blowdown" path or an exit, when it is desired to remove liquid from the reactor.

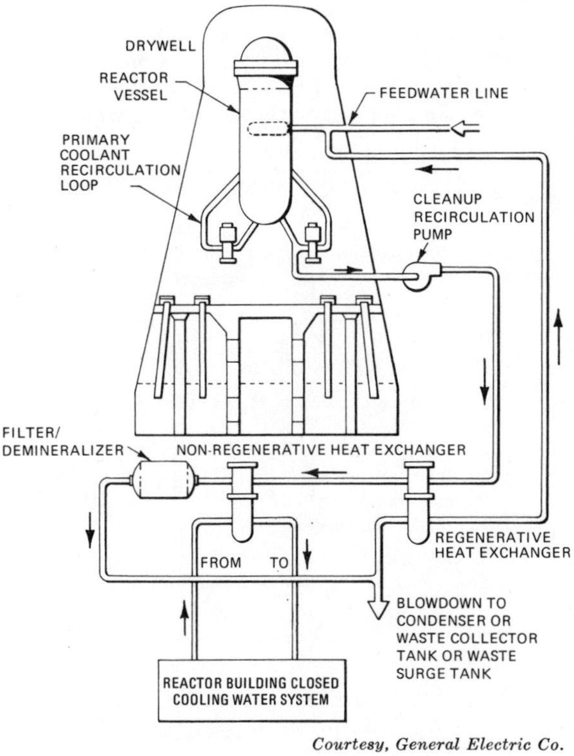

Courtesy, General Electric Co.

Fig. 3-11. Reactor water cleanup system.

3.6.2 *Safety Systems—Standby Liquid Control System*. The standby liquid control system is a redundant reactivity control system furnished in direct conformance with Criteria 26 and 27. The system injects neutron poison as boron (in the form of sodium pentaborate solution), in sufficient quantity to shut the reactor down in any credible situation. The General Electric Co. has chosen to use boron in the pentaborate form because it is much more soluble than the boric acid form. Solution temperatures do not necessarily have to be higher than ambient to prevent crystallization; warmer temperatures are used to decrease the total quantity of solution required for a given boron addition. If boric acid were used, however, there would be no choice—the tank, pumps, piping, etc., would all have to be heat traced and kept warm at all times. Sodium pentaborate is made directly in the poison storage tank by mixing boric acid and borax in stoichiometric proportions with demineralized water.

As seen in Fig. 3-12, the system has two positive displacement pumps and is isolated from the reactor by two parallel, explosive valves. All other valves in the main line (the heavy line in Fig. 3-12) are locked open. The explosive valves each have two electric firing squibs supplied by two different circuits. The pumps can be tested at any time by pumping poison solution from the storage tank to the test tank and then returning it to the storage tank. Just before a shutdown, with the storage tank valved off and the lines cleared of poison solution, the system from the

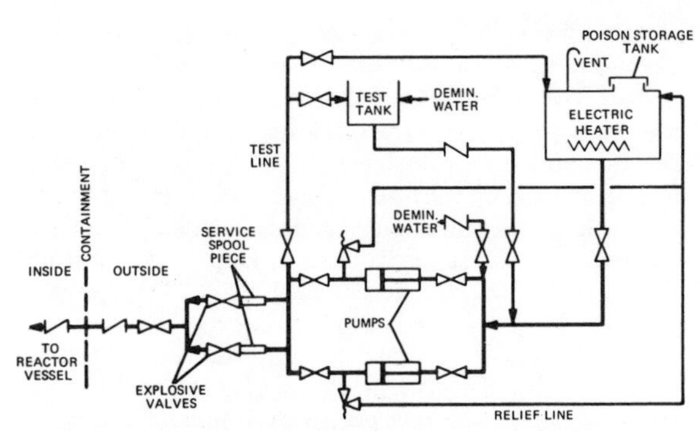

Fig. 3-12. Schematic arrangement of the standby liquid control system.

pumps to the reactor vessel can be tested with demineralized water. This can only be done immediately before a shutdown because the explosive valves must be removed from the line for servicing. Piping, valves, and equipment are all of stainless steel or carbon steel, lined with a corrosion resistant coating. Check valves are used for containment isolation.

Vessels, pumps, valves, and piping conform to Class 2, Section III of the ASME Code. All equipment, piping, supports, etc., are designed as Class I seismic systems.

If this system were ever actually used, the pentaborate solution would be removed by flushing, followed by polishing with the reactor water cleanup system. However, to date, there has been no actual use of this system in any commercial installation except for routine test purposes.

High Pressure Core Spray System (HPCS). The high pressure core spray system is provided to make up water lost in any break or rupture of the primary system and to prevent the core from overheating. Water is sprayed directly onto the core, using the core spray sparger shown in Fig. 3-2. The water supply is taken from the condensate storage tank as shown in Fig. 3-13, and when it is exhausted, the supply is switched over to the pressure suppression pool. This pool is cooled by the RHR system, as described under the containment cooling mode. The system can operate against the entire pressure range from normal primary system pressure on down. A test line, not shown in Fig. 3-13, permits testing of the pump at any time by recirculating from the pressure suppression pool through the pump, and back to the pool. This single pump in the system is independent of all outside power sources, having its own diesel generator set.

Pump and piping are built to Class 2 requirements of Section III of the ASME Code. The entire system is designed and installed to Class I seismic requirements. This system, together with the low-pressure core spray system, the automatic blowdown system, and the RHR LPCI mode, fulfill all of the requirements of Design Criterion 35.

Plants designed prior to 1969 had two systems in place of this single system. The high-pressure end of emergency core cooling was covered by a High Pressure Core Injection System using the same water supply as the HPCS but entering the vessel through the feedwater connections. This system had a single, steam-turbine-driven pump the same as the RCIC pump. The low-pressure end of the cooling was covered by a Core Spray System using the pressure-suppression pool as the water supply. It had two 100 percent subsystems energized by two electric-motor-driven pumps. Only one of the pumps could operate at a time in a normal power-failure situation.

Low Pressure Core Spray System. A low pressure core spray system (LPCS) consisting of a single pump and piping is furnished as a backup to the LPCI mode of the RHR system and is shown schematically in Fig. 3-14. It takes its water supply from the pressure suppression pool and also has a connection to the condensate storage tank. Its water is sprayed directly onto the fuel elements from a ring header inside the reactor vessel. The system comes on automatically on signals of high containment pressure or low reactor water level. System delivery into the reactor vessel is withheld automatically by check valves until the pressure difference between the reactor and containment drops to lower than the LPCS design pressure. The pressure-suppression pool is cooled by the RHR system as described under the containment cooling mode. A test line, not shown in Fig. 3-14, permits testing of the pump at any time by recirculating from the pressure suppression pool, through the pump, and back to the pool. The pump power is supplied through one of the emergency

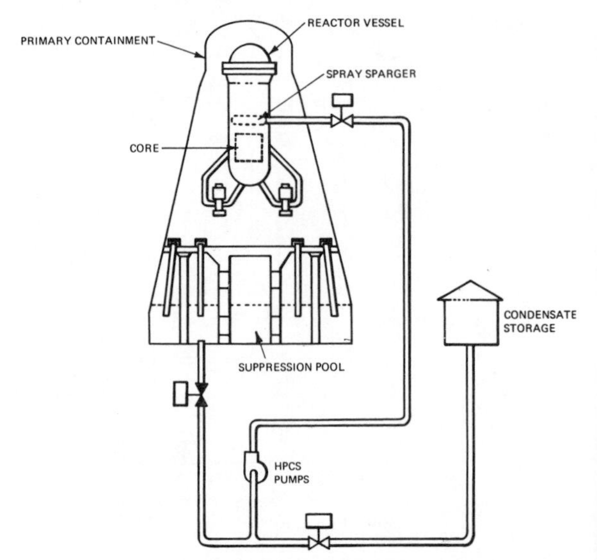

Courtesy, General Electric Co.

Fig. 3-13. High-pressure core spray system. Pump has its own variable-cycle, diesel generator power supply.

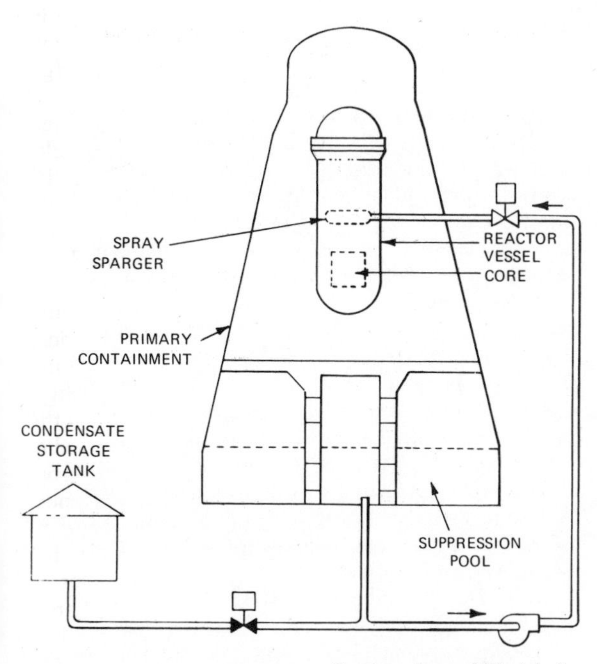

Courtesy, General Electric Co.

Fig. 3-14. Schematic arrangement of the low-pressure core spray system.

buses so the pump can be supplied from the emergency diesel generators if the plant should be isolated from all external power sources.

Pumps, valves, and piping conform to Class 2 of Section III of the ASME Code.

Blowdown System. The automatic blowdown system is the backup for the HPCS system. The system operates by simply lifting the remote controlled relief valves and blowing steam down into the pressure-suppression pool. This takes place at a pressure lower than the set point of the code safety valves. The main purpose of this system is to reduce the pressure rapidly so that one of the low pressure cooling systems can operate. This system requires three coincident signals to initiate operation:

Low low vessel water level
HPCS system not operating
High drywell pressure

3.6.3 *Residual Heat Removal System (RHR)*—A nuclear reactor cannot be shut down immediately in the same manner that a fossil-fuel steam generator can. Shutting down a reactor core stops the chain reaction, after which the residual fission product gamma activity continues to decay. Power level drops to less than 1 percent in three or four minutes. (An equation for decay heat generation is given in Chapter 12.) However, 1 percent of 3000 MWt (a current 900 MWe rating) is still a large quantity of energy, more than 10^6 Btu per hour that must be removed. The RHR system, shown in its entirety in Fig. 3-15, serves the following purposes:

1. Removes decay heat and sensible heat from the reactor during normal shutdown and refueling
2. Restores reactor-vessel water level and maintains it during a loss of coolant accident so that sufficient cooling is provided to the core to prevent melting of the fuel cladding
3. Removes heat added to the pressure suppression pool water during hot standby operation or during operation of the reactor core isolation cooling system. This same mode of operation can be used to cool pressure suppression water after an accident.

Each of these objectives is achieved using the same equipment—pumps and heat exchangers—but with different flow paths. There are actually four RHR pumps and two RHR heat exchangers. In the event of a power failure the RHR pumps are automatically switched over to emergency power. Vessels, pumps, valves, and piping conform to Class 2, Section III of the ASME Code. All piping, equipment, and supports are built to Class I seismic criteria because the system is needed for safe shutdown of the reactor.

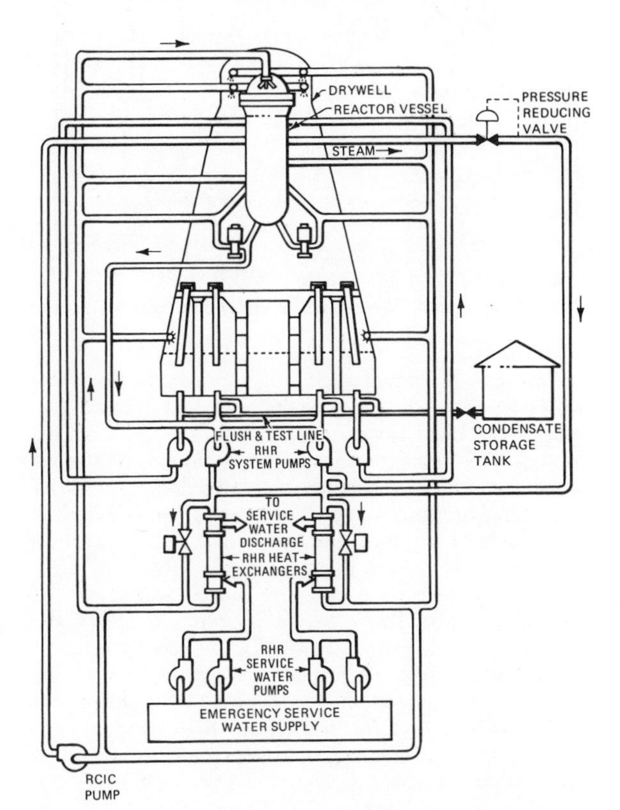

Courtesy, General Electric Co.

Fig. 3-15. Complete diagram of the residual heat removal system.

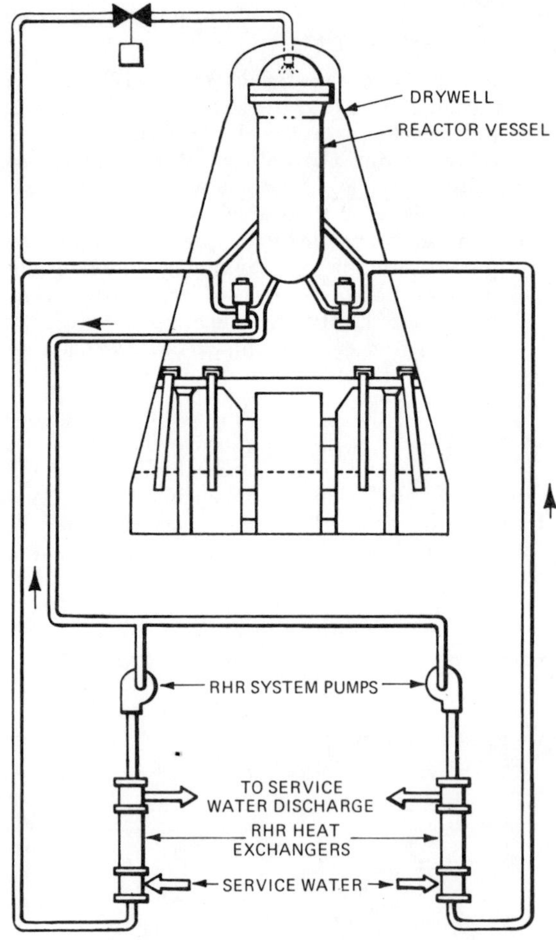

Courtesy, General Electric Co.

Fig. 3-16. RHR system in the shutdown cooling mode.

Because of the diversity of service requirements the capacities of the pumps and heat exchangers match the most severe duty for each. The heat exchangers are designed for rapid cooldown of the reactor for normal refueling and have 200 percent of core safety requirements. The pumps are selected so that three of the four units can flood the core in the low pressure coolant injection mode after the maximum credible pipe break; the fourth unit adds the necessary redundancy to the system. In 1971 the fourth unit was moved to the LPCS (see page 32).

Isolation valves are provided at connections to the primary system together with automatic interlocks which will protect the RHR from high primary system pressures. Four separate pipelines are supplied for flooding the core, as shown in Fig. 3-15, and these provide the redundancy necessary for emergency duty. The equipment is installed in separate areas and emergency pipelines have physical protection against missiles and other types of mechanical damage.

RHR Shutdown Cooling. In this mode of operation, shown in Fig. 3-16, both heat exchangers can be used together with two pumps for each heat exchanger. The system is started after the reactor vessel pressure is below 200 psia. At this time, and until the reactor is depressurized, only one half of the system is operated for shutdown cooling. The other half is held ready for emergency use in the low-pressure coolant injection mode. After depressurization, the second half also operates in the shutdown cooling mode to increase the reactor vessel cooling rate. From core shutdown until the shutdown cooling system is started, reactor decay heat is removed by steam generation and by by-passing the steam around the turbine into the main condenser. As has been seen in the preceding description of operation, from the point of view of reactor safety, the shutdown cooling system actually has two 100-percent capacity loops, backed up by at least two different power sources.

A side connection terminating in a spray nozzle inside the reactor vessel serves to cool the vessel head during the cooldown period.

RHR Hot Standby. The RHR system in the hot standby mode operates with the reactor core isolation cooling system (RCIC) to cover the full pressure range when the core is isolated from the condenser. The decay heat generates high-pressure steam which, in turn, lifts the power operated relief valves (not the code safety valves) on the main steam line and the steam is then discharged to the pressure-suppression pool. When the pressure drops low enough, the steam is diverted to the RHR system from the pressure suppression pool and the RCIC system. In this situation, the RHR heat exchangers act as condensers using the service water as condensing water. The condensate, as shown in Fig. 3-17, is returned to the reactor vessel via the RCIC pump.

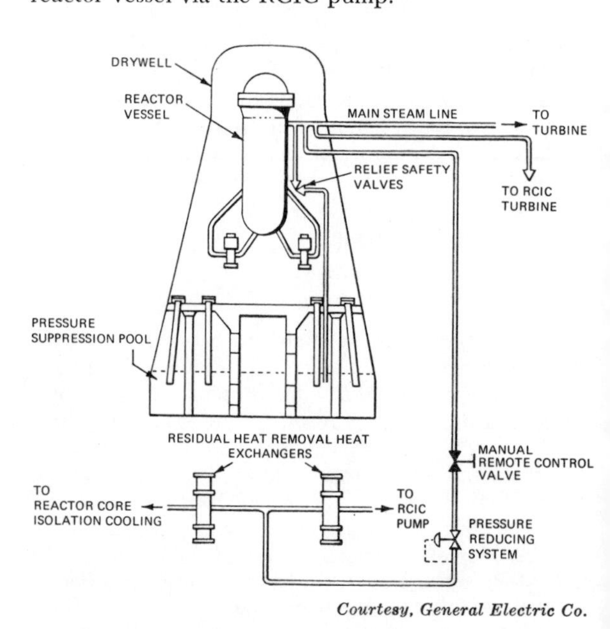

Courtesy, General Electric Co.

Fig. 3-17. RHR system in the hot standby mode.

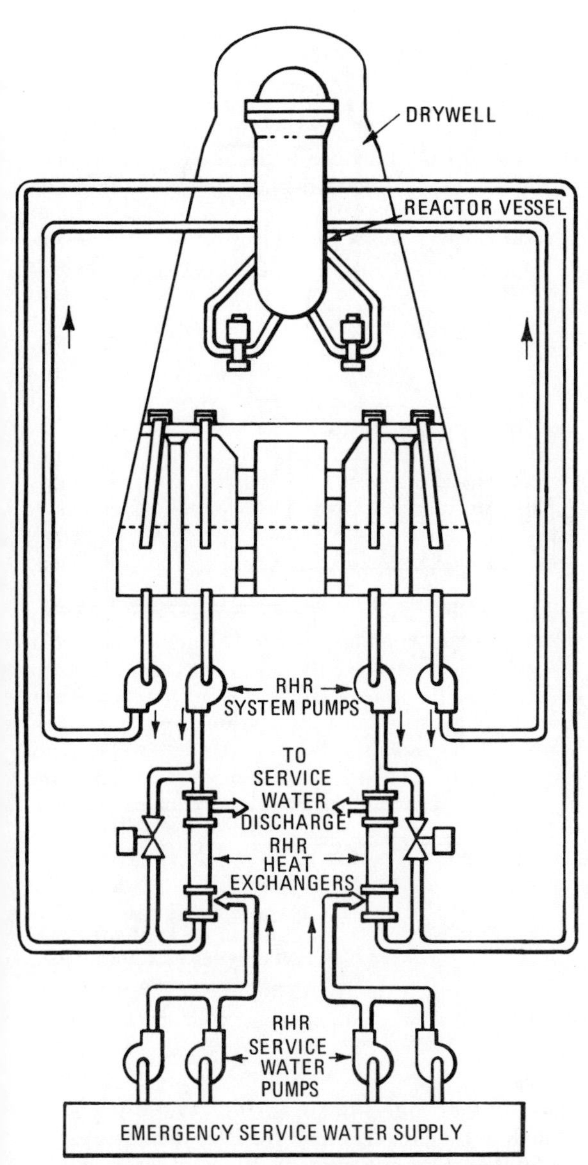

Courtesy, General Electric Co.

Fig. 3-18. RHR system in the low pressure coolant injection mode.

As can be seen, the hot standby mode has a pressure-reducing capability to permit it to accept steam before the reactor vessel pressure reaches RHR design pressure. In the hot standby mode, no pumps are required and both heat exchangers are needed to carry the load. This operating mode is a core cooling standby mode, not an emergency mode. It is intended that the reactor be returned directly to load from this situation.

RHR Low Pressure Coolant Injection Mode. This is an emergency mode of operation rather than a standby mode. The purpose of this mode of operation is to restore reactor water level after a loss of coolant accident. The system restores water level to at least half the core height and then maintains it at this level. The system, as shown in Fig. 3-18, takes its suction from the pressure suppression pool. Since the containment structure is watertight, a closed loop is created using the pressure suppression chamber as a sump. System capacity requires the use of three of the four RHR pumps during the level restoration portion of the operation, after which, only one pump is required to maintain this level. Instrumentation selects an undamaged flow path from the two that are available. Heat can be rejected to the service water system during this mode of operation by going through, rather than around, the RHR heat exchangers.

RHR Containment Cooling Mode. The containment cooling mode acts as both a standby cooling system and an emergency system. Fig. 3-19 shows that this system can recirculate to cool the pressure-suppression pool or cool containment by spraying cooled suppression pool water directly into the con-

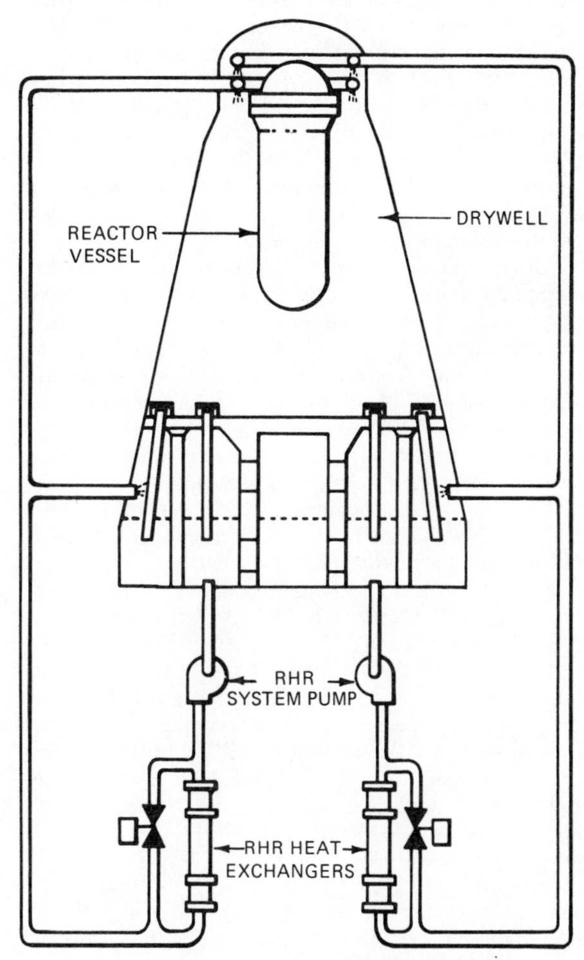

Courtesy, General Electric Co.

Fig. 3-19. RHR system in the containment cooling mode.

tainment atmosphere. The mode acts as standby cooling in concert with the RCIC system. The main purpose at this time is to limit suppression pool temperature during an RCIC or hot standby situation so the pool may still perform its pressure suppression duties, if required.

After an accident, the sprays are used to assist in controlling containment pressure by condensing steam and cooling noncondensable gases. In either situation, the RHR system has two 100-percent capacity systems.

3.6.4 *Reactor Core Isolation Cooling System (RCIC)*—The RCIC is a standby core cooling system used when the reactor is isolated from the turbine due to closure of the main steam isolation valves, turbine throttle valve, or turbine stop valve. It is not associated with any accident. In an RCIC situation the reactor has scrammed, and steam—generated by decay heat—has lifted the power operated relief valves discharging steam to the pressure suppression pool. The RCIC system shown in Fig. 3-20 using a 100-percent capacity pump maintains reactor water level using the water supply in the condensate storage tank. The pump is operated by a turbine using the steam generated in the reactor before it passes into the pressure suppression pool. A connection is provided to the pressure suppression pool so that it can act as a backup water supply, if necessary. The RHR containment cooling mode is available to control pressure suppression pool temperature should it become necessary.

All controls for this system are operated from the station batteries. Thus, with the steam turbine driven pump, this system is completely independent of any external power source. To insure proper pump operation, the pump is installed below the condensate storage tank and alongside the pressure suppression pool. Pump and turbine are carbon steel. If the system should fail to operate, the reactor operator can start the high pressure core spray system.

The RCIC system maintains the water level in the reactor high enough to prevent automatic initiation of the emergency core cooling system.

Pumps, valves and piping are built in conformance with Class 2 of Section III of the ASME Code.

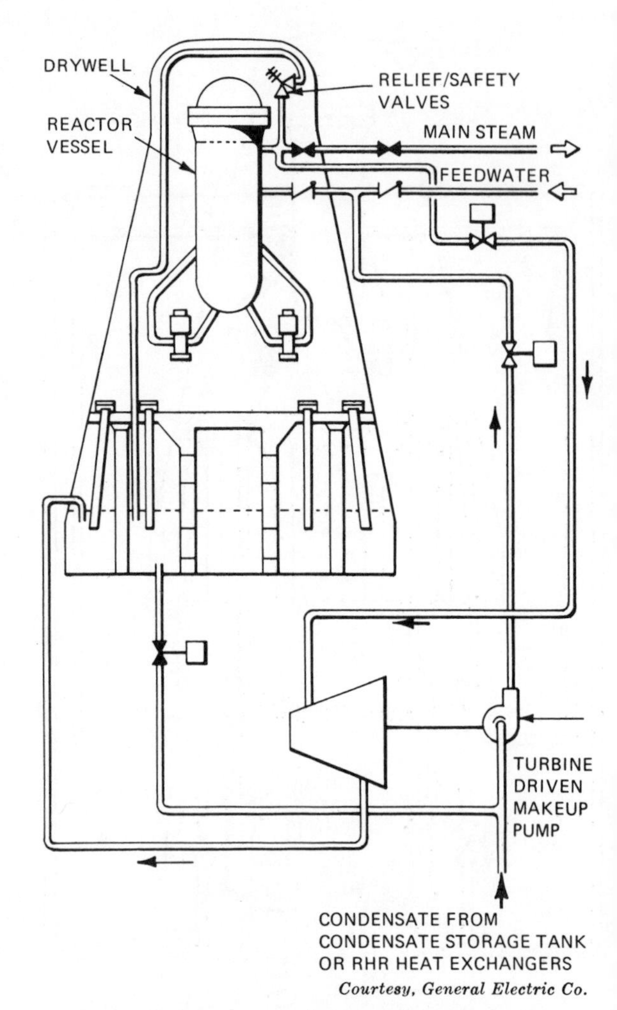

Courtesy, General Electric Co.

Fig. 3-20. Reactor core isolation cooling system. Decay heat generated steam is used to drive the turbine.

3.6.5 *Fuel Storage Pool Cooling System*—This system removes the decay heat generated by spent fuel elements after removal from the reactor during the cooling off period before shipment back to the fuel reprocessor. The cooling capability of this system operates in parallel with the shutdown mode of the RHR system while the vessel is open during refueling. The system contains a filter demineralizer to maintain water purity and clarity. Vessels, valves, pumps, and piping conform to Class 3, Section III of the ASME Code. The system, as shown in Fig. 3-21, supplies both the reactor well and the fuel storage pool through discharges at a point near the bottom of the pool. Water leaves the pools through skimmer arrangements at pool surface, overflows to a surge pool, and then enters pump suction. Water is passed through or by-passed around the filter demineralizers, as required. Piping penetrates the pool walls with a siphon break high up, so that if a pipe shears off, external to the pool, spent fuel shielding will not be jeopardized. Pumps, heat exchangers, and filter demineralizers are installed as two 50-percent units. There are no spare units. Pumps are sized to pump one total pool volume in twenty-four hours. A cross connection between this cooling system and the RHR systems permits the RHR to act as a standby unit and be ready to furnish additional cooling when necessary.

3.6.6 *Closed Loop Cooling System*—The closed loop cooling system is an inhibited, demineralized water system acting as a heat sink for equipment in the reactor building. It, in turn, dissipates its heat to the service water system. A typical system, showing the equipment serviced and their building locations, is shown in Fig. 3-22. This system acts as a buffer be-

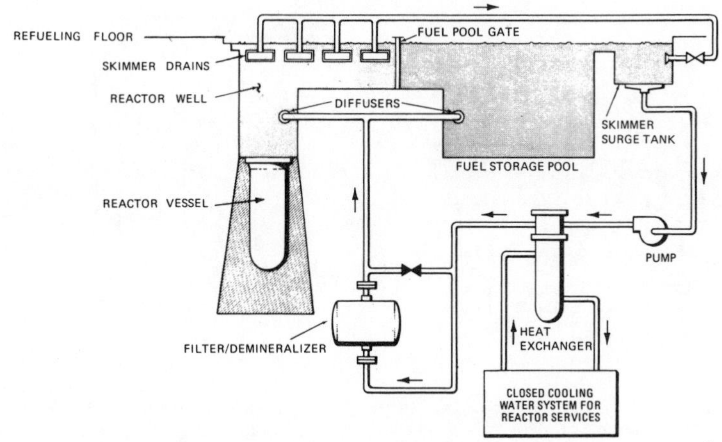

Courtesy, General Electric Co.

Fig. 3-21. Fuel pool cleanup and cooling system.

tween radioactive systems and the environment, so that two failures, in tandem, are required to release activity. Depending on the site location, the closed loop heat exchanger may see salt water in the tubes. If this is true, good salt water practice, described in Chapter 8, must be followed. Facility licensing records indicate that a single loop is acceptable for both the reactor building and the turbine building. In this system the number of units is optimized for economy and then a single spare unit is added. Thus, if the heat load requires three heat exchanger shells, a fourth shell would be included and the piping would permit each shell to be withdrawn from service for cleaning, without interfering with operation. The system is designed in conformance to Class 3, Section III, of the ASME Code.

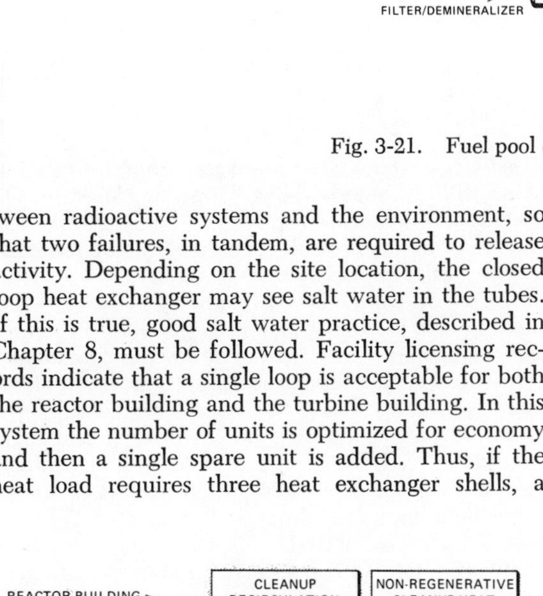

The system is continuously monitored for radioactivity. In a normally operating plant this system has no activity; presence of activity is evidence of an equipment failure in one of the components served. The system pumps and piping are carbon steel and the heat exchanger is compatible with the final heat sink fluid. This system is arranged so that on loss of normal power it can be manually switched over to emergency power, usually from diesel generators, and restarted.

3.6.7 *Service Water System*—The service water system is the final heat sink for all plant heat loads except the main condenser. The water in it is usually of poor quality and contains the normal solids found in lakes, rivers, oceans, etc. The system always requires some type of strainer or screen. No specific direction can be given for materials or velocities because they will vary from site to site. The best practice is to research the experience of existing installations in the area, and to test the water. Corrosion tests should be run on aerated turbulent water for as long a period as possible. In addition to its primary duty as a heat sink, the system serves as a jockey pump for the fire system, a backup firewater supply, and for emergency reactor water flooding.

The system is comprised of two banks of pumps, each totaling 100 percent duty. Typically, there are four service water pumps and four booster pumps as shown in Fig. 3-23A, but as shown in Fig. 3-23B, recent designs do not have the RHR booster pumps nor the pressure control valves. Radiation monitors are furnished instead. All pumps are needed for normal operation but only one service water pump and two booster pumps are needed for an emergency shutdown. All pumps are arranged for manual switch-

REACTOR BUILDING
CLEANUP RECIRCULATION PUMP COOLERS
NON-REGENERATIVE CLEANUP HEAT EXCHANGER
FUEL POOL COOLING HEAT EXCHANGER
RHR PUMP COOLERS
SAMPLE COOLERS
CLEAN SUMP COOLERS
REACTOR RECIRCULATION PUMP MOTOR COOLERS
REACTOR RECIRCULATION PUMP COOLERS
DRYWELL COOLERS
REACTOR BUILDING COOLING WATER HEAT EXCHANGERS
DRYWELL (CONTAINMENT)
TURBINE BUILDING COOLING SYSTEM
SERVICE WATER PUMPS
COOLING WATER SUPPLY

Courtesy, General Electric Co.

Fig. 3-22. Auxiliary closed loop and service water cooling systems.

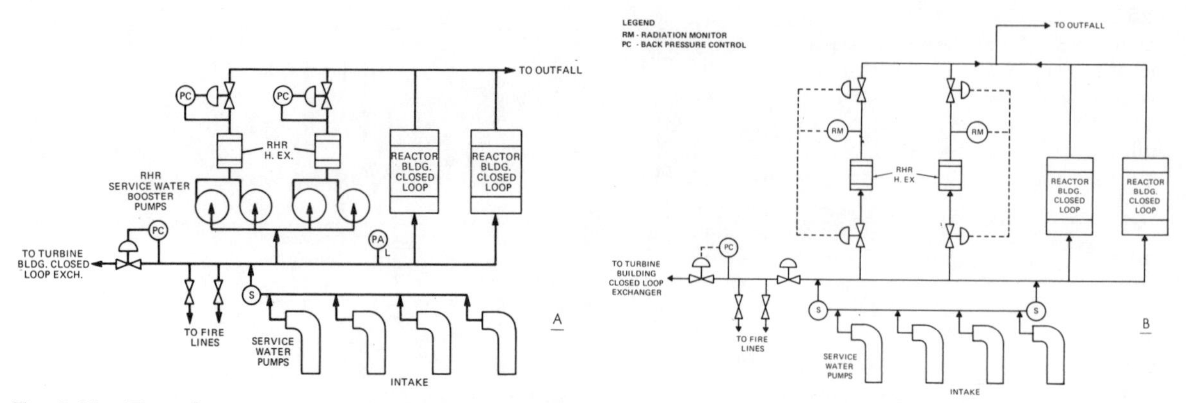

Fig. 3-23. Typical service water cooling system. As seen in B, recent designs do not have the RHR booster pumps nor the pressure control valves. Radiation monitors are furnished instead.

over to emergency power so that the operator can select the ones he wants.

As can be seen, there is no secondary barrier between primary water in the shell and environmental water in the tubes, in the RHR heat exchangers. In units designed prior to 1969, Fig. 3-23A, booster pumps raised the pressure of the service water to the RHR heat exchangers so that if a leak should occur, service water will leak in rather than have primary water leak out. Note that the pressure to be surmounted must also include any contribution by containment pressures. Back-pressure control valves on the outlet of each RHR heat exchanger guaranteed exchanger pressure. In 1969, and later, the service water was monitored for radioactivity as shown in Fig. 3-23B. At any indication of activity, the circuit is isolated and the second circuit is put into service. If service water system pressure drops in an emergency, the turbine building supply is closed off and its essential loads are supplied by other means. The system is designed to the same codes and standards as the closed loop system.

3.6.8 *Standby Gas Treatment System*—The standby gas treatment system is provided to remove gaseous fission products from either the containment structure or the reactor building, in the unlikely event of an accident. The system contains parallel trains of particulate and iodine filters preceded by demisters, air heaters, and prefilters. A typical system is shown in Fig. 3-24. Upon actuation, the reactor building is sealed and both fans start. The system is also connected to containment, a sealed structure. Both fans operate to reduce building internal pressure to a subatmosphere pressure—minus

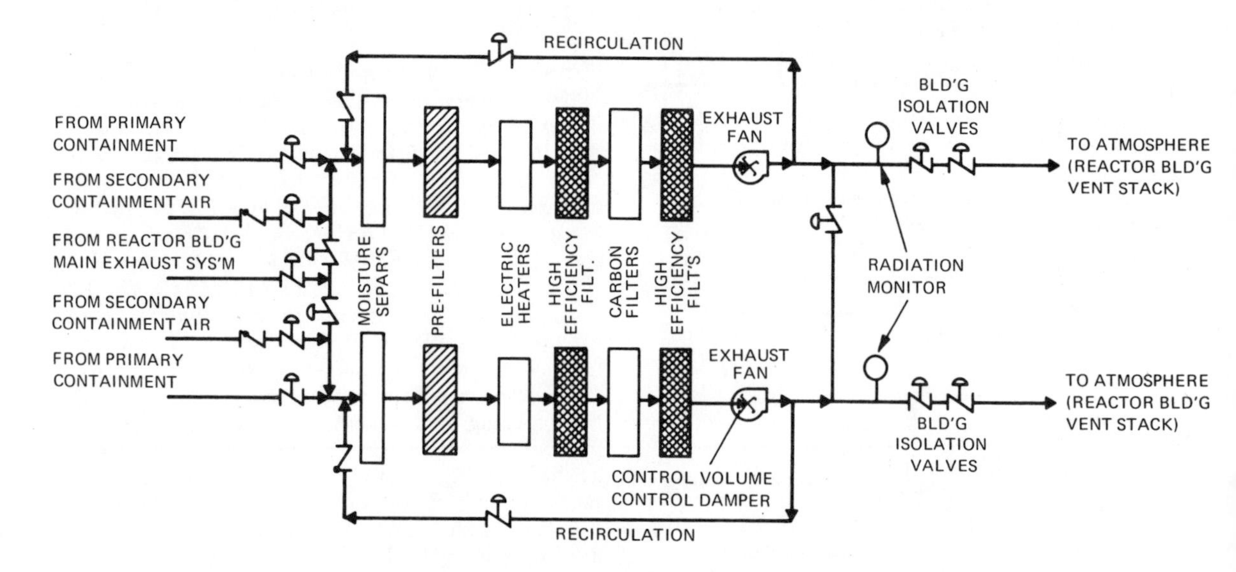

Fig. 3-24. Schematic arrangement of a standby gas treatment system. System processes gas from both the containment building and the reactor building.

0.25 inch of water is a representative design level. Upon reaching design level one fan is shut off and the other operates to hold the building at the negative pressure. Each fan is designed to exhaust the inleakage to the building under vacuum conditions. Building inleakage must be estimated, however a typical value is 100 percent of the building volume per day.

The system is designed for automatic and manual startup. Manual initiation is from the control room; automatic initiation is from high drywell pressure, low reactor water level, or high radiation in the refueling floor exhaust duct. Note that even if the signal originates in a sealed containment structure, the reactor building is sealed and the system starts. Particulate removal is by HEPA filters (99.97 percent, or more, of 0.30 micron particles removed). Iodine is removed by activated charcoal beds with a removal efficiency in excess of 99 percent. All the equipment is Seismic Class I, and usually, each equipment train is housed in a Class I structure for further protection. The demister and air heater are provided to insure that no water droplets reach the filtering equipment. A cross connection, not shown on Fig 3-24, must be provided between filter trains to remove the decay heat from the adsorbed iodine in the inactive charcoal bed. The radiation detectors in the refueling floor exhaust duct must be located so that the radiation in the duct is detected and the reactor building sealed, before the contaminated air reaches the isolation valves. Isolation valves with closing times of 3 or 4 seconds are required. The use of electronic instrumentation together with a pneumatic valve operator practically eliminates the reaction time of the control system.

3.6.9 *Ventilation*—Ventilation and cooling in containment is accomplished by water-cooled fan coil recirculating units. Ambient conditions during operation are dictated solely by the needs of the electrical gear. The air-cooling systems have no safety functions for containment cooling after an incident. If the recirculating pump motors are air-cooled, they must be given special consideration to insure sufficient cooling. Particular attention must be given to the air flow patterns to insure proper cooling of the reactor vessel cavity.

3.6.10 *Condensate Storage*—Condensate storage is placed in its normal location in the feed water cycle before the preheaters and it has connections into the engineered safety features systems. The tank itself has design features that provide a water supply for the emergency core cooling systems at all times. If, for any reason, this supply is actually used it must be replaced before any additional capacity is restored. This capacity, which runs between 100,000 and 200,000 gallons, must be determined in conjunction with the reactor vendor.

For best results the storage tank ought to be corrosion resistant. The lowest priced tank available at the present time is of carbon steel lined with epoxy or vinyl. The vessel should be constructed to an API code since Section VIII of the ASME Code does not go lower than 15 psig; it would impose an unnecessary cost burden to make this vessel carry the 15 psig pressure rating required by Section VIII when it can be atmospheric. Piping connections to and from the tank are designed in accordance with ANSI Standard B31.1.0. Tank, piping, supports, etc., are Seismic Class II because the tank is a backup water supply to the pressure-suppression pool.

3.7 Control

The BWR is operated as a constant pressure variable-flow device. The recirculation loop operates to take advantage of the negative reactivity created by the void space inside the steam bubbles. The basic control sequence, starting with the plant on load at equilibrium, is the following:

The power dispatcher sends a signal to increase power. This signal resets the recirculation flow controller and increases the recirculation flow rate. The higher fluid velocity through the core sweeps the steam bubbles out at a faster rate, which reduces the core void space and results in increased reactivity and higher energy production. Increased production of steam bubbles continues until a new equilibrium power level is reached. The increased steam production results in an increasing pressure in the reactor vessel. The rising pressure is detected by the pressure controller which signals the turbine throttle-valve to open further, thus reducing the pressure to the control point again. In order to reduce reaction time to a load change, at the same time the flow controller receives its signal, the pressure controller is temporarily reset to a slightly lower pressure. Water in the reactor flashes, making more steam available for the turbine almost immediately. As the power level in the reactor increases to the load demand, the pressure setting is returned to normal. To decrease power, the same sequence of events takes place but in an opposite direction.

Reactor power level control is achieved by a combination of recirculation flow-rate and control rods containing boron carbide. During plant startup and initial operation, control rod pattern and flow rate are optimized for the anticipated power levels, to give best fuel economy. Thereafter, these settings are used for control. Both the control rods and the flow rate can be controlled either automatically or manually. Normal operation, with only plus or minus a few percent as the power fluctuation range, uses "flow rate" as the control device, so that both the neutron flux pattern and the power distribution in the core are fixed. The control range available, using only flow-rate variation, is 35 percent of the set point. Reactor scram is accomplished by a rapid insertion of all control rods. In the BWR, all control rods are identical and none are built as "regulating" rods as opposed to "safety" rods. In accordance with Design Criterion 26, a backup sodium pentaborate injection system is provided to shut the reactor down if anything unexpected takes place in the primary shutdown system.

3.8 Reactor Protection

The reactor uses a "one-out-of-two taken twice" system of protection. There are two trip channels in the system and both of them must trip, to trip the reactor. In Fig. 3-25, parameter 1 is monitored by four separate channels: 1A, 1B, 1C, and 1D. Channels 1A and 1B are in trip channel R1, while channels 1C and 1D are in trip channel R2. In order to de-energize the scram, both R1 and R2 must trip. Hence, either channel 1A or 1B must trip together with either Channel 1C or 1D, in order to trip the reactor. As can be seen, the system is selective and two of the correct channels must trip.

The following parameters have inputs to the protection system:

a. High pressure in reactor containment—This indicates a possible rupture of the primary system.

b. Low water level in the reactor vessel—This scrams the reactor before it drops low enough to damage the fuel elements.

c. High reactor pressure—This indicates an abnormal situation in the reactor. Trip value is set high enough so that normal pressure transients do not actuate the channel.

d. High neutron flux—This is a direct measure of power output.

e. High liquid level in the scram discharge receiver—High levels will prevent or interfere with the high speeds required in scram situations.

f. Fast closure of the turbine control valves—Pressure increases in the reactor cause power excursions from the increase in reactivity due to the sudden decrease in bubble voids. This scram controls the magnitude of the power excursion.

g. Closure of the turbine stop-valve—This is the same category as "f," above.

h. Main steam line isolation-valve closure—This is in the same category as "f" and "g," above. Scram is initiated at 10 percent valve closure.

i. High activity in the steam line—Abnormally high levels of activity in the main steam line trip the reactor and close the isolation valves, to confine the activity to the containment building.

j. Loss of plant auxiliary power—This causes the scram valves to open, allowing the scram.

k. Manual trip.

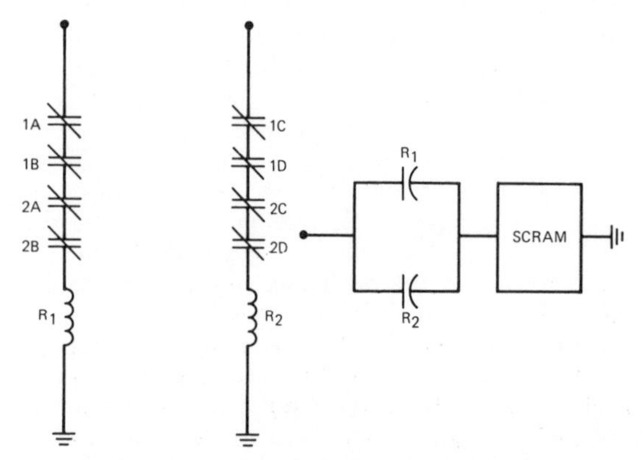

Fig. 3-25. Reactor protection-system scram circuitry illustrating the "one-out-of-two taken twice" design.

All protection system circuitry is designed to trip in the unenergized condition. Hence, during normal operation, all circuits are energized and circuit failure is equivalent to channel trip. This implements a fail-safe philosophy such that any protection system circuit or component failure places the channel in the safest condition.

A "trip element," referred to earlier, is actually a complete instrument channel starting with the parameter transducer. Thus, the reactor water level trip will have a single set of nozzles and manifolds on the reactor vessel, but the manifolds will have four differential pressure liquid-level transducers, each with its own instrument channel ending in a trip contact in the protection circuitry. Each individual sensing channel can be tested during reactor operation by tripping the channel without scramming the reactor because of the "one-out-of-two taken twice" design.

3.9 Plant Loading

The plants designed today are normally base-loaded and are not intended to follow load for the first twenty years. After that, it is expected that newer plants will produce power at lower cost and so the units built now will follow the load. The load changes that the plant can accept without tripping are:

(1) A 20-percent reduction of initial power in the first minute after the reduction starts
(2) A positive or negative ramp change of 1 percent of full power per minute at any power level.

3.10 Refueling

The boiling water reactor is refueled once a year in an operation scheduled to take from 12 to 18 days, depending on the reactor size. Normal fuel management replaces 1/4 to 1/3 of the core annually, and moves 25 percent of the core to different locations. (See Chapter 1 for the considerations involved in fuel management.)

The sequence of steps in refueling a BWR is as follows:

1. Shut down from power.
2. Depressurize the vessel by blowing down, then fill the complete vessel with condensate to increase the rate of cooling.
3. Simultaneously with Step 2, purge the interior of the drywell and the pressure-suppression

chamber. Remove the concrete shielding above the top of the drywell.
4. Remove the top cover of the drywell, then install the drywell flange surface protector.
5. Lower the water level inside the reactor pressure vessel to just below the parting flange surface.
6. Unbolt the pressure vessel head and remove it. Install flange surface protector. Both the drywell head and the pressure-vessel head are stored in designated areas on the refueling floor.
7. Remove the steam dryers and store them in the dryer-separator storage pit.
8. Fill the dryer-separator storage pit and the reactor refueling well with condensate and equalize the level with the fuel storage pool. Remove the gate between the fuel storage pool and the reactor refueling well. The procedure, up to this point, is estimated to take 68 hours.
9. Remove fuel, shuffle fuel, seek out failed assemblies if necessary, install new fuel, service control rods and in-core flux monitors as required.
10. Check the core, verify physics calculations, and run required tests. Steps 9 and 10 require from 100 to 200 hours, depending upon the reactor size.
11. Re-install steam separators and close off the fuel storage pool. Drain the reactor refueling well and the separator-dryer storage pit.
12. Decontaminate the reactor refueling well.
13. Install the steam dryers.
14. Remove the vessel flange surface protector and reinstall the vessel head and appurtenances. Hydrostatically test the primary system.
15. Remove the drywell flange surface protector and replace the drywell head. Replace drywell shielding.
16. Start up and check all monitors during the approach to full power. Steps 11 through 16 are estimated to require from 123 to 130 hours.

Fuel is purchased as fuel bundles, described previously. Common practice is to have 25 to 35 percent spare channels. (Irradiated channels are reused. See Fig. 3-3.) Prior to refueling, new fuel bundles are assembled into channels and a fuel assembly is prepared. New material may be assembled by contact operations; assembling new fuel bundles into recycle channels is done remotely, under water—in the fuel storage pool, which has special equipment for the purpose. Actual movement of fuel is directed from the control room by a licensed operator manipulating the fuel grapple, an assistant, and a responsible supervisor. Continuous voice communication is maintained throughout refueling between the control room and the refueling party.

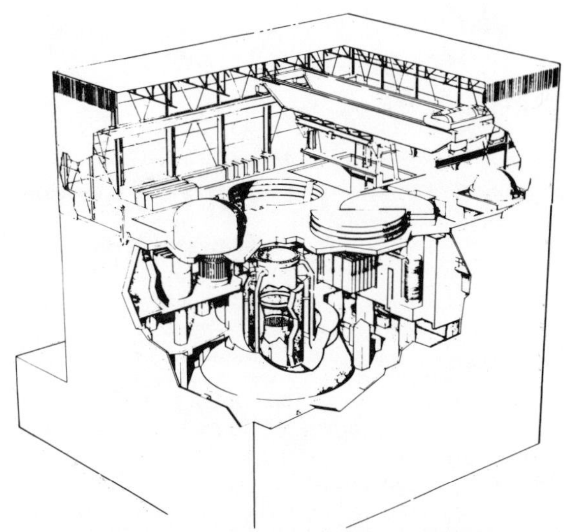

Courtesy, General Electric Co.

Fig. 3-26. Typical BWR reactor refueling area.

A typical view of a plant open for reloading is shown in Fig. 3-26.

Spent fuel bundles are stored in the fuel storage pool to permit their activity and heat production to decay to acceptable levels before shipping them out to a reprocessing plant.

3.11 Containment

The BWR has always been built with a single containment concept—a relatively small drywell mounted above and connected to a pressure suppression pool. Figures 5-1 and 5-2 illustrate the arrangement. In the concept, steam generated by a primary system failure is released in the drywell which is connected to the pressure suppression pool by a series of downcomers so that the downcomers are under a head of approximately 10 ft of water. Thus, when the steam pressure exceeds the 10 ft of water, the steam is bubbled through the downcomers into the pressure suppression pool where it is condensed, relieving the pressure. The pressure suppression pool is cooled by the RHR system in the containment cooling mode. Design pressure for containment is 50 to 60 psig. Containment volume is one- or two-hundred-thousand cubic feet rather than one- or two-million cubic feet, as in the PWR.

3.12 Radioactive Waste System

Radioactive waste systems were supplied by the reactor vendors in the period 1958 to 1968. After that, the vendors divorced themselves from radioactive waste system supply. At the present time the vendor supplies criteria information only: What is produced? How much? Activity level, etc. A complete discussion on these systems will be found in Chapter 11.

CHAPTER 4

High Temperature Gas Cooled Reactors

4.1 General

This type of reactor is the youngest in the United States, in terms of operating history. Its cousins in England, the gas-cooled Magnox and Advanced Gas Reactor, have accumulated operating histories that match or exceed those of the American water reactors. In this design, heat generated in a carbide-base fuel is transferred to compressed helium (700 psia); the helium, in turn, transfers the heat energy to water in a once-through steam generator. The reactor is reflected and moderated by graphite. The design takes advantage of the thermal characteristics of the helium, carbide, and graphite, to develop temperatures of 1450 degrees F in the coolant, and to produce steam at the modern turbine conditions of 2500 psig, 955 degrees F and 1000 degrees F reheat. The thermal cycle in shown in Fig. 4-1.

This reactor, a converter, is the intermediate step between the uranium fissioning reactors and future breeder reactors. Its fuel is a mixture of uranium-235 and thorium-232. During the reaction, the thorium absorbs a neutron and is transformed to U-233, which takes part in the fissioning process. The next logical step in the development of this plant is a transformation from a closed-cycle system generating steam for a turbine, to an open-cycle system working with a gas turbine. Work is being done in this area in Europe at the present time. It is to be expected that at such time as this becomes technically feasible, it will prove to be more efficient than any thermal cycle in use today.

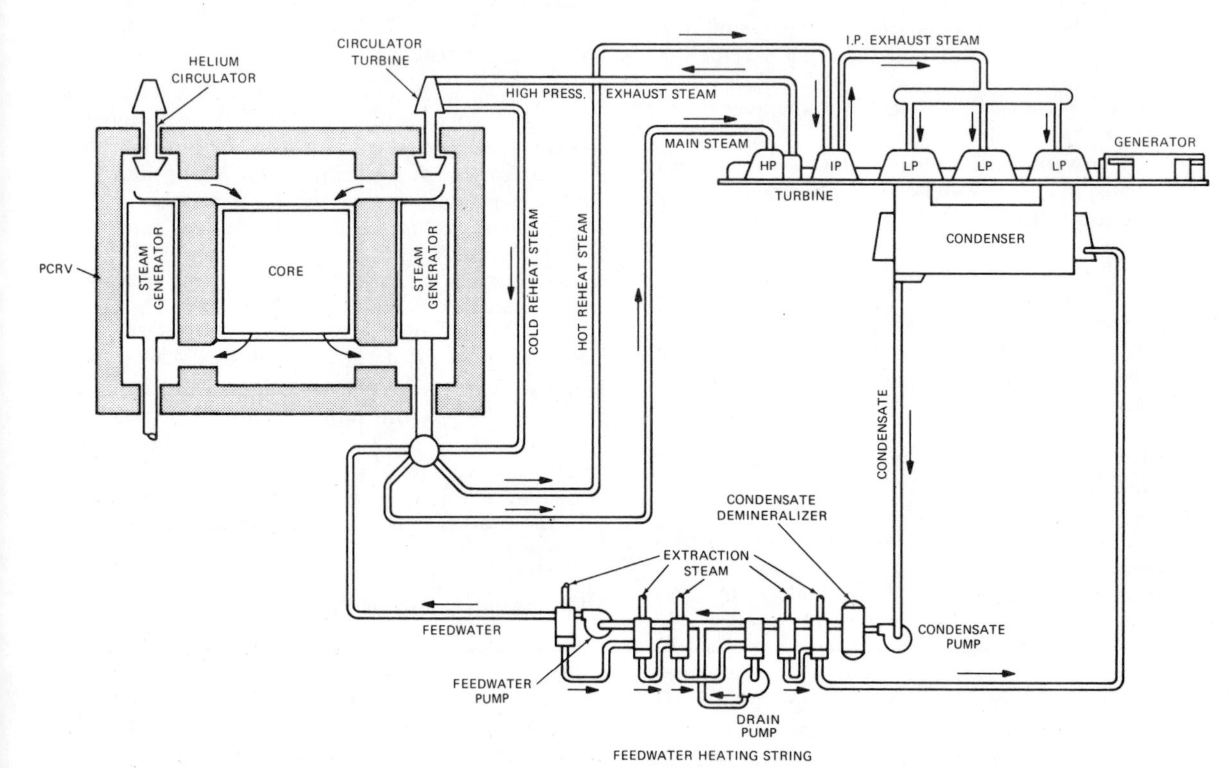

Fig. 4-1. Thermal cycle for the High Temperature Gas Cooled Reactor. The six-flow turbine and five feedwater heaters are illustrative only.

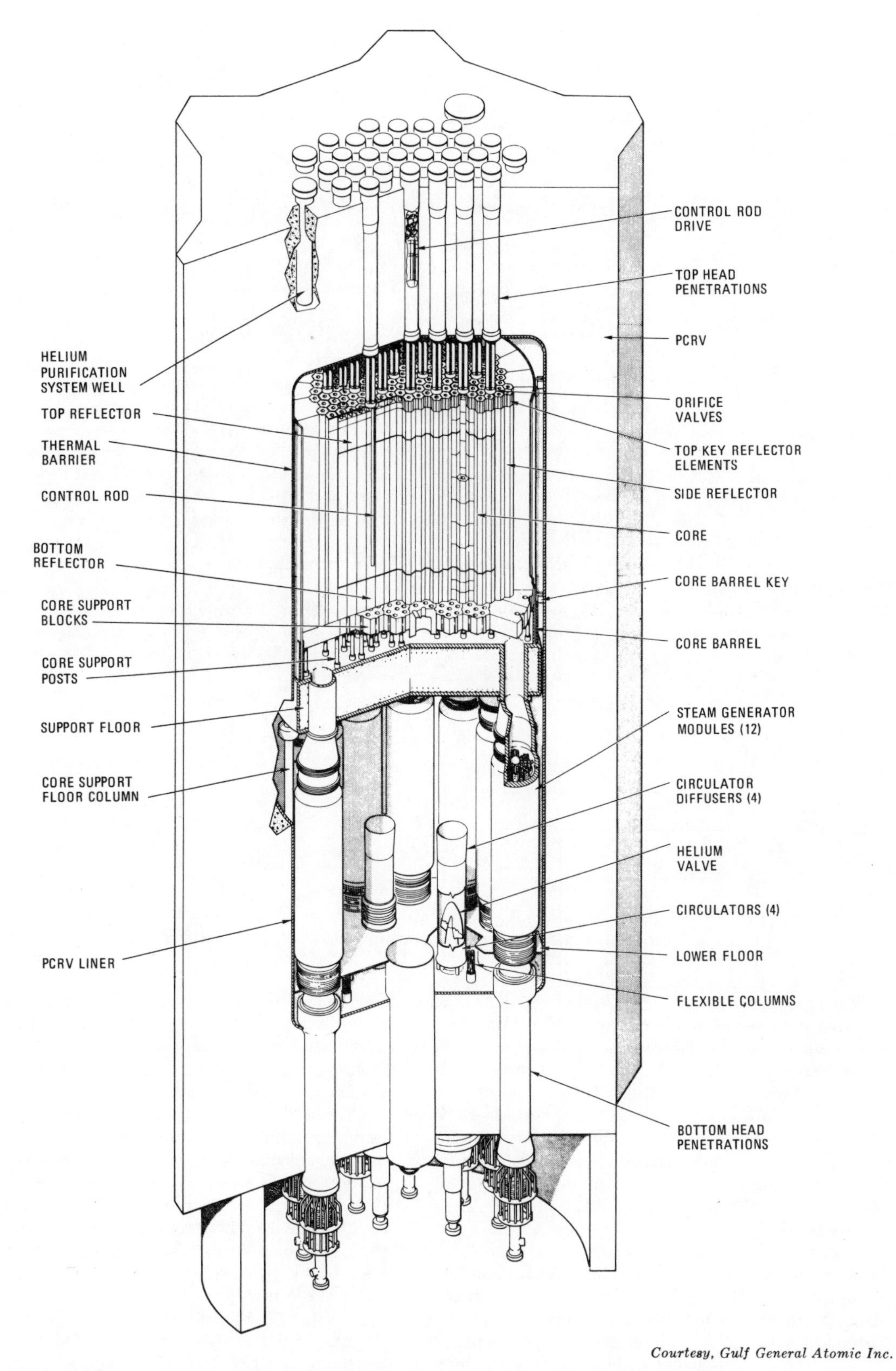

Courtesy, Gulf General Atomic Inc.

Fig. 4-2. General arrangement of the reactor within the PCRV at Fort St. Vrain.

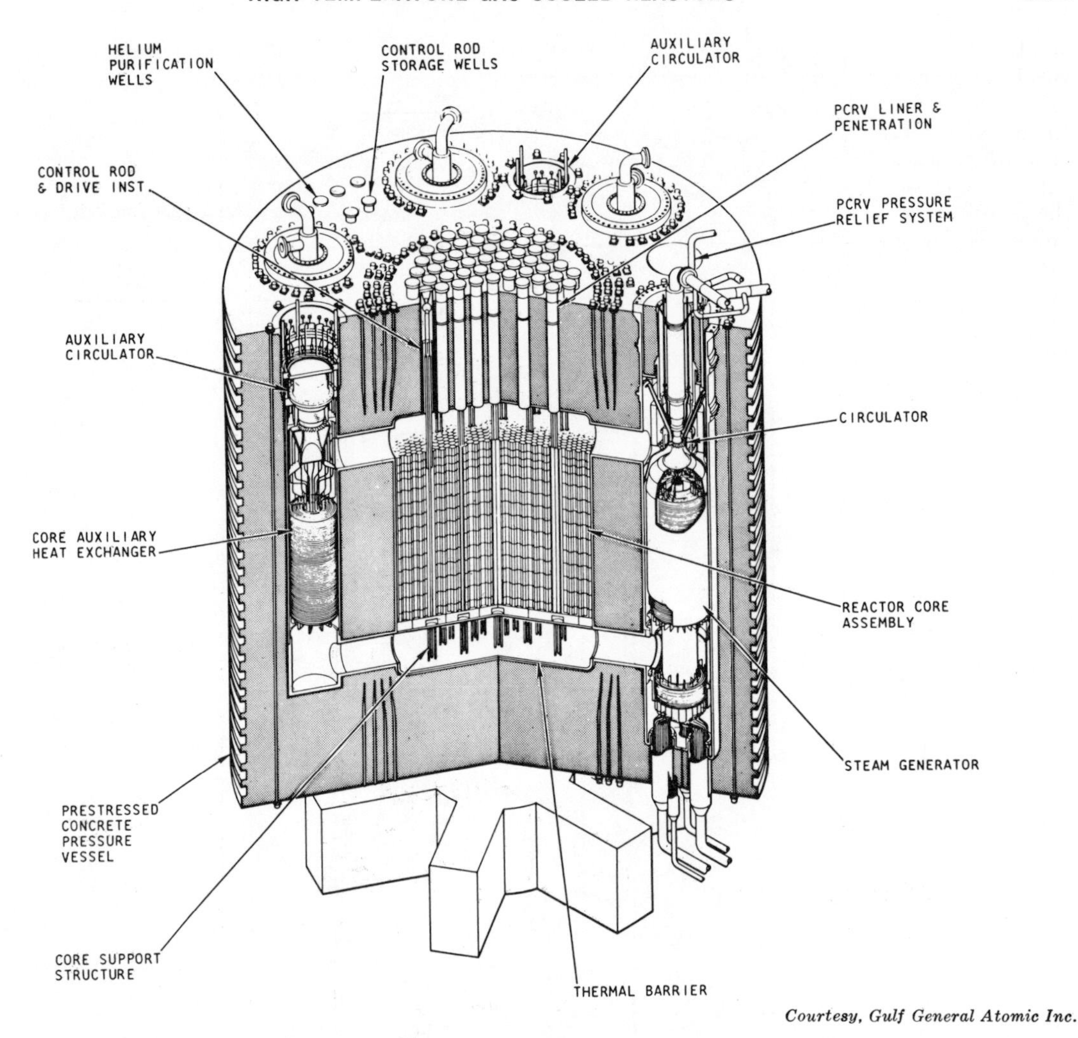

Courtesy, Gulf General Atomic Inc.

Fig. 4-3. General arrangement of the reactor inside the PCRV for the proposed 1100 MWe plant. Note the equipment rearrangement and general design simplification.

This reactor is marketed by Gulf General Atomic, Inc. At the present time there is one 40 MWe prototype plant at Peach Bottom, Pa., that has been operating since June, 1967. A 330 MWe demonstration plant, the Fort St. Vrain Nuclear Generating Station, near Platteville, Colo., is almost finished. Fort St. Vrain thermal rating is 837 MWt from the reactor. Gulf General Atomic, Inc., is presently offering the power industry a reactor plant rated at approximately 1100 MWe and 2783 MWt. The first sale of two units was announced in September, 1971. Overall plant efficiency for these plants runs from 39 to 41 percent, the same as a modern fossil-fuel station, as compared to 31 or 33 percent for a water reactor plant. With the mandatory implementation of the National Environmental Protection Act this increased efficiency of the gas-cooled reactor will assume more and more importance. In the proposal evaluation stage of a project, a capital credit for this efficiency differential will be of the order of five- to fifteen-million dollars, depending on the heat-rejection system required by the site.

One of the most important characteristics of this design is that a complete primary heat system—reactor, circulators, and steam generators—is housed inside a prestressed concrete reactor vessel (PCRV). A steel lining on the inside surface of the vessel cavity forms an impermeable membrane. The complete reactor assembly for Fort St. Vrain is shown in Fig. 4-2. Figure 4-3 is an assembly of the proposed 1100 MWe unit.

In Fig. 4-2 the steam generators and circulators are placed below the core so that the coolant path is relatively complicated. Cold helium flows from the circulators, up through the annulus, between the core barrel and the PCRV liner, turns 180 degrees and flows down through the core, and through the steam generators into a discharge plenum. The helium then passes through the circulator to complete the circuit.

Note the rearrangement of equipment in Fig. 4-3. The steam generators and circulators have been moved up, outboard of the core. The flow path is a simple loop and reactor construction is simpler. Helium flows down through the core and up through the steam generators and circulators. There are six steam generators and circulators in the 1100 MWe unit instead of the four circulators and twelve steam generators in the Fort St. Vrain plant.

4.2 Buildings

The Fort St. Vrain facility has one building making up its nuclear island. This is a multistory building with the prestressed-concrete reactor vessel in the center, surrounded by all of the necessary reactor auxiliary systems, in much the same fashion as the boiling water reactor. A recent ruling by the AEC requires a separate containment building for the PCRV so that the 1100 MWe units will be built with

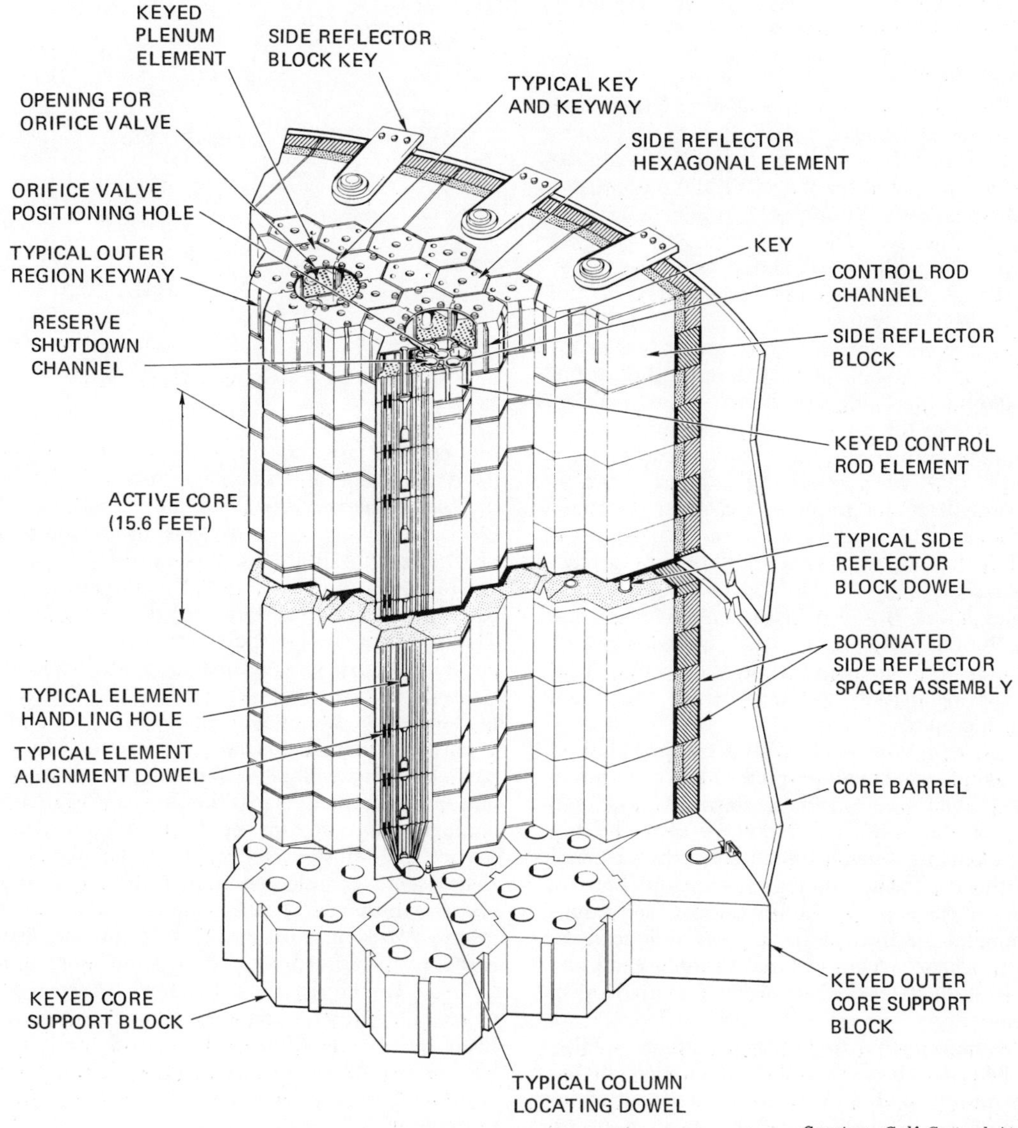

Courtesy, Gulf General Atomic Inc.

Fig. 4-4. Core arrangement for Fort St. Vrain. Compare this with the core arrangement for the 1100 MWe unit in Fig. 4-26.

a reactor containment building and a reactor service building. The reactor service building will house all the auxiliary systems, spent fuel storage facilities and the hot service facility. The control room, relay rack room and cable room, D.C. power supplies, instrument air and control room air-conditioning may be housed in their own building or in a separate wing of a larger structure, whichever is more convenient.

4.3 Safety Classes

This reactor concept has not yet been sufficiently produced commercially to have had any work done in the safety area. As the first two 1100 MWe units progress, the definitions of safety classes and their relationships to existing codes, ought to evolve.

4.4 Fort St. Vrain

All of the system and component descriptions that follow describe the power station at Fort St. Vrain. At the end of the chapter will be found the modifications that are proposed for the 1100 MWe plant.

Control rods enter through the top penetrations, as shown in Fig. 4-2. These same penetrations are used for refueling. Similar blind cavities in the top of the PCRV are used to house components of the coolant purification system.

The graphite reflector-moderator pieces are keyed together in an arrangement that permits sliding between adjacent blocks due to thermal growth differences, but retains the necessary geometry of the core. The complete core is surrounded by a carbon-steel core barrel that gives structural support, controls thermal growth of the graphite, and assists in channeling the gas. The inner face of the PCRV liner is provided with a ceramic heat insulation to protect the carbon-steel liner.

The circulators, installed in the bottom head, are individually removable from their imbedments. A large, center access penetration in the bottom head is used for installation and removal of the steam generator modules.

4.4.1 *Reactor Core*—The core is comprised of an assemblage of machined graphite blocks within a cylindrical, steel core barrel, as shown in Fig. 4-4. All parts of the core are graphite except the top plenum elements which distribute the coolant through the core. The outer sections (top, bottom, and sides) of the core are slightly poisoned to reduce the thermal load in the core barrel, core support floor, and the top liner structure. Flow through each element is controlled by the orifice valve at the top of the column.

Each column has six fuel elements, shown in Fig. 4-5, stacked one above the other. Each element has locating dowels and sockets to insure continuous coolant passages through a complete stack, and each has 210 fuel holes and 108 coolant holes. Some elements have a burnable poison—boron carbide—to assist in reactivity control. Above and below each fuel column, are pure graphite reflector blocks with the same shape and same coolant flow channels as the fuel elements. A wall of pure graphite reflector blocks surrounds the active core inside the poisoned (boronated) graphite, as shown in Fig. 4-4.

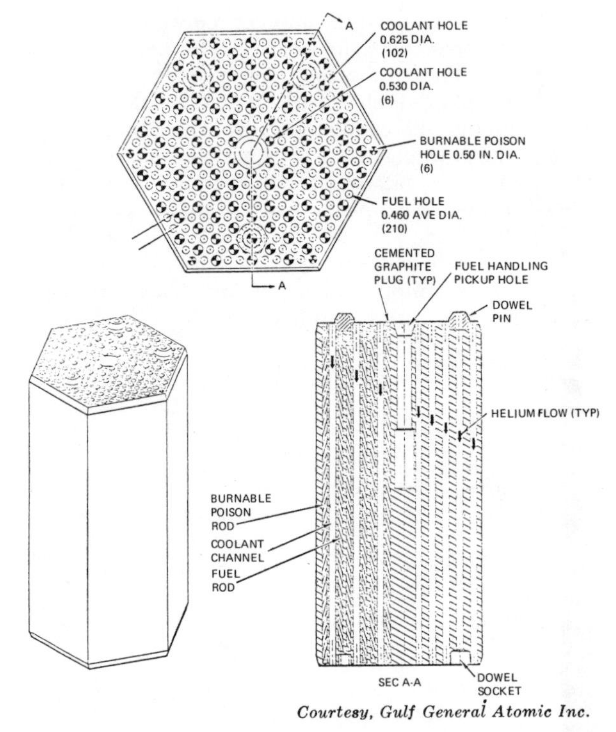

Courtesy, Gulf General Atomic Inc.

Fig. 4-5. Graphite fuel element.

The fuel is a mixture of fully enriched (93.5 percent) uranium-235 carbide and thorium-232 carbide. The individual fuel particles are coated to make them impermeable to gaseous fission products. Fuel is in the form of two types of particles. The first type is a fissile particle, about 200 microns in diameter, with 1 part U-235 and 4.25 parts thorium. The second particle is fertile, about 350 microns in diameter, and has only thorium. Each fuel rod in Fig. 4-5 is a homogeneous mixture of both fissile and fertile particles. In this reactor some of the higher energy (fast) neutrons released from fissioning U-235 convert the thorium-232 to fissionable uranium-233. The uranium-233, in turn, takes part in the fissioning reaction, releasing heat and creating more uranium-233 to continue the chain reaction. At the beginning of element life the chain reaction is supported entirely by uranium-235; at the end of life only about 25 percent of the chain reaction is supported by uranium-235, the rest is supported by fissioning uranium-233.

Fuel is in the core for six years with 1/6 of the core replaced every year. The fuel is not shuffled (moved to a different core location) in this reactor.

Courtesy, Gulf General Atomic Inc.

Fig. 4-6. The control rod drive mechanism of Fort St. Vrain.

After the core has reached equilibrium a typical refueling has the following approximate breakdown:

Fresh U-235 (fully enriched)	4.9%
Recycled U-235	0.1%
Recycled U-233	2.2%
Thorium-232	92.8%

Note the relative independence of U-235, as compared to the light-water reactors.

4.4.2 *Control Rods*—In Fig. 4-4, one can see the control rod element surrounded by fuel elements. Each control rod element has a coolant channel, two holes for control rods and a hole, closed at the bottom, for the reserve shutdown system. The two control rods are driven by a common drive package—a cable and winch arrangement—shown in Fig. 4-6, that scrams with gravity force alone. The control rod, itself, consists of a series of annular cylinders mounted on a metal spline attached to a cable and arranged to permit relative motion between adjacent sections. The rod section consists of two, concentric stainless-steel cylinders capped at both ends; the annulus is filled with a graphite boron composite containing 40 percent boron. The drive packages control the motion of the control rods.

The backup shutdown system consists of a hopper of boron carbide granules over each control assembly. The system operates by breaking a rupture disc with gas pressure, and letting the boron carbide fall into the reserve shutdown hole in the control element. The granules can be recovered afterward by use of a vacuum cleaner.

4.4.3 *Prestressed Concrete Reactor Vessel (PCRV)*—The reactor vessel is constructed at the site of concrete with an internal, mild steel impermeable membrane. The concrete is reinforced with conventional reinforcing bars and high-strength tendons. A cross section of the PCRV is shown in Fig. 4-7. The vessel has three groups of prestressing tendons: Longitudinal tendons, passing from top to bottom, provide longitudinal compression. Circumferential tendons provide hoop compression, and head tendons provide head compression. The tendons are installed in imbedded conduits and are hydraulically tensioned after the concrete is completely cured. The tendons are grouted in place after tensioning.

All components or pipe lines are installed in special imbedments. Each gas circulator imbedment is designed so that the circulator has two seals in tandem and the circulator is removable through the imbedment. The steam generator module (See Fig. 4-8) imbedment does not permit complete removal through the imbedment; however, a large access penetration is provided in the center of the bottom

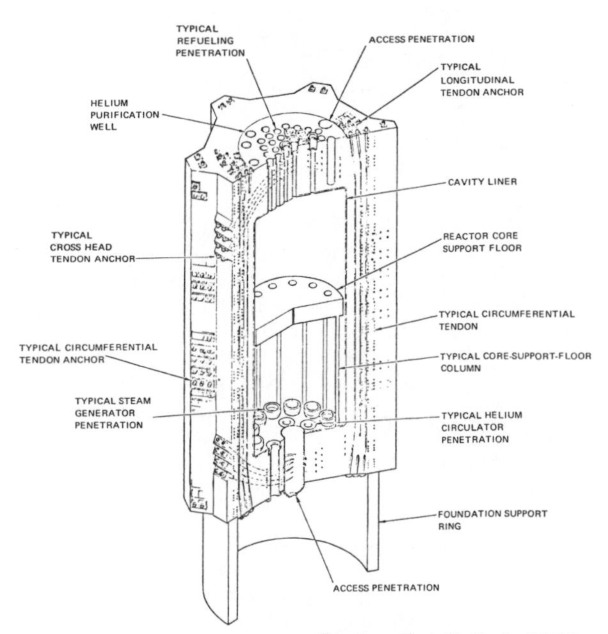

Courtesy, Fort St. Vrain PSAR

Fig. 4-7. Cross section of the prestressed concrete pressure vessel.

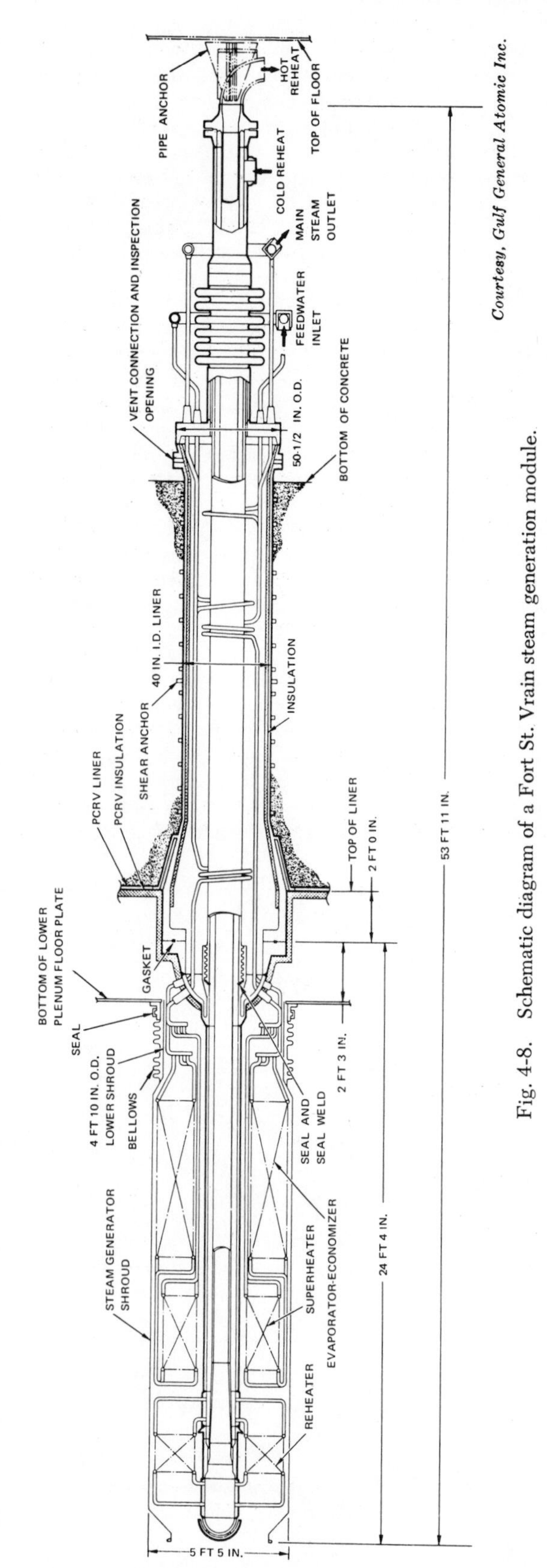

Courtesy, Gulf General Atomic Inc.

Fig. 4-8. Schematic diagram of a Fort St. Vrain steam generation module.

head for removal of a module if it should ever become necessary.

4.5 Primary Cooling

The primary cooling loops are comprised of gas circulators, steam generation modules, and flow passages designed into the reactor assembly, as shown in Fig. 4-2. In the current 330 MWe Fort St. Vrain unit, cooling is divided into two loops, each with six steam generation modules and two gas circulators. The six steam generation modules are grouped together and called one steam generator. System pressure varies from 700 psia at the circulator discharge, down to 686 psia at circulator suction. System temperature-rise averages 700 degrees F through the core —from 750 degrees F to 1450 degrees F. System pressure is constant and load following is achieved by variation of both temperature rise and mass flow-rate, as subsequently described in the control sequence. The total helium inventory in the cooling cycle is about 7400 lbs. A comparison of the energy available for release in an accident in this gas system, with the example in Chapter 5, indicates that there is approximately one-fifteenth the energy available here.

4.5.1 *Steam Generation Modules*—Each steam generation module, shown in Fig. 4-8, has an evaporator-economizer-superheater section and a reheater section. Steam conditions at the superheater outlet are 2400 psig and 1000 degrees F; reheater outlet conditions are 585 psig and 1000 degrees F. The tube bundles of the steam generator are made of helically wound tubes; a single, continuous tube runs from the feedwater inlet header through the economizer, evaporator, and superheating sections, in turn. The reheater section is a series of tubes fastened into two concentric headers. Hot helium flows through the generator entering at the reheater end and leaving at the economizer end. Steam and water flow countercurrent to the helium through the economizer, evaporator, and reheater sections, but co-current through the superheater section. Details of these sections are shown in Fig. 4-9.

Pressure in the reheater is lower than primary coolant pressure, while pressure in the main steam portion of the module is higher than that of the primary coolant. Materials of construction are carbon steel, low-alloy steel, or Incoloy 800 depending on operating temperature.

4.5.2 *Gas Circulators*—Gas circulators are single stage axial compressors driven by a steam turbine, as shown in Fig. 4-10. Exhaust steam from the high-pressure turbine drives the circulator before entering the reheater. Emergency power is provided by a water-driven pony turbine on the same shaft, or by the main turbine, with steam from an auxiliary boiler. The circulator has two, tandem, labyrinth-shaft seals with purified buffer helium injected between the two seals. Clean helium injection insures that any leaking helium will not be contaminated. Bearings are a

Courtesy, Gulf General Atomic Inc.

Fig. 4-9. Details of a steam generation module showing the economizer, evaporator, and superheater sections.

hydraulic hybrid type with both hydrostatic and hydrodynamic capability. Two of the four circulators are used for shutdown cooling. Circulator speed varies during shutdown with gas density. As the system pressure decreases, circulator speed increases to maintain adequate mass flow until a maximum of 10,200 rpm is reached while refueling. Each circulator has a shut-off valve to prevent backflow, as shown in Fig. 4-11. Figure 4-12 shows a full-size Fort St. Vrain prototype gas circulator being installed at a test facility.

4.6 Auxiliary Systems

4.6.1 *PCRV Cooling*—The PCRV concrete requires protection against thermal damage from reactor heat; its temperature must not exceed 150 degrees F. To insure this, cooling tubes, divided into two systems, are fastened to the concrete side of the steel membrane by continuous fillet welds. Each of the systems is capable alone of controlling concrete temperature within safe limits. The systems are arranged so that adjacent tubes do not belong to the same system and a tube failure will not endanger the concrete. A typical arrangement is shown in Fig. 4-13. Both of these systems are provided with demineralized water by the reactor plant closed loop cooling system described a little later. The PCRV cooling systems are backed up by the plant fire-water system. Provision is made to heat the water should it become necessary to minimize concrete temperature variation during a prolonged shutdown.

4.6.2 *Steam-Water Dump System*—Each steam generator has its own dump system, as shown in Fig. 4-14. The sequence of events in a dump is: First, to close off the feedwater supply while simultaneously tripping the reactor; then second, to open both parallel dump valves. Each valve has sufficient capacity for the procedure; the second valve is provided as a backup to the first. Each valve is powered from dif-

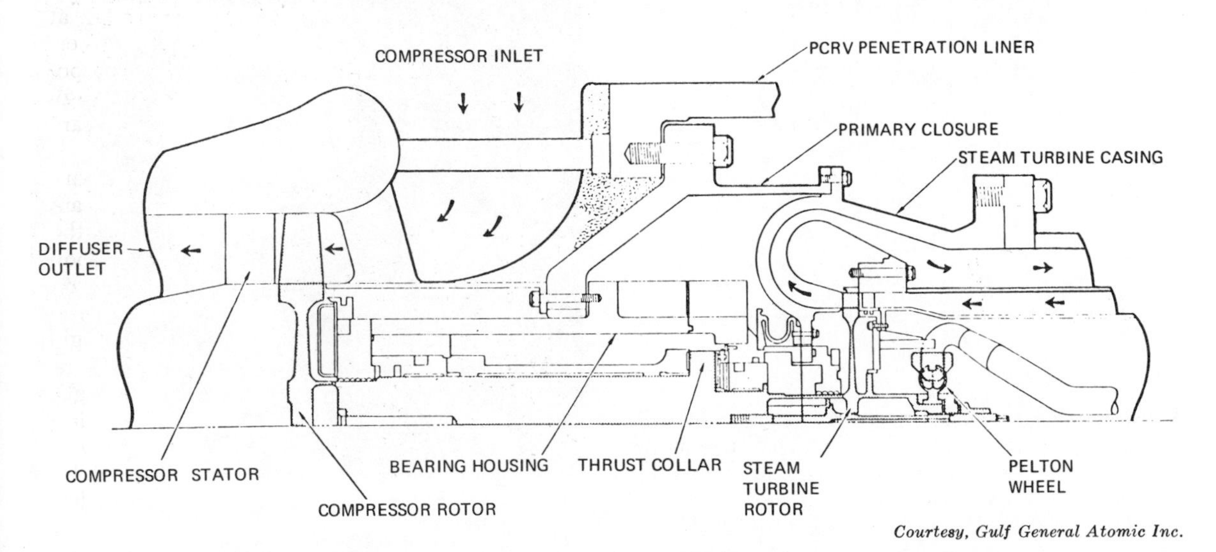

Courtesy, Gulf General Atomic Inc.

Fig. 4-10. Fort St. Vrain gas circulator showing both the main steam-drive turbine and the emergency water-drive turbine.

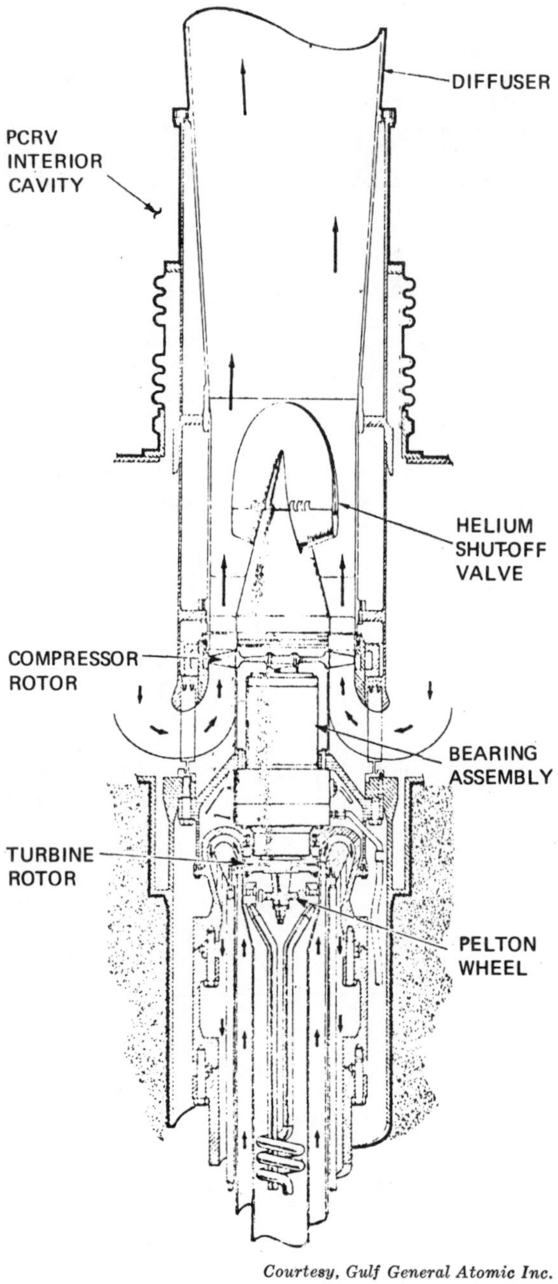

DIFFUSER
PCRV
INTERIOR
CAVITY
HELIUM
SHUT-OFF
VALVE
COMPRESSOR
ROTOR
BEARING
ASSEMBLY
TURBINE
ROTOR
PELTON
WHEEL

Courtesy, Gulf General Atomic Inc.

Fig. 4-11. Schematic arrangement of the Fort St. Vrain gas circulator.

ferent sources so that power failure in one valve will not affect the second valve. If the reactor trip-dump was initiated by high moisture, the intact steam-generator loop will continue operating for reactor cooldown. If the trip was "high pressure" or "reactor high-moisture" initiated, the reactor operator will restart the intact steam-generator loop for cooldown, as soon as it is identified. The heat energy produced by the core during the short "no flow" period can safely be absorbed by the reactor materials.

4.6.3 *Shutdown and Emergency Cooling*—Shutdown and emergency cooling are handled by the main cooling-system components. One gas circulator can circulate sufficient coolant to remove decay heat although two are used. One economizer-evaporator-superheater or one flooded reheater can transfer the required decay-heat load. Each helium circulator can be driven by either the steam turbine or the water turbine. Feedwater is available either from electric-motor-driven feed pumps or a steam-turbine-driven feed pump. The motor-driven feed pumps supply driving water for the circulator turbine in addition to feedwater. Any time backup electric power is not available for the feed pumps, an auxiliary boiler is started and placed on standby until such time as backup electric power is restored. The auxiliary boiler has the capability of supplying both the feed pump and the gas circulator. Additionally, the heat capacity of the reactor core itself can absorb up to four hours' worth of decay heat energy without exceeding normal fuel operating temperatures.

By using main system components, or components run at regular intervals during normal operation, there is a greater degree of certainty of system opera-

Courtesy, Gulf General Atomic Inc.

Fig. 4-12. A full-size Fort St. Vrain prototype gas circulator being installed at a test facility at Valmont Power Station. All four circulators will be tested at this facility.

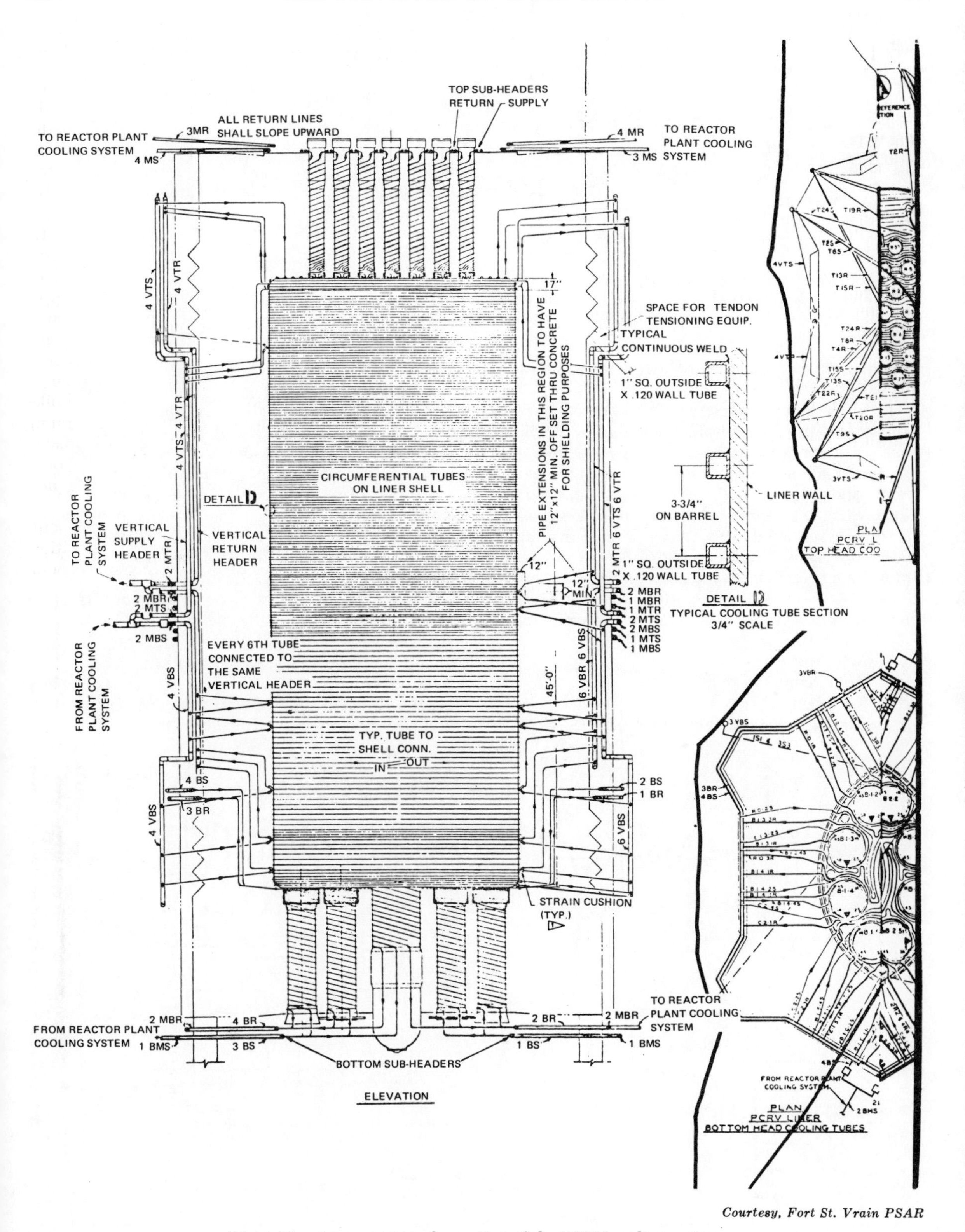

Courtesy, Fort St. Vrain PSAR

Fig. 4-13. Arrangement of a portion of the PCRV cooling system.

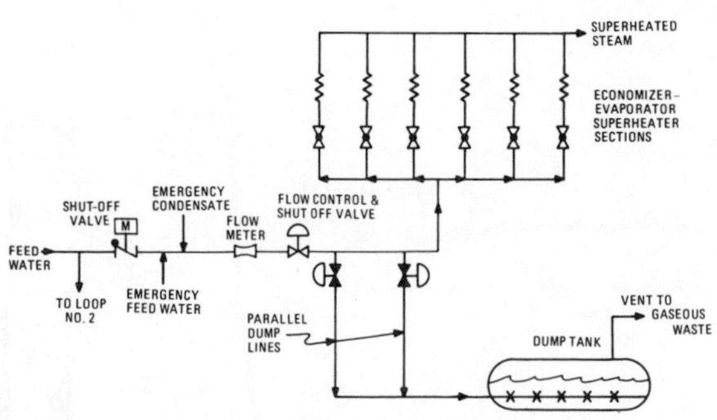

Fig. 4-14. Schematic diagram for the steam water dump system for one loop. Fort St. Vrain can dump both loops; the 1100 MWe design can dump only two of six loops.

tion than if the system is only operated on test at regular intervals.

4.6.4 *Helium Purification*—The helium purification system provides purified helium for circulator seals and the top PCRV seals. It purifies helium after the initial fill at plant commissioning, before it is pumped to storage, and after a boiler-tube failure. The system also collects active impurities passing through it, although this is not one of its design objectives. The system has two, full-size processing trains; one is operating, while the other is in activity decay, regeneration, or standby. The process, as shown in Fig. 4-15, consists of the following steps:

1. Exit from process
2. High temperature filter/adsorption
3. Gas cooling
4. Drying (molecular sieve)
5. Cooling to cryogenic temperatures (−295°F)
6. Low-temperature adsorption (−320°F, charcoal)
7. Regenerative heating with gas of Step 5
8. Post filtering
9. Heating
10. Hydrogen and tritium removal (titanium sponge, +720°F)
11. Regenerative cooling with gas of Step 9
12. Filtering
13. Cooling
14. Return to process.

A purification train operates for six months, after which it is taken out of service for regeneration, and the standby train is placed on stream. The contaminated train is held for two months to permit decay of active impurities before regenerating. During this period the low-temperature adsorber is continuously held at −320 degrees F to hold its sorbed contaminants and remove isotopic heat. At the end of the two months, the train is regenerated in a day and put on standby until needed. The single hydrogen removal train is on stream for about a month and is then also regenerated in a day. In regeneration, all impurities are passed into the gaseous waste system for final processing and disposition.

All the equipment is Class C, Section III of the 1968 ASME Code, with the same pressure rating, 845 psig, as the PCRV. Valves are Class II of the Pump and Valve Code, and piping is Class II of ANSI B31.7. Materials can be carbon steel except for the cryogenic applications which can be austenitic stainless steel or 9 percent nickel steel.

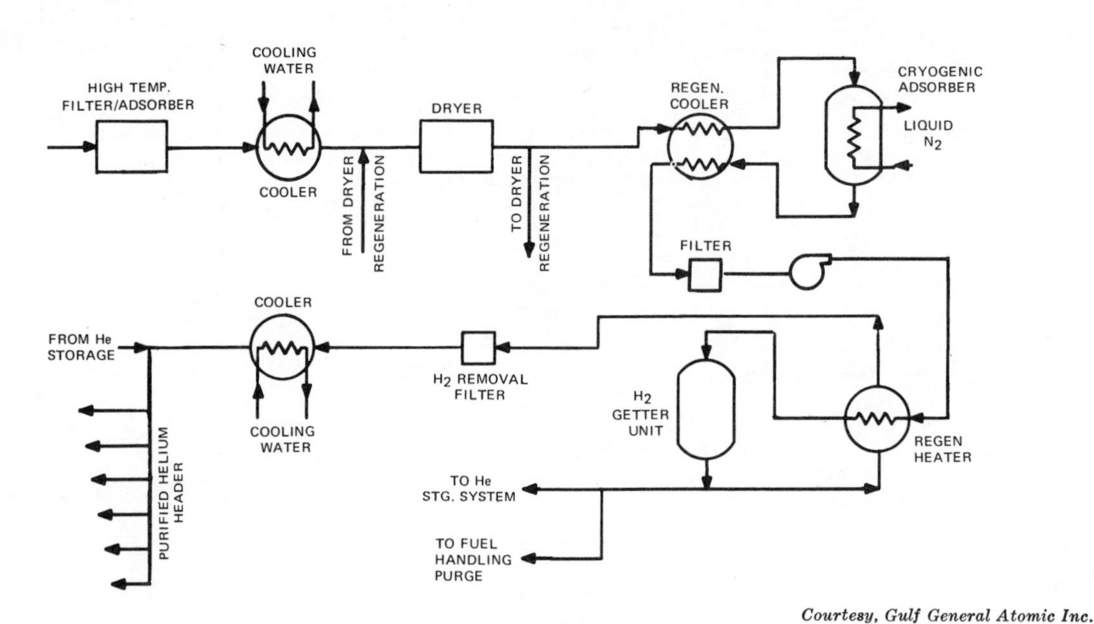

Courtesy, Gulf General Atomic Inc.

Fig. 4-15. Simplified schematic of the helium purification and hydrogen removal system.

The purification and hydrogen-removal equipment is installed in penetrations or cavities in the top of the PCRV. This minimizes pipe runs and provides physical protection for the equipment. Pipelines leaving and entering the wells have isolation valves at the wells. These valves fulfill the same function as the containment isolation valves in a water reactor.

The plant is designed to deliver a constant volume of gas regardless of system pressure. Since the plant can operate from atmospheric pressure up to full helium-system pressure, mass flow varies with pressure.

4.6.5 *Helium Storage*—The helium storage system provides storage volume for plant helium inventory during depressurization, furnishes a supply of high-pressure purified helium for plant use, and provides means of pumping down the primary system as well as pumping it up. The system schematic is shown in Fig. 4-16. During normal operation the storage tanks are "empty" and the pressure tanks are at 1100 psig. All helium enters the storage system either from a trailer or through the purification system; thus no activity ever enters the system. All bulk transfers between the storage system and the primary system are done in two distinct steps. The first step is pressure equalization at a controlled flow rate followed by pumping with the transfer compressors. Depressurizing the primary system down to 12 psia for refueling takes about 12 hours; repressurizing takes about half the time. To avoid the possibility of an accident (pressure excursion) during refueling, the high-pressure supply tanks are bled into the primary system during depressurization.

The compressors are multistage reciprocating units with an oil adsorber downstream. System construction is welded carbon steel. Vessels conform to Section VIII of the ASME Code and valves and piping conform to ANSI B31.1.0.

4.6.6 *Liquid Nitrogen System*—The liquid nitrogen system is provided in the plant to furnish the −320 degrees F temperature required by the low-temperature charcoal adsorbers in the helium purification system and a source of cold, dry gas for use in the primary coolant moisture monitors. The system is a simple one consisting of nitrogen recondensers (compression units plus condensers), storage vessels, and an interconnecting piping system. Although the system has three recondenser units (shown in Fig. 4-17) whose combined capacity is 125 percent of

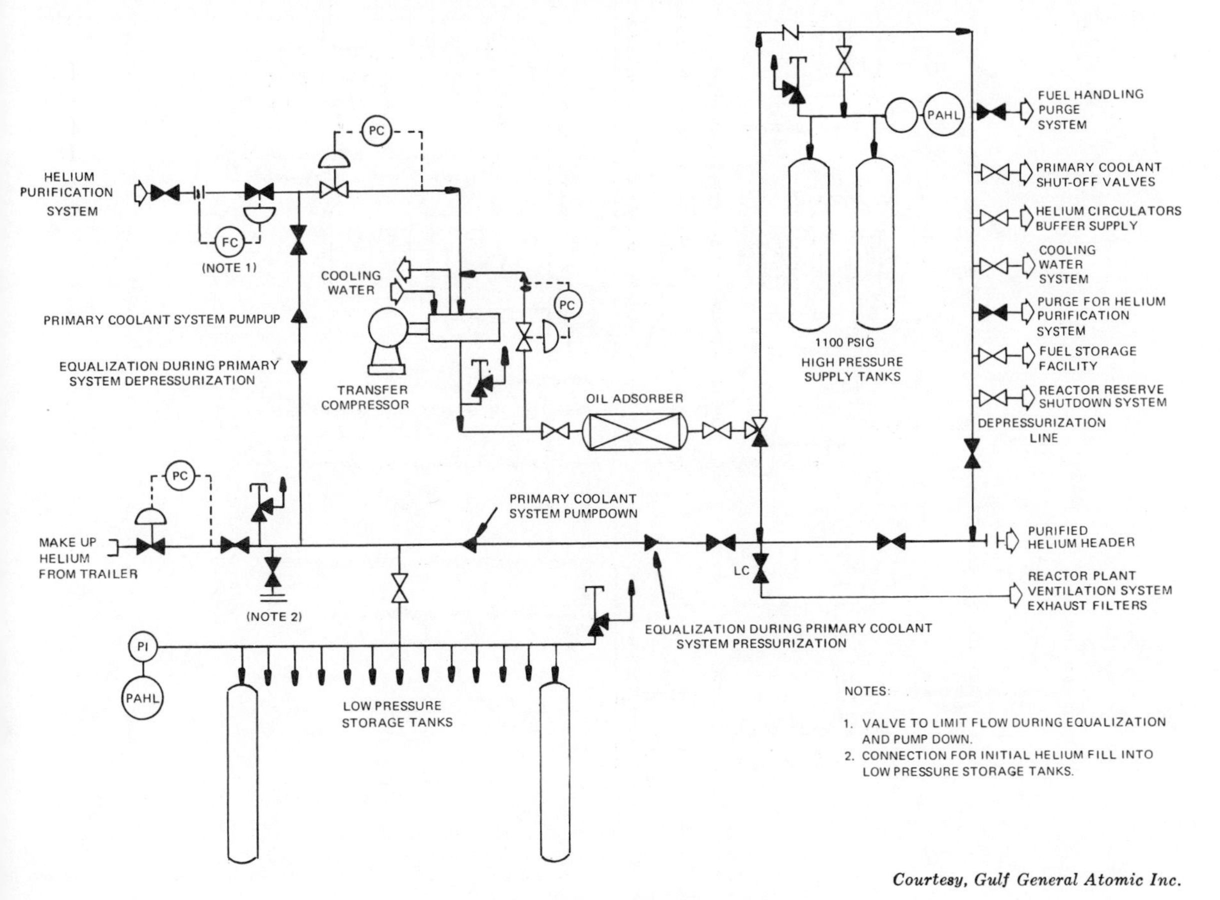

Courtesy, Gulf General Atomic Inc.

Fig. 4-16. Helium storage system.

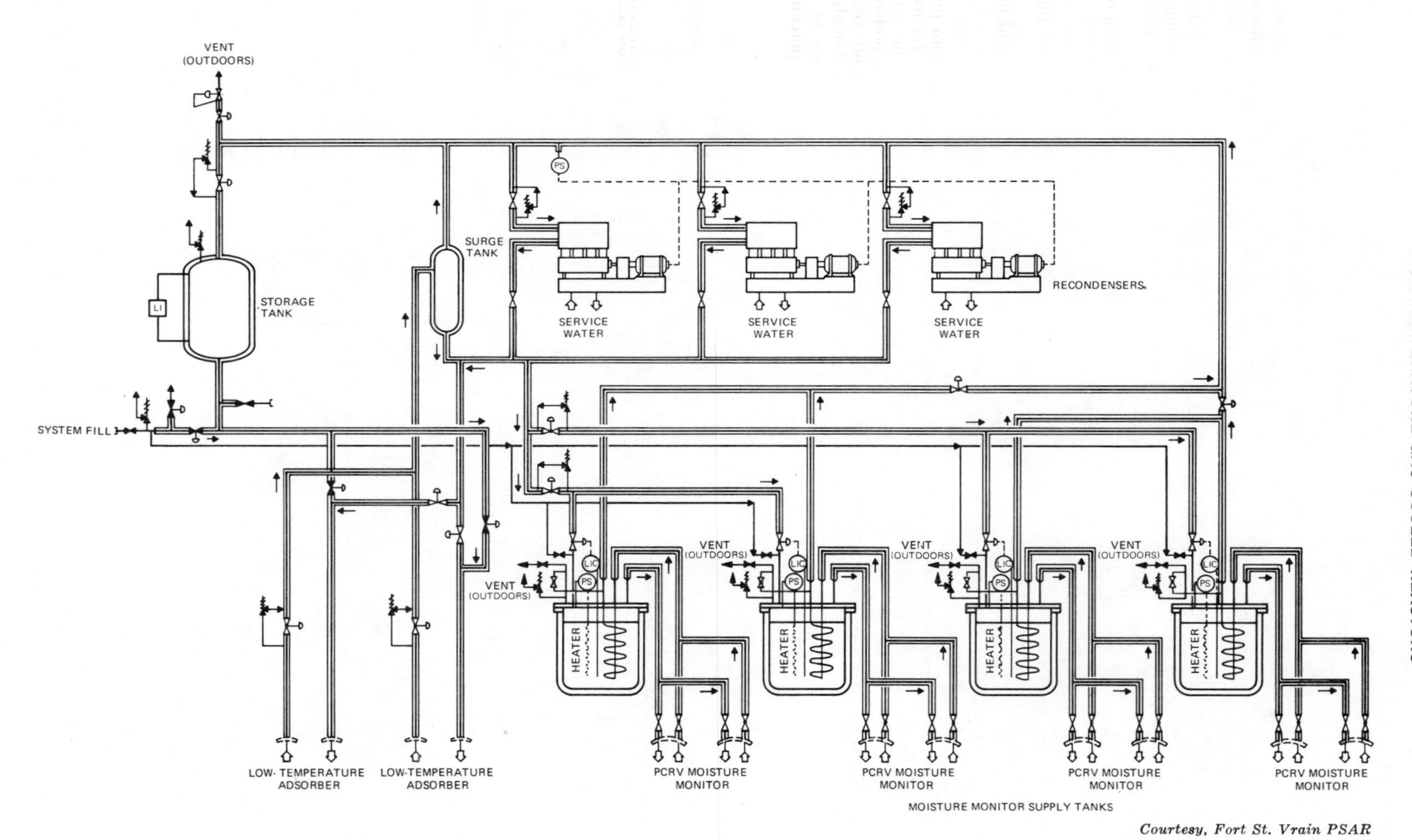

Courtesy, Fort St. Vrain PSAR

Fig. 4-17. Schematic diagram of the liquid-nitrogen system.

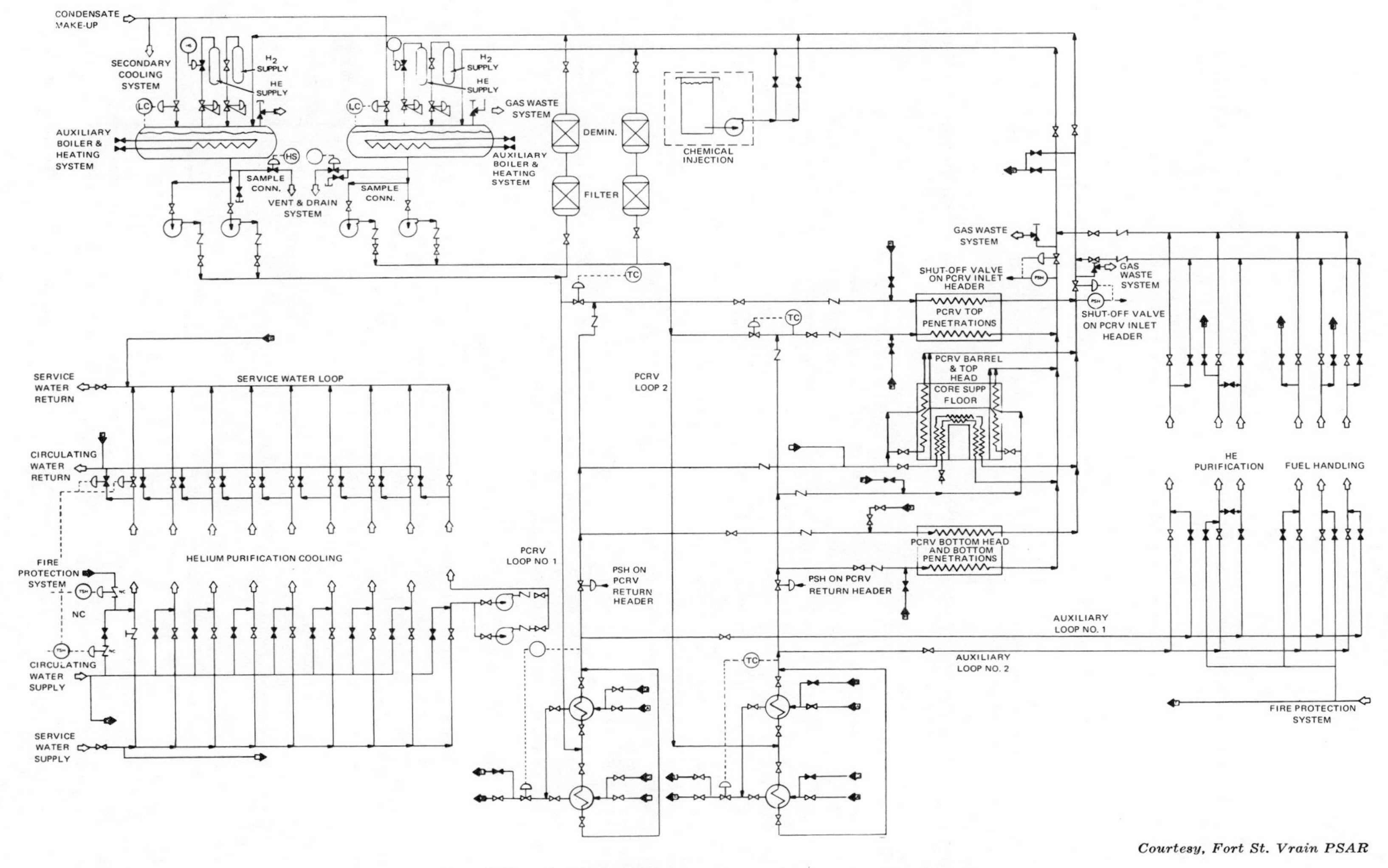

Fig. 4-18. Auxiliary cooling water systems in a reactor plant.

requirements, a liquid-nitrogen storage tank with some 25 hours-of-refrigeration capacity is provided for emergency service. In this time period, additional liquid nitrogen can be trucked in to keep the plant operational.

Operation is straightforward. The recondensers and tanks are on the operating floor so that all liquid flow is by gravity. Cold gas with liquid droplets returns from the adsorbers to the recondensers through a surge tank which removes the liquid. The gas is recondensed, and on exiting from the recondensers joins with the liquid from the surge tank. The liquid nitrogen then flows back to the adsorbers to be re-evaporated.

The moisture-monitor supply tanks provide cold gas to the monitors. The warmed gas returns through the coils in the supply tanks for precooling, and then passes to the suction of the recondensers. Liquid level in each supply tank is controlled automatically by a level controller.

Standard cryogenic construction is used in the system. All equipment and piping use vacuum jacket insulation. Any equipment and/or piping which can be isolated between two closed valves is equipped with a relief valve to relieve the pressure that develops as the trapped liquid nitrogen evaporates. This is shown in Fig. 4-17.

4.6.7 *Circulator Auxiliary System*—The circulator auxiliary system supplies buffered helium for injection, bearing water and emergency drive water for the water turbine. Helium is injected into the zone between the seals, and flow splits evenly; fifty percent leaks into the primary system and fifty percent leaks out and is drained as a water-gas mixture with bearing water. The wet gas is separated in a high-pressure separator, recompressed in a water piston compressor, dried to −100 degrees F dew point, and recycled to the circulator. Make-up gas is furnished to the buffer helium circuit from the helium purification system.

Bearing water is condensate and the inventory is supplied from the feedwater system. The water is cooled, filtered, and injected into the bearings by two separate systems. Bearing drains and separated water from the helium-water drain are collected and recycled.

All major equipment in both the buffer helium subsystem and the bearing water subsystem—pumps, compressors, heat exchangers, and filters—are duplicated.

Bearing water circulating pumps and buffer helium recirculators are on two separate emergency buses to ensure that failure of a single power supply does not shut down the plant.

Water-turbine power water is supplied by the feedwater system, backed up by the fire system and the condensate system. Backup water for the bearings is supplied by the condensate system.

4.6.8 *Reactor Plant Cooling Water Systems*—Reactor plant cooling water systems, shown in Fig. 4-18, are comprised of three systems; viz:

1. Two closed demineralized-water loops (100 percent capacity each) serving the PCRV, core support structure, helium purification fuel storage, fuel purge, and the liquid nitrogen system.
2. The service water system serving the water turbine coolers, helium transfer compressors, buffer helium and circulator bearing coolers, radwaste pumps and compressors, the closed demineralized-water-loop coolers, and the booster service water system.

Water in the service water system is used on a "once-through" basis and is discharged to the outfall.

Each closed loop system is pumped through two 50 percent coolers, then through the heat loads, in parallel, and back to the system surge tank. The water is subject to radiolytic decomposition as it passes through the PCRV coolant tubes. The tank, therefore, is pressurized with hydrogen to 2 psig, to permit the hydrogen concentration in the water to build up and control dissociation.

Each closed loop has two 100 percent capacity pumps. The PCRV-liner cooling tubes and the fuel storage facility are backed up with water from the fire protection system. All other closed loop heat loads are backed up with water from the circulating system.

All the pumps are supplied from the essential power bus. Piping is in accordance with ANSI B31.1.0 and vessels conform to Section VIII, Division 1, of the 1968 ASME Code.

4.6.9 *Reactor Building Ventilation*—This reactor design does not have a containment building as does

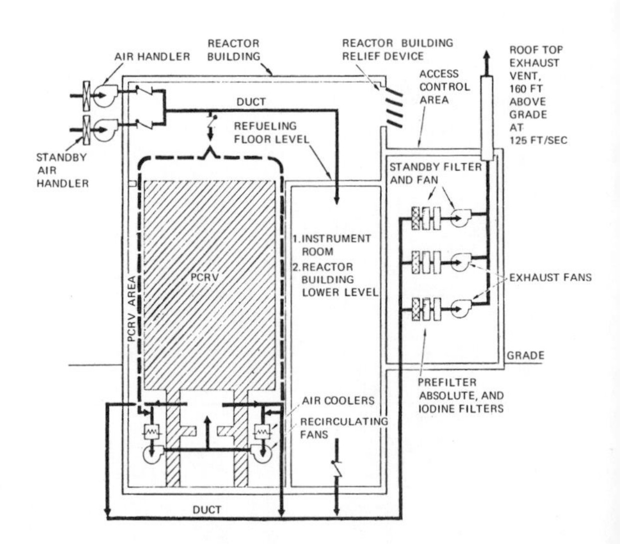

Fig. 4-19. Air-flow paths in a reactor building ventilation system.

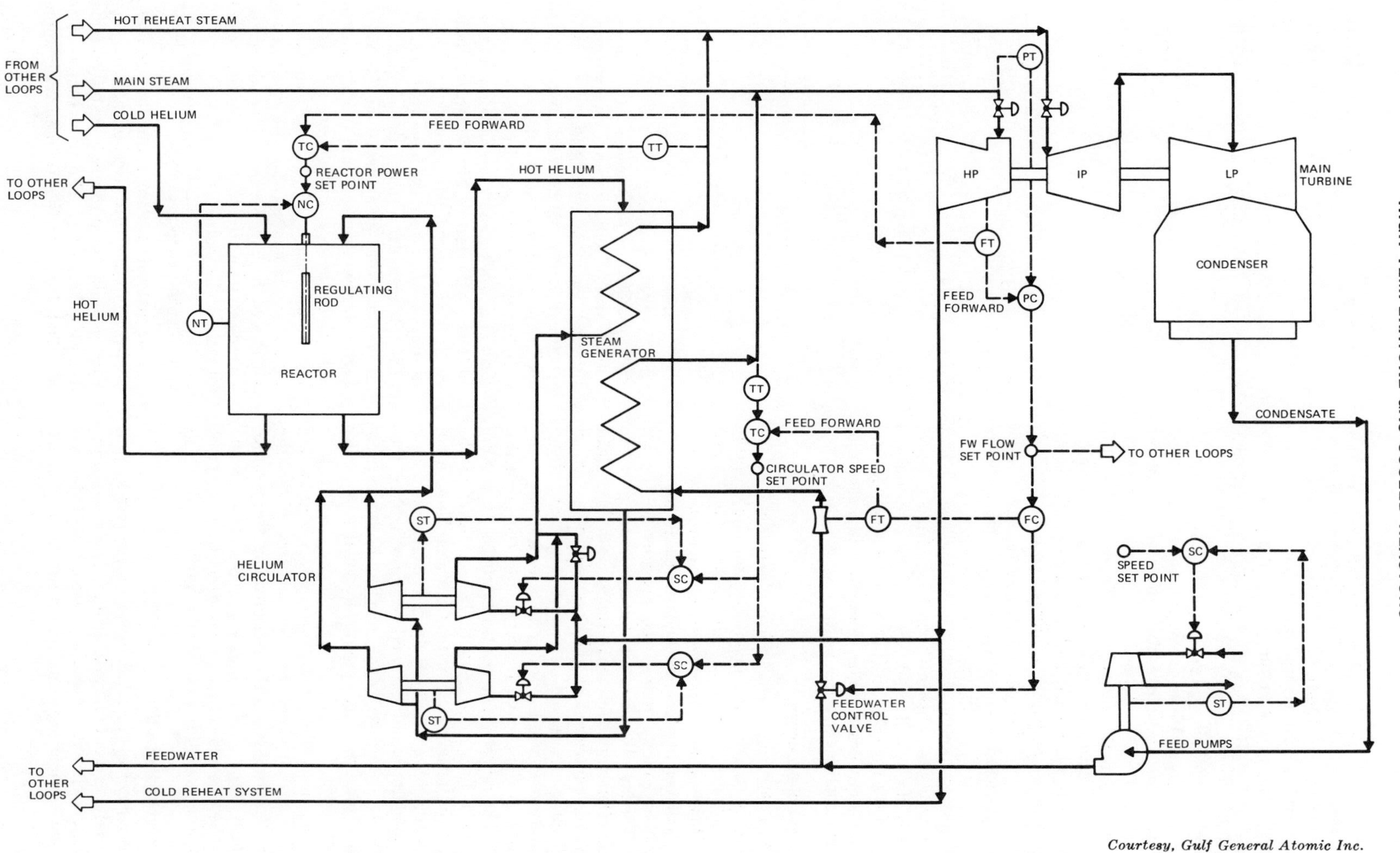

Courtesy, Gulf General Atomic Inc.

Fig. 4-20. Simplified schematic of Fort. St. Vrain control system. The "feed forward" loops are anticipatory signals to reduce overall plant-reaction time.

a water cooled reactor. The building uses conventional materials. It has restricted leakage features through the use of appropriate joint designs and sealing materials. The ventilation system is a "once-through" type with separate supply and exhaust systems, as shown in Fig. 4-19. The supply system has two 100 percent capacity subsystems. The exhaust system has three 50 percent capacity subsystems. Illustrated in Fig. 4-19, all exhaust air is passed through absolute filters and charcoal (iodine) filters before it is released to the atmosphere. Radioactive waste vent gases join the exhaust air downstream of the filters. The exhaust stream leaves the vent at a velocity of 125 ft per second and mixes with the atmosphere. The combination of filtering, velocity, and height is such that at no time are the people of the surrounding area endangered by airborne activity.

Calculations are made to determine the highest temperature attainable in the building from a vessel leak or a steam leak, and all ventilation equipment is designed to operate in that ambient temperature. Fans are supplied from the emergency buses so that they may continue to operate after loss of main power.

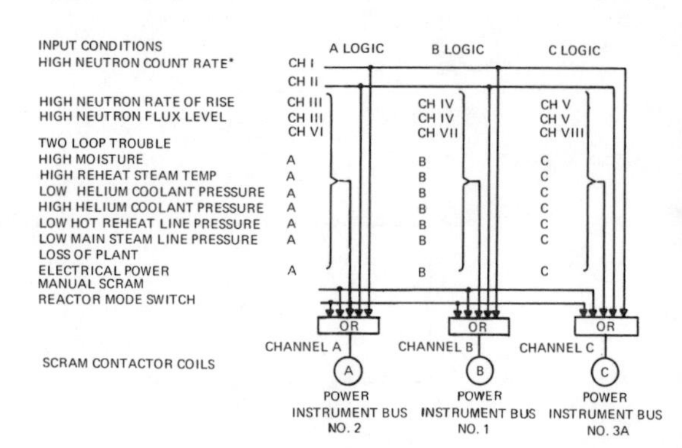

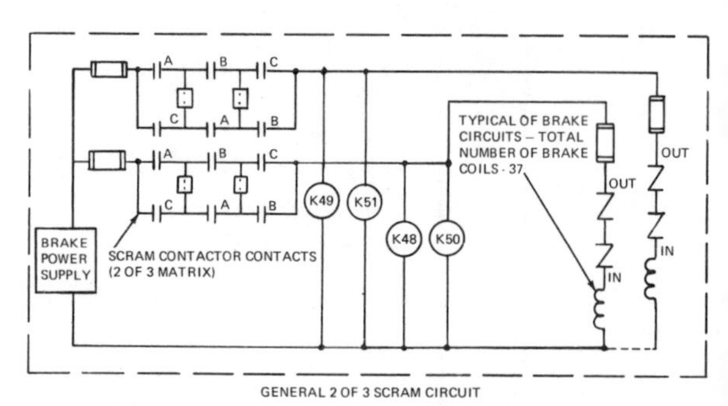

Courtesy, Fort St. Vrain PSAR

Fig. 4-21. Reactor protection circuitry. Top of figure shows parameters having input to the protection system. Bottom of figure shows the scram circuit matrix.

4.7 Control Sequence

In this plant the reactor follows the turbine. A load demand increase signal opens the turbine throttle. As the steam flow increases, the pressure at the exit of the first turbine stage decreases. A pressure controller detects this change and increases the feedwater flow rate to restore the pressure to the set point. The increased flow rate lowers the temperature of both the main superheat steam and the reheat steam; and individual temperature controllers detect the deviation from each set point. The main steam temperature controller reacts to increase the mass flow-rate of the helium by increasing the circulator speed. The reheat steam controller reacts by moving the control rods, increasing the power level of the reactor and raising coolant temperature. Thus, more heat is transferred through the steam generator, and steam conditions are returned to normal. Note that unlike the pressurized water reactor, both temperature and mass flow rate are manipulated in a power level change.

As shown in Fig. 4-20, anticipatory signals are provided to cut down time of response. Both the pressure controller and the reheat-steam temperature controller get their anticipatory signal from steam flow. The main steam temperature controller gets its anticipatory signal from feedwater flow.

4.8 Reactor Protection

The reactor is protected by a fail-safe trip system that uses "two-out-of-three" logic. There are nine parameters that can trip the reactor: high neutron flux, low primary coolant pressure, high primary coolant moisture, high primary coolant pressure, high reheat steam temperature, low superheated steam pressure, low hot reheat header pressure, loss of plant electrical power system, and "two loop coolant trouble" (there are only two coolant loops). Each parameter has three measurement channels terminating in a set of alarm contacts, as shown in Fig. 4-21. As can be seen at the top of Fig. 4-21, one fault in channel A, B, or C will trip the channel. In the lower portion of Fig. 4-21 it can be seen that two of the three channels must trip (open the contacts) in order to de-energize the scram brake on the rod drives. The high neutron count rate and the high neutron rate-of-rise (short period) contacts are active during startup or refueling, and are by-passed at normal power. The reactor mode contacts are the equivalent of an ON-OFF switch. The roman numerals on the neutron channels indicate that there are eight neutron detection channels.

4.9 Coolant Loop Protection

The helium coolant normally operates with less than 1 ppm moisture. Higher moisture levels lead to the possibility of a graphite-water reaction releasing H_2, CO, and CO_2. The only source of moisture is a failed tube in the economizer-evaporator-super-

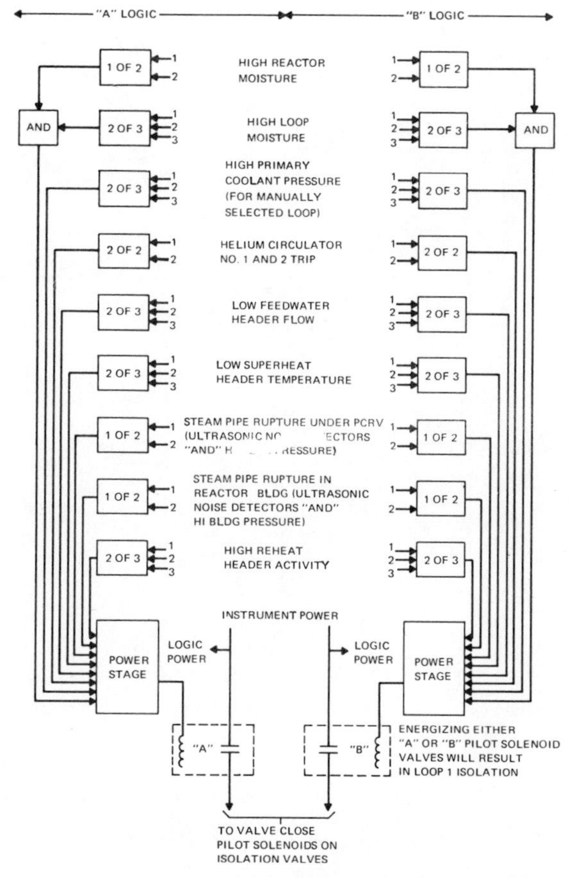

Courtesy, Fort St. Vrain PSAR

Fig. 4-22. Primary coolant loop protection system. The illustration shows one loop only. A single input to either power stage will shut down the loop.

heater section of a steam-generator module. Presence of moisture in the coolant will also increase the total pressure of the coolant and since reheat steam, as was previously noted, is lower in pressure, helium will leak into a failed tube. Hot reheat steam which normally has only a background activity level is continuously monitored; therefore any increase in activity above background gives evidence that there has been a tube failure.

To protect the reactor and the equipment, each loop has its own shutdown system, shown in Fig. 4-22. Shutting down of a single loop represents one-half of the "two-loop trouble" trip described in the reactor protection system. The following parameters may trip at any time the reactor is on:

1. Reactor moisture high
2. Loop moisture high
3. Primary coolant pressure high
4. Circulators tripped
5. Hot reheat header activity high
6. Steampipe rupture (under PCRV or in reactor building)
7. Superheated steam header temperature low

The parameter listed below is activated only when the reactor is at power:

Feedwater header flow low

High moisture level in one loop shuts that loop down and dumps the water into dump tanks. High primary coolant pressure or moisture dumps the water in both loops. These loop trips also shut down the reactor.

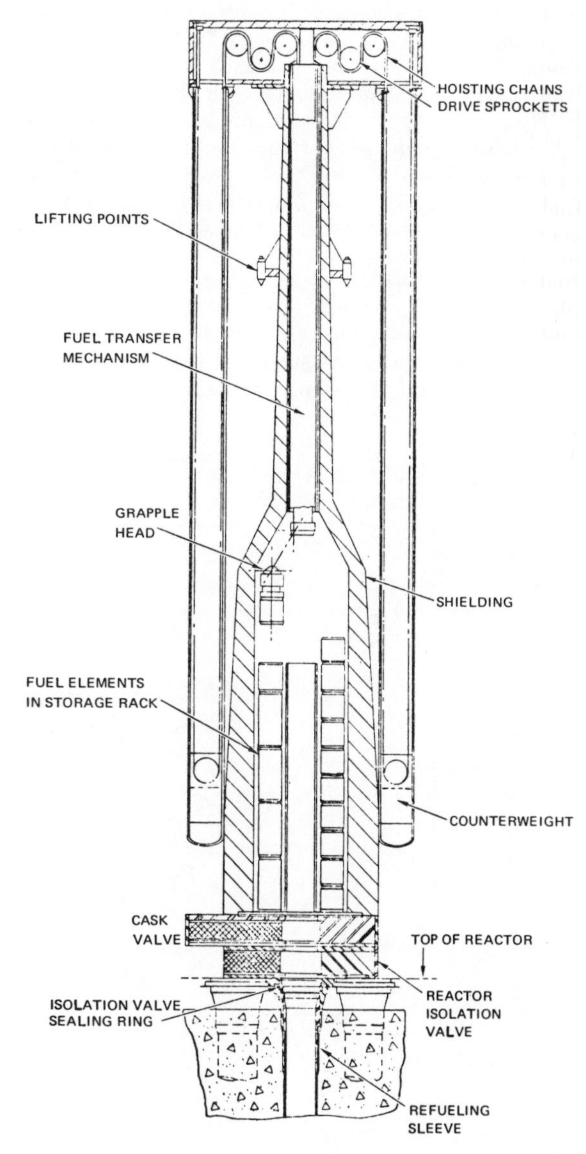

Courtesy, Fort St. Vrain PSAR

Fig. 4-23. Fuel handling machine.

4.10 Plant Load Control

The plant can follow load variation automatically over the range from 25 to 100 percent of full load. Maximum normal rate of load change is 5 percent per minute.

4.11 Fuel Handling

Fuel is manipulated in a fuel handling machine that is placed on top of a special seal valve. Briefly, a control and orifice assembly, including the two control rods out of the control element, is transferred to a special storage well using an auxiliary transfer cask. Spent fuel is removed from the core by the fuel handling machine shown in Fig. 4-23; then new fuel, previously placed in the machine, is installed. The fuel handling machine changes all the fuel in one fuel sector without moving from its position over a control rod element. One fuel sector is comprised of one hexagonal control rod element plus the six fuel elements that surround it. All refueling is accomplished with helium temperature at 250 degrees F and 12 psia.

The fuel is stored in wells, each of which has four water-tight compartments, as shown in Fig. 4-24. The fuel is stored in helium (at 12 psia) in the inner compartment at a temperature less than 750 degrees F. The inner wall has two independent sets of cooling tubes, fed from the booster service water system. Each set of cooling tubes has adequate capacity to guarantee safe fuel cooling. The annular space between the double wall is filled with a granular material which acts as both a biological shield and a heat sink. Heat is removed from the spent fuel elements by convective flow of the helium and is transferred to cooling water in the tubes.

Each compartment, as shown in Fig. 4-24, is equipped with a closure identical to that on the reactor.

The auxiliary cask, shown in Fig. 4-25, handles helium purification equipment as well as taking part in the refueling procedure.

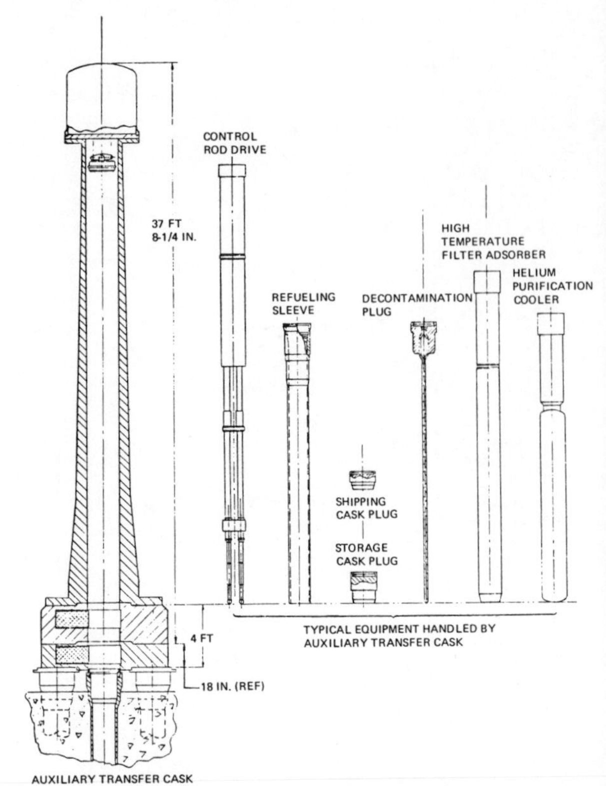

Courtesy, Fort St. Vrain PSAR

Fig. 4-25. Auxiliary transfer cask.

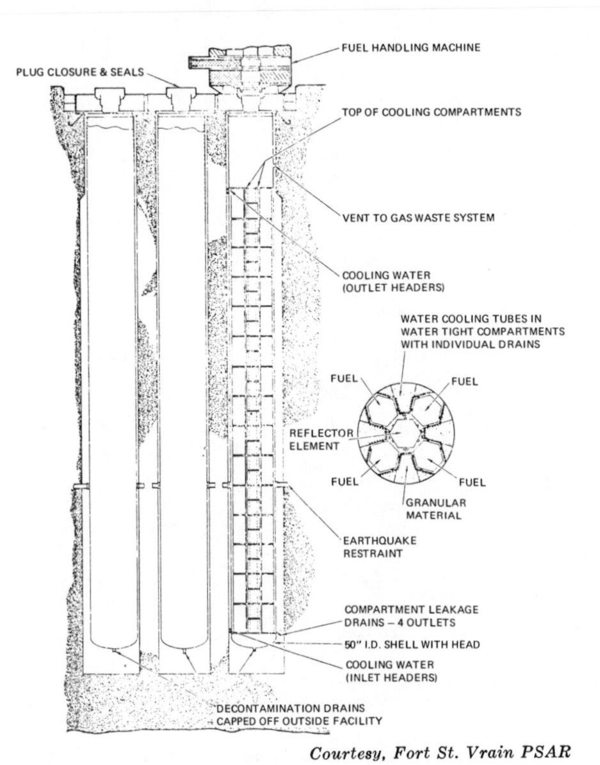

Courtesy, Fort St. Vrain PSAR

Fig. 4-24. Spent fuel storage well.

4.12 Radioactive Waste

Radioactive waste problems with the gas cooled reactor are generally less demanding than with a water-cooled reactor. The primary coolant, helium, is inert and does not produce any corrosion products. The PCRV cooling system has limited exposure to neutron irradiation but does develop some irradiated corrosion products; this system is purified by mixed-bed demineralization which produces a solid waste. Other solid wastes are generated during equipment decontamination and fuel handling in the form of rags, floor paper, rubber gloves and shoe covers, etc. Liquid wastes are produced during decontamination operations or as a result of a failed tube in a steam generator. Gaseous wastes are continuously produced in the plant vent system and the spent fuel purge system, and every six months when a helium purification train is regenerated. The gaseous waste system uses filtration (high efficiency and activated charcoal), compression, and time-delay tanks to meet federal discharge requirements. It is interesting to note that the "zero release" gaseous waste-system concepts (refrigerated adsorption) described in Chapter 11 were incorporated in the Fort St. Vrain facility several years earlier than in the water plants.

For any future gaseous waste systems for a High Temperature Gas Cooled Reactor, the goal should be a "zero release" system.

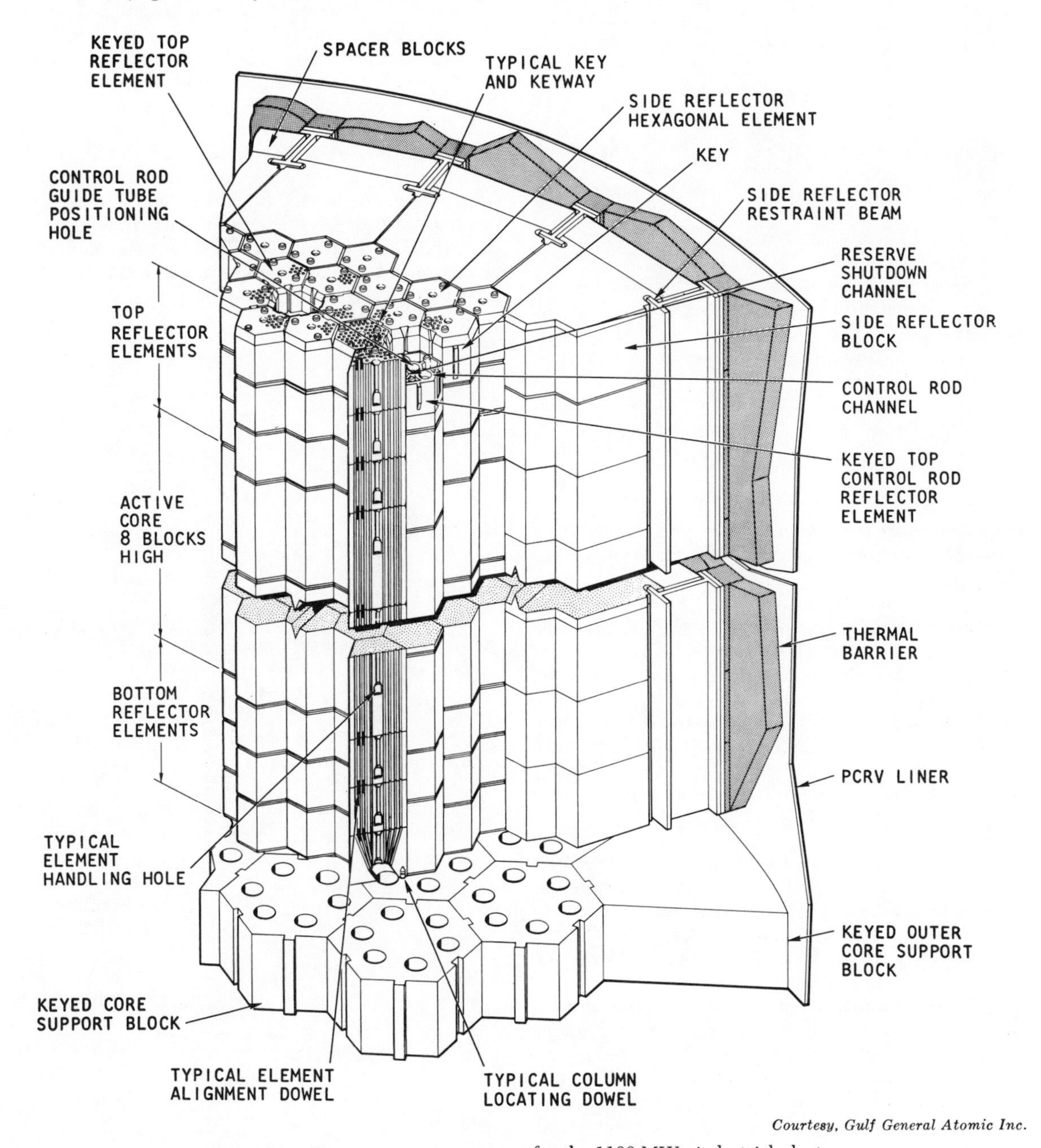

Courtesy, Gulf General Atomic Inc.

Fig. 4-26. Reactor core arrangement for the 1100 MWe industrial plant.

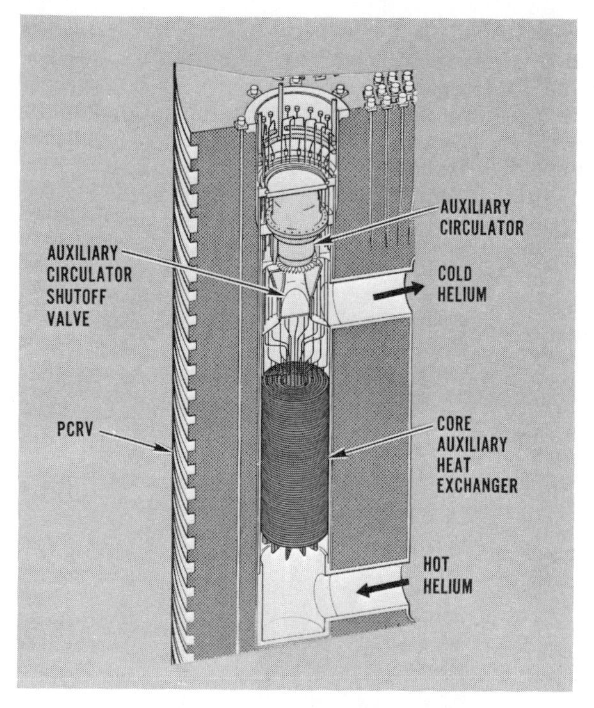

Courtesy, Gulf General Atomic Inc.

Fig. 4-27. Auxiliary cooling loop for the 1100 MWe industrial unit.

This plant has a "hot service facility" in which decontamination of equipment is performed. It is provided with shield windows and manipulators so that remote or contact decontamination and maintenance can be performed. Equipment access is through a hatch in the ceiling so that the reactor building crane can serve the facility. The facility is provided with hot water, dry air, and a special decontamination solution fluid system. The decontamination solution system has provision for particulate filtration, fluid reuse, and discharge to the liquid waste system. It is intended that the refueling mechanisms will be serviced in this facility.

4.13 1100 MWe Unit

Economic studies by Gulf General Atomic Inc. indicated that an economic size for the power industry would be about 1000 MWe. They have, therefore, developed a unit that produces 2783 MWt and reaches 1100 WMe with an overall efficiency of approximately 39½ percent.

Although the major systems were defined at Fort St. Vrain, the industrial unit concentrated on design simplification and increased safety. Compare the core design and support of Fig. 4-2 and 4-3. In Fig. 4-2, the core, supported on relatively long columns with two false floors to direct coolant flow, is surrounded by a steel core barrel so that cool gases can pass up the annulus between the core barrel and the PCRV liner. In Fig. 4-3, columns, false floors, and the core barrel are gone and the structure is shorter and more stable. It is easier to conform to Seismic Class I requirements in the new design than in Fort St. Vrain. Core arrangement has been simplified as indicated by a comparison of Fig. 4-26 and Fig. 4-4.

Referring again to Figs. 4-2 and 4-3, note how the circulators and steam generators have been rearranged. Each circulator-steam-generator combination has its own individual loop. Coolant flow is thus more uniform and can be more easily controlled. Each circulator or steam generator can be removed from its own penetration; the large central access hole of Fig. 4-2 is no longer necessary.

In Fort St. Vrain, emergency and shutdown cooling is accomplished solely with the speed variation of the main circulators. Each turbine also is equipped with an auxiliary waterwheel should drive steam not be available. In the 1100 MWe unit the main circulators may be used when steam is available. There are, however, three separate, complete cooling loops,

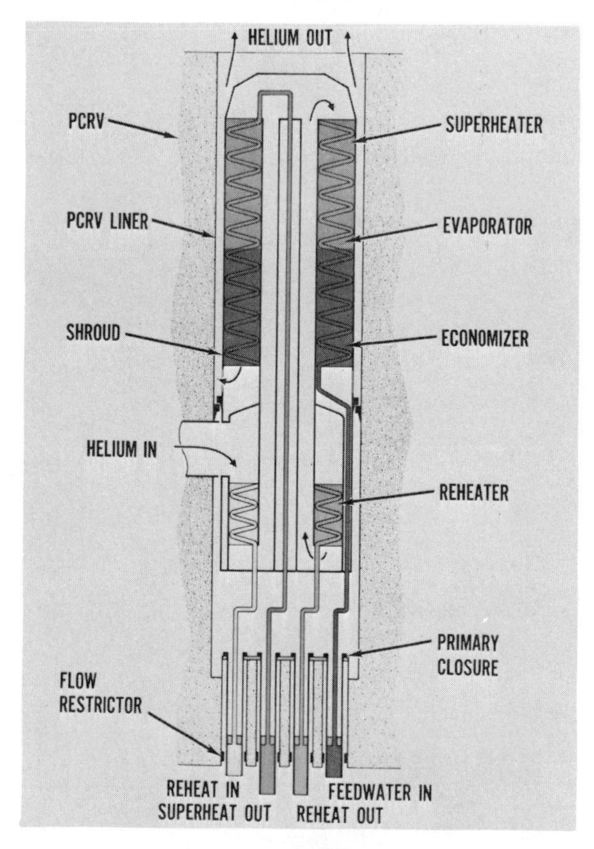

Courtesy, Gulf General Atomic Inc.

Fig. 4-28. Diagram of the new steam generator module. Six of these units are used in the 1100 MWe unit.

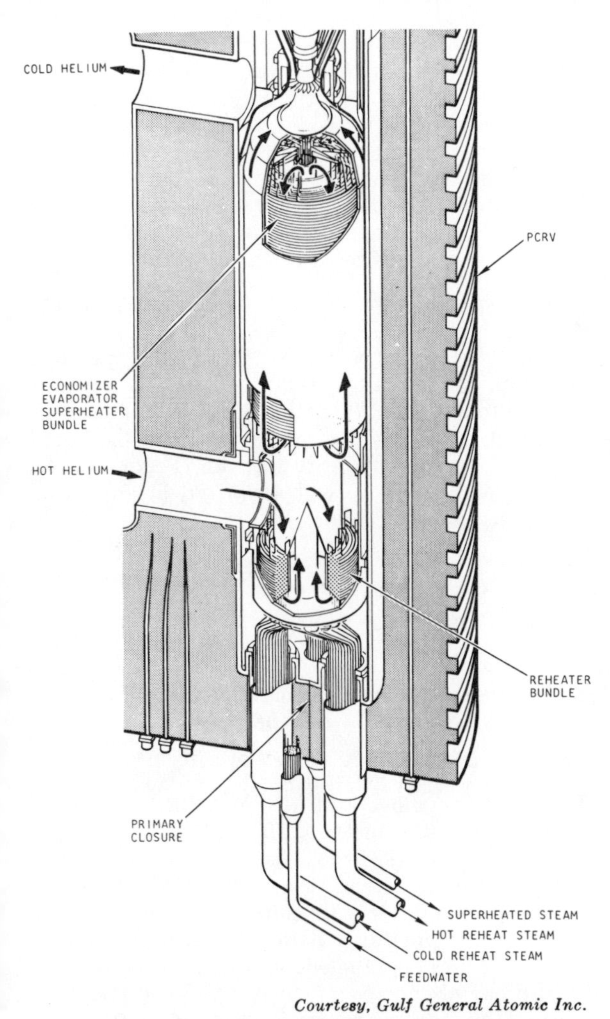

Courtesy, Gulf General Atomic Inc.

Fig. 4-29. Proposed arrangement and gas flow path of the steam generator module of Fig. 4-28.

driven by variable-speed electric motors for emergency service. Each loop, as shown in Fig. 4-27, has both a circulator and a water-cooled heat exchanger; however, only two of the three loops are required. Core heat absorption capability in an emergency is essentially the same as in Fort St. Vrain.

A major change has taken place in the containment concept. In Fort St. Vrain each penetration was sealed to the PCRV with a double closure and a leaktight building was erected around the PCRV. In the industrial design, penetrations have a single closure and the PCRV has been surrounded by a containment vessel. A design-basis accident has been postulated and containment has been designed to accept the incident. The accident is based on reactor depressurization through an opening of 100 sq ins. The addition of the containment vessel is an added safety feature because the PCRV is the same, highly redundant vessel as in Fort St. Vrain.

The steam generator modules are redesigned and the new design is shown in Figs. 4-28 and 4-29.

There have been additional, relatively minor changes made but they are not described here. What should be remembered is that the changes have all been in the direction of greater safety and simpler design. The plant has not yet been subjected to the test of licensing and detail design but it is inevitable that further evolution will take place during design of the first unit.

Refueling has been modified to decrease refueling time. The refueling machine has been redesigned so that transient storage of fuel elements being handled is in a separate fuel-transfer cask. Thus, fuel can be transferred in and out of the reactor from one transfer cask, while fuel is being placed in storage out of a second cask. This redesign recognizes the cost of down time to a major production unit. Additionally, the fuel cycle is now a four-year cycle with ¼ of the core replaced at each refueling.

CHAPTER 5

Containment

5.1 General

In a nuclear installation, containment is defined as the process or procedure of restricting to sharply defined volumes, the distribution of those radioactive materials which are intimately involved in nuclear fission. In the normal installation there are three concentric containment systems and all three of these systems must be breached before any activity can be released to the surroundings.

In a water reactor the first, or innermost of the systems, is the fuel pin. The fissionable material, normally in the form of a spherical or cylindrical pellet of uranium oxide, is sealed into a Zircalloy-4 tube about 0.400 in. to 0.500 in. in diameter. Each tube or pin has sufficient gas space in it to contain all the radiolytic gases evolved from the fuel during its active life.

The fuel pin, in turn, is positioned in a fuel lattice inside the reactor vessel. The reactor vessel is part of the primary coolant system, all of which forms the second containment barrier. The primary coolant system comprised of the reactor vessel, pumps, piping, pressurizer, and steam generator (for a PWR) is located within the third barrier called the containment structure or containment vessel. For a modern PWR 1000 MWe station this structure can be as large as 140 ft in diameter and 200 ft high. It is built as a pressure vessel and may be capable of containing an internal pressure as high as 60 to 65 psig.

Initially, reactor facilities housed low-power (up to 10 MWt) reactors installed in water pools. This arrangement was such that much of the energy released in an excursion was absorbed by the pool water and relatively small amounts of steam were generated to create an internal pressure. In that generation of reactors, containment design was of the order of 5 to 8 psig. This could be economically contained in a building of reinforced concrete with an elastomeric coating on the inside as the impermeable membrane. The structure was easy to construct and could be of any convenient economic shape; in Hamilton, Ontario, for example, a seventeen-sided structure was built.

With the advent of Shippingport the situation changed. Temperatures and pressures increased so that the amount of energy stored in primary systems increased two or three orders of magnitude. A typical, present-day PWR system operates at 2200 psia and 580 degrees F with subcooled compressed water. A typical 600 MWe plant contains approximately 11,000 ft^3 of water or 660,000 lbs. In the operating condition the enthalpy of the compressed liquid is 585.9 Btu per lb. When the pressure boundary is breached, gage pressure goes to zero and the liquid can only contain 180 Btu per lb. Therefore, the amount of primary coolant vaporized in an unconfined situation would be:

$$\frac{\text{excess energy per lb} \times \text{total lbs in inventory}}{\text{latent heat of vaporization per lb}} = \text{lbs of vapor}$$

$$\frac{(585.9 - 180) \times 6.6\text{x } 10^5}{970.3} = 2.76\text{x } 10^5 \text{ lbs vaporized}$$

Since the specific volume of atmospheric steam is 26.8 ft^3/lb, this vapor would occupy:

$$26.8 \times 2.76 \times 10^5 = 7.42 \times 10^6 \text{ ft}^3$$

However, containment structures actually have between 1.5×10^6 and 2×10^6 cubic feet of free volume; thus the specific volume of the steam generated has to be between 5.42 ft^3/lb and 7.25 ft^3/lb which corresponds to a steam-pressure range of 80 to 60 psia. The preceding illustration is oversimplified because it uses only the energy stored in the coolant. It also has inaccuracies in that additional energy is released by the core; energy is absorbed by the structure itself, the atmosphere in the structure, and by cooling systems of various types. It must also be remembered that the building starts with 14.7 psia of air whose pressure increases according to the "perfect gas" laws as the air comes to the same temperature as the steam. Actually, the excess energy is less because the water and steam come to equilibrium at the final pressure in the building. At 60 psia the enthalpy of saturated water is 262 Btu per pound which is less available excess energy than assumed in the first calculation. The actual calculation is an iterative process on a computer which assumes a final condition, calculates the energy available to produce this condition, and finally, checks the total pressure and volume of the steam produced and the heated air against the free volume available in the containment structure.

At 60 psia the excess energy is 585.9 − 262 or 223.9 Btu per lb, and the water vaporized is:

$$\frac{223.9 \times 6.6 \times 10^5}{915.5} = 1.62 \times 10^5 \text{ lbs.}$$

In a structure with 1.75×10^6 ft^3 of free volume, the specific volume of the saturated steam will be:

$$\frac{1.75 \times 10^6}{1.62 \times 10^5} = 10.8 \text{ ft}^3/\text{lb}$$

This corresponds to a steam pressure of 39 psia. If the next iteration is performed at a pressure of 55 psia, it will be seen that the difference between the assumed final pressure and the calculated final pressure is less than the difference when the assumed pressure was 60 psia. Hence, for this example, if the utility requires a 1.75×10^6 ft^3 free-volume building, the actual final pressure will be 57 psia steam pressure plus 19.7 psia air pressure, for a total pressure of 76.7 psia, or 62 psig. The building design pressure should be 71 psig so that there is a 15 percent safety margin in the design. In this instance, it would probably be cheaper to build a slightly larger building and reduce the design pressure to about 60 psig, than to build the smaller building rated at the higher pressure. Normally, containment design pressures are in the range of 40 to 65 psig, while design temperatures are assumed to be the saturated steam temperatures corresponding to these pressures.

5.2 PWR Containment

The PWR structures are vertical, flat-bottom cylinders with hemispherical or ellipsoidal domes as seen in the scale model of Fig. 5-1. When design pressures and temperatures increased so that reinforced concrete could no longer do the job economically, designers then turned to steel, a material which served as both the strength member and the impermeable membrane. At the beginning, there were no contractors to construct these structures, and plant designers turned to the builders of large field fabricated tanks and vessels, who were accustomed to constructing free-standing steel structures and who were able to adapt to the requirements of this construction. Since the structural strength required to build the free-standing structure also provided sufficient strength for the design pressure load, the first structures utilized concrete for its shielding properties only. With the growth of the unit size from 100 MWe to the current 1,000 MWe units, multiple-loop PWRs appeared and containment structures approached the 140 ft-diameter size. Now the free-standing structure needed additional strength and the designers again started to use concrete for its strength, while using the steel structure, basically, as the membrane. The concrete strength design uses post-tensioned tendons for three basic reasons. Each tendon is an individual entity and the failure of one or two tendons would not result in as catastrophic a situation as would one failure in a steel structure. The system of individual tendons achieves the beneficial redundancy of design in the structure, that is built into the reactor safety channels. The second reason is the ease of inspection and maintenance; by installing each tendon in a

Courtesy, Chicago Bridge and Iron Co.

Fig. 5-1. Model of a steel PWR containment structure with an air space between the vessel and the concrete shield structure. The figures of the men are in the correct scale.

conduit, it can be inspected at regular intervals. The third reason is that by the use of post tensioning, more efficient use is made of the concrete and thereby less concrete is used.

In 1968 and 1969, utilities building PWRs started to question the need for paying the extra cost for constructing a freestanding steel membrane and following it up with the construction of a post-tensioned concrete structure. The pros and cons of the situation are still not too clearly defined. Some owners feel that if the steel membrane and the concrete could go up together, one lift at a time, the steel could be lighter, there would be only one charge for scaffolding, and the net effect would be a cost saving and a shorter construction period due to concurrent steel and concrete work. With the improvement of field heat-treating methods for plates and the increased cost of interest during construction, two plants in the 800 to 1000 MWe range have tried yet another technique which is said to be economical. A free-standing steel structure was built first of 1¾-inch plate. A separate concrete structure was built outside the steel structure using slip forms. The concrete structure was built in three weeks. Emergency systems to handle radioactive iodine were sized to handle leakage from the containment structure into the annulus between the steel and the concrete. The only valid conclusion then, is that each time a unit is planned an economic evaluation of all designs must be made, to arrive at a correct selection for the individual case.

5.3 BWR Containment

The boiling water reactor uses a differently shaped containment structure. The containment system, as shown in Fig. 5-2, is best described as an inverted light bulb (the drywell) on a torus (the suppression chamber). The two structures are connected by a series of parallel pipelines comprising the drywell vent system. While the normal PWR containment is dry and depends on volume for pressure limitation, the BWR carries a water inventory in the pressure suppression chamber. The top head of the structure is removed during refueling. In operation, the steam generated in an accident is conducted through the vent system and released below the water surface in the suppression chamber, where it is condensed. A typical drywell has an upper cylindrical section 35 ft in diameter, atop a spherical section 65 ft in diameter. Thus, although the free volume of this containment structure is of the order of 260,000 ft^3 as compared to a PWR containment volume of 2,000,000 ft^3, the design pressure is about 56 psig, the same approximate design pressure as the much larger PWR structure. The reason for this much smaller BWR containment structure having the same design pressure for an equivalent energy release, is the condensing action of the pressure suppression chamber. The drywell is enclosed in concrete, for shielding purposes.

In 1969, General Electric Co. revised its product line and offered a new containment concept consisting of a frustum of a cone on a cylinder, as shown in Fig. 5-3. In this concept the reactor vessel is up near the top of the cone so that the parting line of the reactor pressure-vessel head is level with the part-

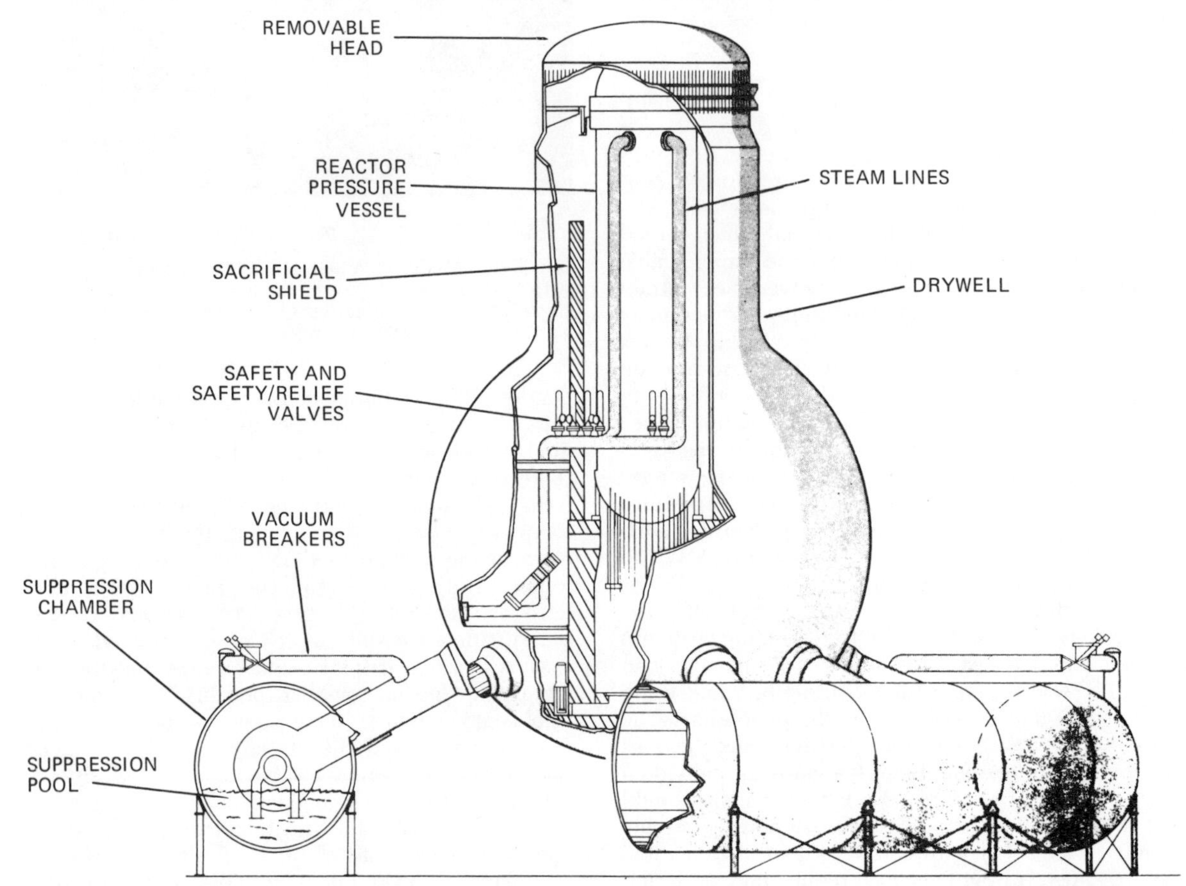

Courtesy, General Electric Co.

Fig. 5-2. The original, all steel "inverted light-bulb and torus" containment design for the Boiling Water Reactor.

Courtesy, General Electric Co.

Fig. 5-3. Over-and-under design for the Boiling Water Reactor Containment. Note the RHR system pumps just outside of the pressure suppression pool at the bottom level.

ing line of the head of the drywell, as shown in Fig. 5-3. The lower portion of the structure, below the floor in Fig. 5-3, is the pressure suppression chamber. The drywell vent pipes are the vertical pipelines starting at floor and going down. This concept is called the "over-and-under design."

The "over-and-under design" can be built as a steel vessel with a concrete shield, a steel membrane in a prestressed-concrete vessel, or a steel membrane in a reinforced-concrete vessel. The simpler contours of this design, as compared with the inverted light bulb and torus, result in a significant economic saving in the cost of the "over-and-under design." An additional advantage of this design is that it permits the BWR owner to perform an economic evaluation of the three construction designs to permit him to select the most economic alternate for his site.

5.4 High Temperature Gas Reactor

Containment for the high temperature gas reactor has a slightly different connotation than for a water reactor. The fuel, uranium and thorium carbide, is coated with pyrolitic graphite and dispersed in a graphite matrix. The pyrolitic graphite is the first containment barrier. The second barrier is the reactor pressure vessel, a post-tensioned concrete vessel (PCRV) with a steel membrane. The third containment barrier is a containment building that houses the PCRV, as shown in Fig. 5-4. The entire primary system: circulators, steam generators, core, etc., is contained within the reactor pressure vessel. (See Figs. 4-2 and 4-3 in Chapter 4, for typical reactor structure.) Experience with this type of reactor vessel is limited in the United States, although the British have used this construction in their Magnox and AGR reactor stations. Design pressures for these structures are primary system pressure, about 700 psia. Since the coolant is helium gas, the coolant inventory is not the huge energy reservoir that water is, and accident effects are less severe. As noted previously, a concrete vessel, due to the redundancy of its reinforcing, is not subject to catastrophic failure.

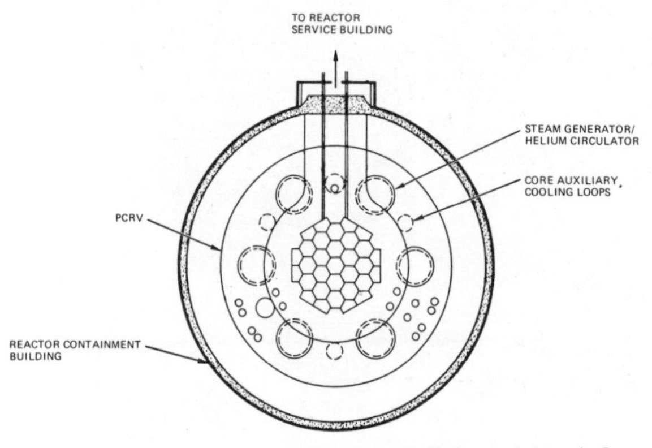

Courtesy, Gulf General Atomic Inc.

Fig. 5-4. Typical arrangement of an HTGR containment structure. Inside diameters will run from 120 to 130 ft.

No containment vessel has yet been built for an HTGR; several studies, however, have been made of the most economical structure, drawing on experience gained in PWR containments. As has already been noted, there are no large energy reservoirs in the reactor and hence, no high pressures anticipated in the event of a PCRV rupture. Therefore, when the structural concrete has been designed to satisfy shielding requirements, it is found that the structure can accommodate up to 15 psig with just a little more reinforcement. A steel liner is then added on the inside, as an impermeable membrane, and the design is complete. The economic evaluation required is to select between a reinforced concrete structure and a prestressed-concrete structure. Present criteria indicate that if an ASME Section III, Class MC vessel is used (which can be done) the cost of the complete structure is exorbitant since the same amount of concrete is needed for shielding anyway. The PCRV for an 1100 MWe reactor is approximately 97 ft in diameter and 88 ft high, plus about a 15-ft-thick support pedestal underneath. The external, circumferential prestress wire winding machine requires about 10 ft of clearance for operation. Vertically, there is a refueling floor above the PCRV upon which the 43-ft-high refueling machine runs; above that is a polar crane. Hence, there is a minimum vertical clearance of approximately 60 ft required above the top of the PCRV. The minimum internal dimensions of the containment must then be 117-ft-diameter by 163 ft to the springline of the dome of the structure.

5.5 Design Standards

Steel containment structures are designed in accordance with Section III, Class B of the 1968 ASME Boiler and Pressure Vessel Code or Section III, Class MC of the 1971 Code. Concrete is designed according to the Building Code Requirements for Reinforced Concrete of the American Concrete Institute. These codes are supplemented by applicable local codes and structural steel codes.

The following types of loading are considered in the design:

1. Pressure and temperature transients caused by the postulated loss of coolant accident
2. Test pressure
3. Dead and live loads including prestress loads
4. Operating temperature gradients in the structure
5. Loads imposed by pipeline rupture
6. External pressure loads
7. Uplift due to buoyant forces
8. Winds
9. Tornados

10. Earthquakes
11. Missiles
12. Temporary loads imposed during construction.

You will note that external pressure loads are also considered. These may occur during a leak rate test or after an accident, while the building is still sealed. It is brought about by a sharply rising barometer that produces an atmospheric pressure higher than the pressure captured within the structure. An increase in barometric pressure of ½ inch of mercury results in an external pressure on the building of 36 lbs per sq ft.

Steel plate for the structure is selected from the list of materials acceptable to the code. The plate is basically structural steel but differs in that its chemistry is more closely controlled. The chemistry control requirement is imposed because the material must have a nil ductility transition temperature (NDT) 30 degrees F lower than the lowest anticipated working temperature. The accepted specification test for this requirement is a vee notch Charpy impact test value of not less than 15 ft lbs. Consider what this criterion means in the Iowa or Nebraska area. Winter outdoor design temperatures are −10 degrees F to −20 degrees F. If credit is taken for a 15 degree F temperature rise through the shielding concrete, minimum design temperature is −5 degrees F for the steel structure and the Charpy impact test is performed at a temperature of −35 degrees F. Allowable working stresses are taken from the ASME Boiler and Pressure Vessel Code, Section III, Class B (1968) or Class MC (1971).

Design criteria for concrete are still confused. There is no single recognized code authority for the design in the same sense that the ASME Boiler and Pressure Vessel Code governs the design of metal vessels. The 1971 edition of Section III of the ASME Code will have a concrete design subsection added sometime after its initial issue. All concrete designs must still be justified to the Division of Reactor Licensing of the AEC on an individual basis. At the present time the American Concrete Institute and the American Society of Civil Engineers are working to produce a code for these structures. The NDT concept has only been applied to anchorages for post-tensioning and tendons; reinforcing bars do not have an NDT requirement. In the absence of governing criteria, general practice is to use reinforcing steel conforming to ASTM Specification A-615; stranded tendons conforming to ASTM Specification A-416 and ¼-inch-diameter tendon wire conforming to ASTM Specification A-421.

5.6 Structural Protection

Internal partitions and floors in the structure are provided for normal building design and for an engineered safety-feature service. Design criteria require that the containment structure be protected against mechanical damage from internally generated missiles and pipe failures. Missiles can be generated by a failure of a control rod hold down or the blank on a spare connection. If a high pressure pipe line shears, the two ends can whip around and cause damage. The containment structure must have sufficient internal devices to protect against these hazards. Externally, the mechanical hazards considered thus far are hurricane originated flying poles and airplanes. The airplane hazard is applicable only if a plant is beneath or very close to a landing or take-off pattern.

5.7 Penetrations

Any entry through the containment structure is defined as a penetration regardless of size or purpose; thus, a 12 ft by 12 ft entry for a truck and a ½-inch-diameter conduit for a pair of signal wires are both "penetrations." A penetration is a complex device; it must not leak; it must accept any load transmitted to it, while at the same time, it must transmit this load into the containment structure in an acceptable manner. All welds used in its construction must be inspected and its design must be such that all seals can be individually tested.

The last requirement means that seal welds are enclosed with a pressure-tight shroud channel so that design pressure can be imposed on the weld at any time. Air-lock doors are provided with double seals so that a test pressure can be applied between the two seals and each one can be individually checked.

Electrical leads are usually grouped together in convenient amounts and passed through a single penetration as shown in Fig. 5-5.

Note that in Fig. 5-5, each compartment and each weld can be pressure tested. The welded shroud

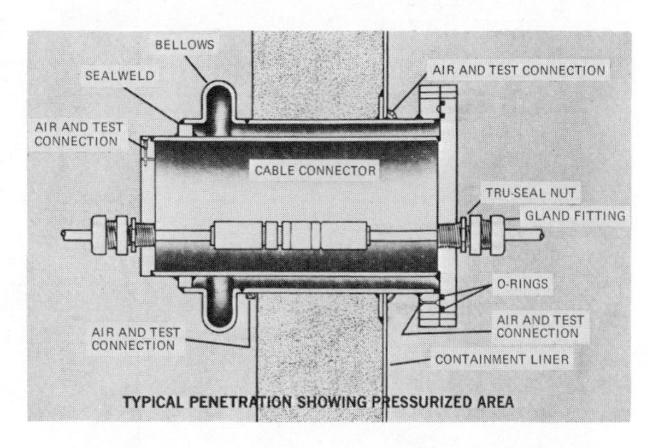

Courtesy, Westinghouse Electric Corp.

Fig. 5-5. Cross section of a typical electrical penetration. Note that every volume can be independently pressure tested.

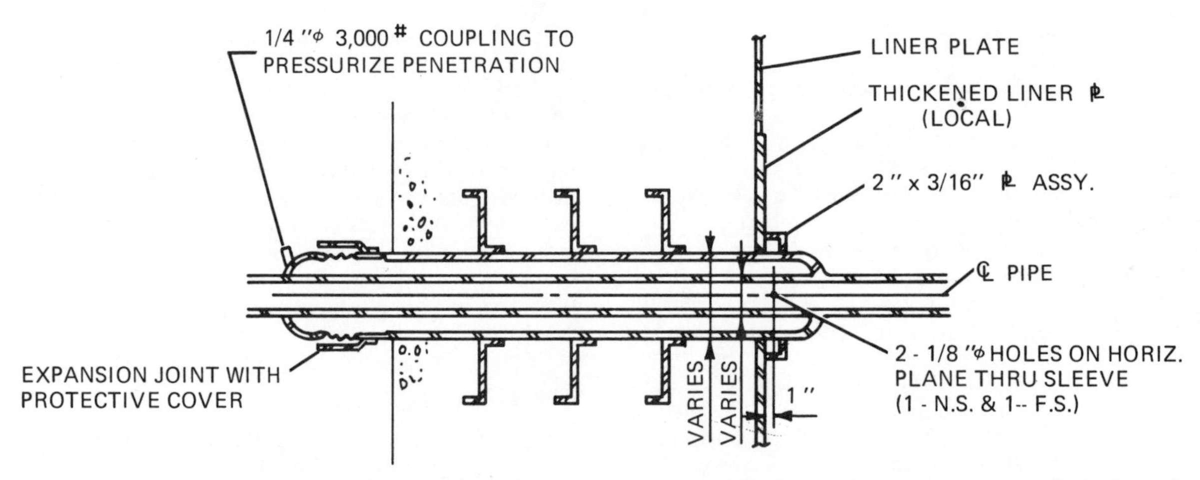

Fig. 5-6. Cross section of a typical cold pipe penetration. In this particular design the pipeline size would be limited by the anchor load that could be accepted by the concrete wall.

channel over the weld seam at the containment liner provides both the test chamber and a second barrier for leakage prevention. The air and test connection into the cable chamber is provided on pressurized water reactor plants only to pressurize the penetration during an over-pressure incident.

Piping penetrations are broken down into two categories—cold penetrations as shown in Fig. 5-6, and hot penetrations as shown in Fig. 5-7.

In designing a hot penetration, the concrete of the structure must be given thermal protection. In the main steam penetration in Fig. 5-7, the pipeline is insulated and a cooling coil is provided. The cooling medium may be recirculated water or air; in either case, all cooling connections are made outside of the containment structure. The cooling coil is usually an embossed coil on the inside surface of the imbedded sleeve and forms an integral part of the sleeve.

Both of the penetration designs are shown with corrugated expansion joints because the designs are

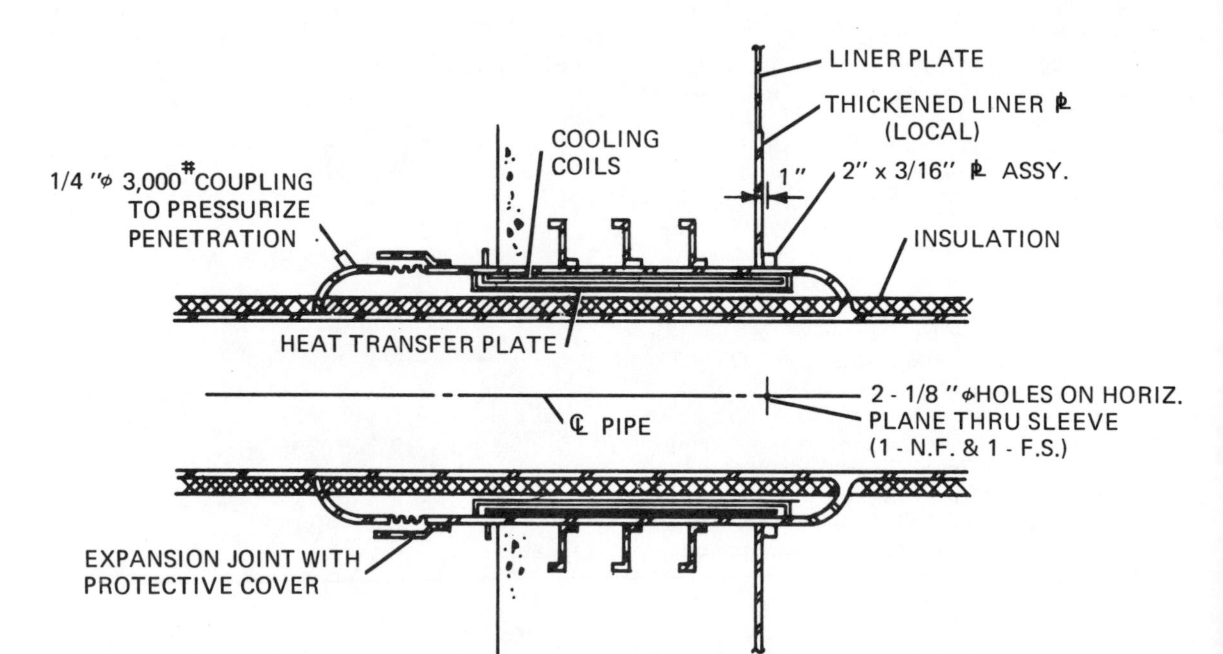

Fig. 5-7. Cross section of a hot pipe penetration. The cooling coils are provided to limit concrete temperatures to acceptable values.

more complex. The expansion joints effectively remove all pipe-transmitted loads from the containment structure. However, the present trend is to try to correlate the pipeline and supporting system design with the containment structure design, eliminating the expansion joint.

In some of the more densely populated areas where pressurized water reactors are installed, the AEC Division of Regulation and Licensing has required that each penetration be pressurized with clean air under accident conditions to insure that any leakage from a penetration is not contaminated. Hence the pressure connection for the electrical penetration and the pressure connection for the annulus in the pipe penetration.

5.8 Penetration Pressurization

The system supplies compressed air to each penetration and operates all the time the reactor is at power. A typical pressurization system is shown in Fig. 5-8. The building is split into several zones, each of which is supplied from its own air receiver. As can be seen in the illustration, the system is backed up with bottled nitrogen for each zone so it is essentially independent of plant power or air. System construction is welded steel and design criteria include Class I seismic requirements. Pressure is normally about 10 psi higher than the maximum anticipated containment pressure. The design conforms to ANSI Standard B31.1.0. The system is so designed that the nitrogen pressure is slightly lower than the air pressure; therefore, the nitrogen is used only if the air supply fails. The supply line to each zone has a flow-restriction orifice so that the normal maximum leakage can be handled, but leakage to another zone is restricted. Each zone is sized to pass twice the allowable leak rate so that double failure can be handled; i.e., allowable leakage both into the building and out of the building at the same time. The air receivers are sized for four to six hours of operation and the backup N_2 cylinder banks have another twenty-four hours' supply. Instrumentation monitors the flow rate and pressure. Abnormally high flow or low pressure will trip an alarm.

5.9 Air Locks

The pentrations previously discussed are fixed types. Entrances for personnel and equipment must

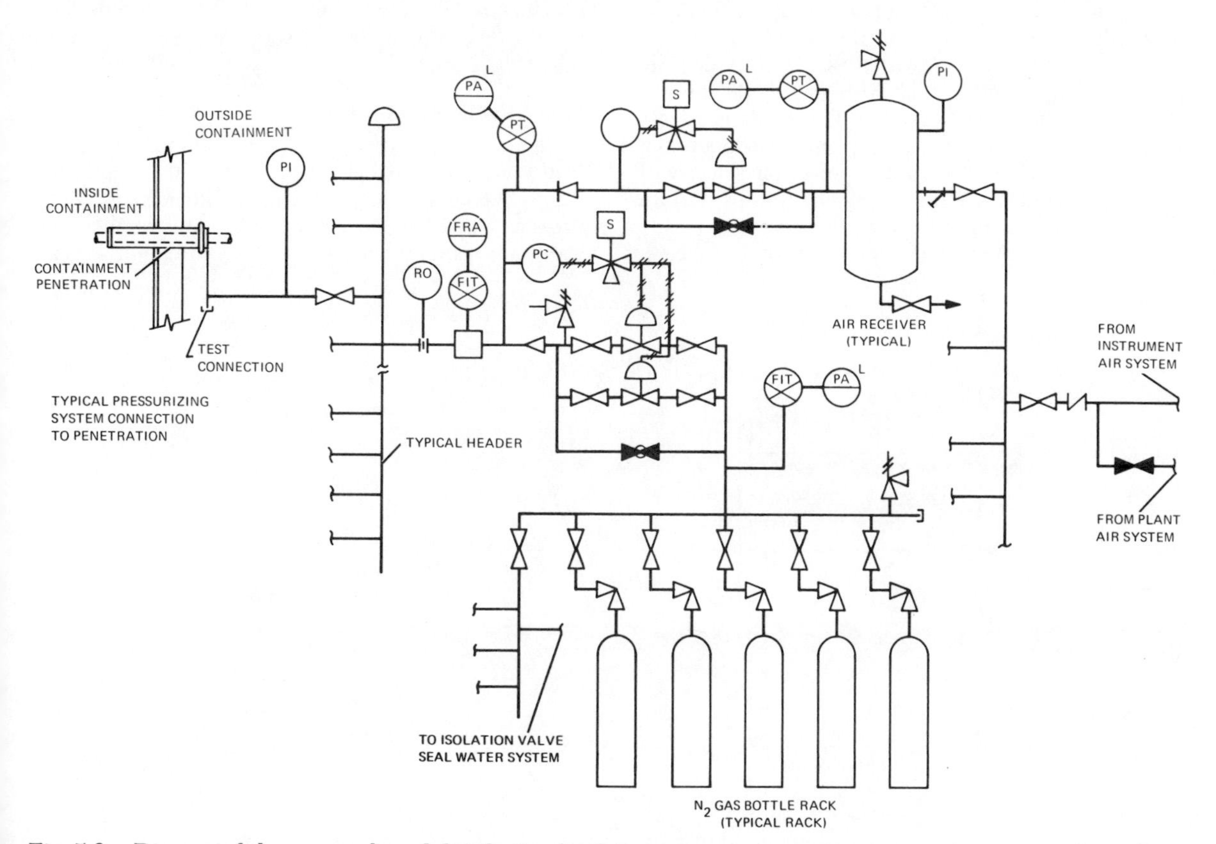

Fig. 5-8. Diagram of the gas supply and distribution for the containment penetration pressurization system for a PWR. It has two independent gas supplies – plant air and bottled nitrogen.

Courtesy, Chicago Bridge and Iron Co.

Fig. 5-9. Assembled personnel lock ready for shipment.

meet the same rigid sealing requirements as the electrical or mechanical penetrations, while serving their primary purpose. Personnel entrances must be available at any time, so typical air-lock construction is used. A chamber is provided with two doors; one opens into the containment structure while the second door opens to the exterior. The doors are interlocked with each other so that only one door may be open at a time. Doors are power-operated with manual overrides in the event of power failure. Figure 5-9 shows an air lock of this type.

The equipment entrances offer a slight choice in design. Size is dictated by the largest assembly which must pass through. For a PWR, this is usually a reactor vessel O-ring; for a BWR, this is the recirculation-pump motor. The reactor vessel and steam generators are installed before the building is closed in and, since removal of this equipment represents an expense in the tens of millions, no provision is made for its removal. The advantage of an equipment airlock and its cost has to be weighed against the lower cost of an equipment hatch that can only be used during plant shutdown periods. The decision has usually been to use an equipment hatch.

Experience has shown only one reliable method of sealing door-type openings—the use of double, inflatable pneumatic seals. In this design, shown in Fig. 5-10A, the two seals are inflated against a flat machined surface, capturing a volume between the seals that may be monitored for leakage. Air is supplied to the seals through an individual local reservoir equipped with alarms and lock-up valves, as shown in Fig. 5-10B. In addition, each seal is monitored and has its own closure valve.

Air locks are built to the requirements of Class B or MC of Section III, of the ASME Boiler and Pressure Vessel Code. Seals are elastomers, usually a neoprene; however, it is expected that the ethylene-propylene terpolymers will supplant neoprene in the next few years.

5.10 Containment Isolation

Of necessity, there are many pipelines that penetrate the containment wall. There are ventilation ducts, feedwater, main steam, demineralized water, process waste, and many others. Some of them are needed to shutdown the reactor safely, others are

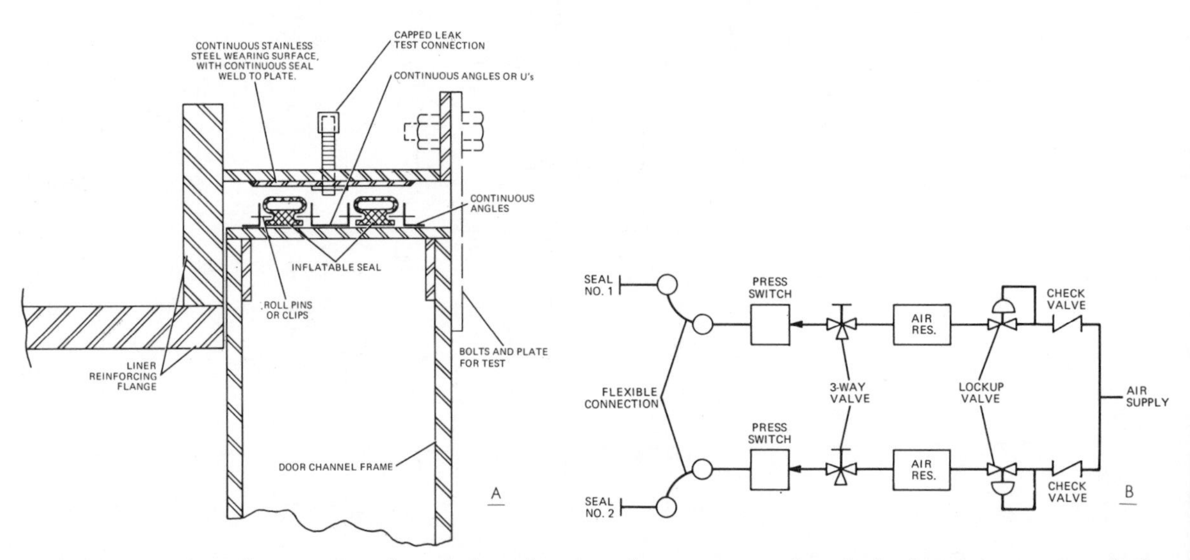

Fig. 5-10. A. Typical cross section through the seal section of a containment door. Seal inflation pressure is set higher than building design pressure. The captured volume between the seals can be pressurized at any time to test the seals. B. Diagram of the air supply for the seals in Part A.

not. All penetrations are isolated on a signal of high building pressure—3 to 4 psig. In addition, all penetrations that open into the building atmosphere—purge connections, sump drain connections, air supplies—close on a signal of high radiation. Valves are actuated by the "engineered safety features" signal just described.

The kinds of fluid penetrations and their isolation valving requirements are defined in Criteria 54, 55, 56, and 57 in Appendix I.

The BWR containment is just large enough to hold the reactor vessel, recirculating loops, pumps and valves, air-handling equipment, and electrical cables, as well as provide adequate access for maintenance and inservice inspection. Therefore, the emergency auxiliary equipment must be in a large building that surrounds the containment structure. In effect, a secondary containment barrier has been created that can handle any leakage from primary containment. The PWR containment structure, however, is much larger, and some of its auxiliary equipment is housed in the containment structure. There is no secondary containment barrier. To make up for the lack of the secondary barrier a PWR may be required to pressurize void spaces in penetrations and provide a water seal on those pipelines which could become leakage paths in an accident. This requirement is evaluated on an individual basis by the Division of Reactor Licensing when an application is made. Initial containment design should provide for any pressurization penetrations that may be required, should DRL require the system.

5.11 Isolation Valve Sealing

This system supplies water to the areas requiring liquid sealing on the high pressure engineered safeguard signal. A typical distribution system is shown in Fig. 5-11. The system is pressurized from the N_2 bottles of the penetration pressurization system, hence, is completely independent of the plant for operating power. All parts of the system are outside of containment. There are two types of sealing: 1. Automatic fluid injection on initiation of system operation for gas lines and low pressure liquid lines; and 2. Manual injection for high pressure liquid lines. Each individual connection requires a check valve to lock-in the seal fluid. In addition, each connection must be individually assessed to determine relief valve requirements in the event of check valve failure and exposure of the seal water system to process line pressure. Normal design pressure for the seal water system is 150 psig. Sealing is accomplished in two fashions, as shown in Fig. 5-12, A and B.

Gate valves over two inches in size are specified with split wedges and the body is pressurized between the wedge halves. In sizes less than two inches, and where gate valves are not used, two valves form the sealing volume. When the gate valves are specified for this service, the specification must include a

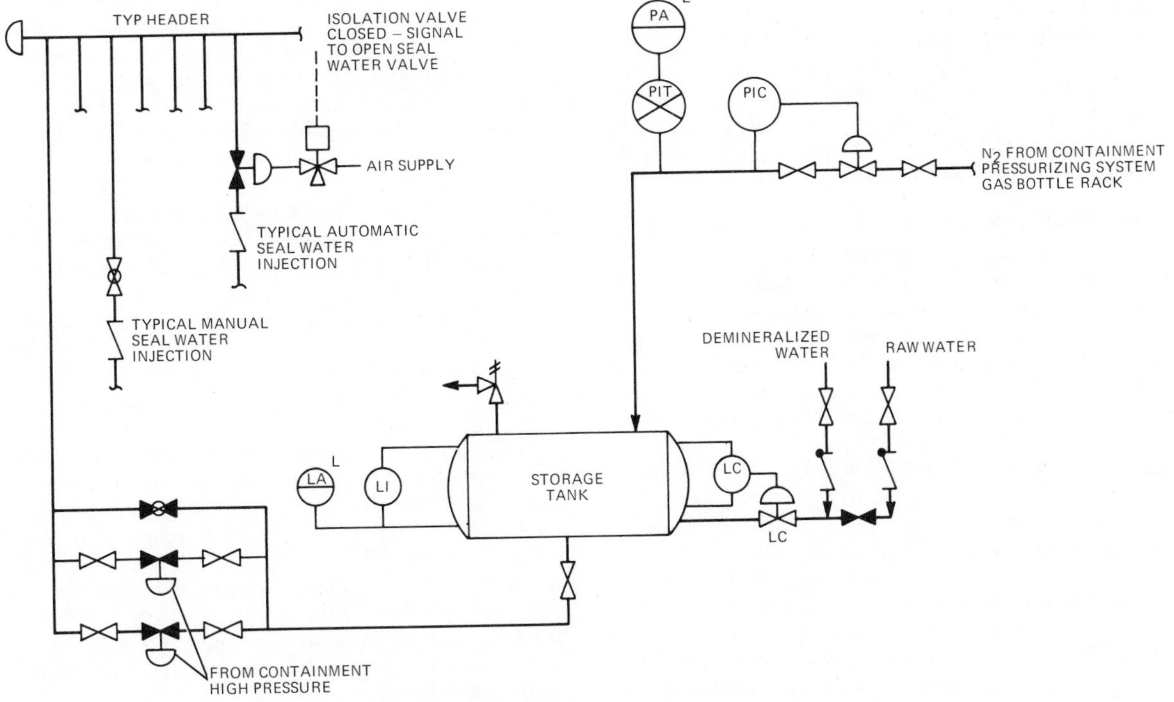

Fig. 5-11. Water system for the liquid seals for containment isolation.

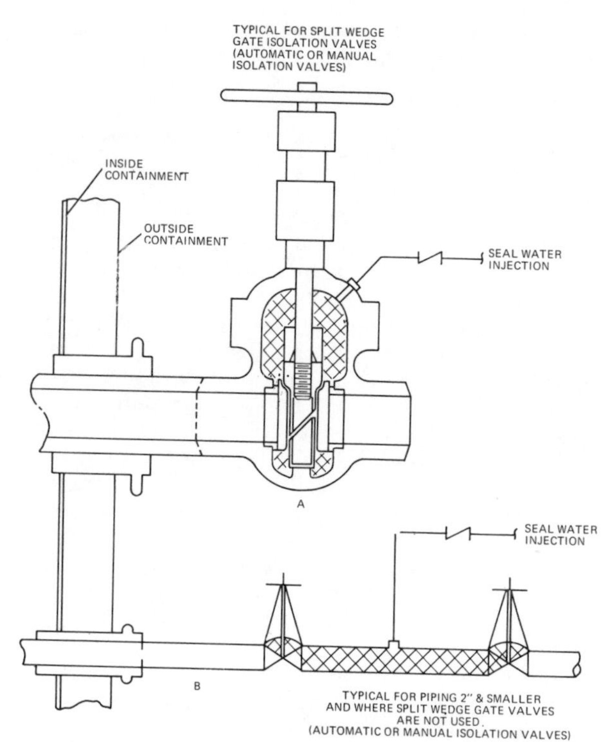

Fig. 5-12. Two methods of providing a liquid seal in a pipeline during isolation.

description of the injection and double sealing together with seat leakage tests from each side and in the injection mode. Globe valves used in the double valve liquid-seal application must have seat tests from both directions because one of the two valves will be required to seal in the reverse direction. System design includes Class I seismic requirements and conformance to ASME Section III, Class 3.

5.12 Ventilation

Ventilation in a containment structure has essentially one purpose—to provide an ambient atmosphere suitable for the equipment. Personnel access to the containment structure is restricted to short periods in a BWR during reduced load operation and in a PWR to certain locations for short periods. Accessible areas are maintained at a maximum temperature of 110 degrees F during operation. Inaccessible areas may go as high as 135 degrees F; the controlling factor is the design of the electrical gear. All areas are maintained at a minimum of 60 degrees F during shutdown periods.

The BWR containment structure is maintained as a closed system during operation; temperature is controlled by duplicate fan-coil units. Until recently the design required inerting containment with nitrogen, which had to be flushed from the structure with filtered outside air before start of refueling. Connections, with isolation valves, are provided from containment into the standby gas treatment system. The nitrogen and purge systems are shown diagrammatically in Fig. 5-13. During accident situations the standby gas treatment system is actuated and heat removal is accomplished by operation of the residual heat removal system in the containment cooling mode. This has been described in Chapter 3.

In a PWR the cooling and ventilation system is more complicated than in the BWR. The system performs the following functions:

a. Maintains the structure at predetermined temperatures, during normal conditions, by the addition or removal of sensible heat.

b. Removes sensible and latent heat under accident conditions to limit temperature and pressure increases.

c. Purges the containment structure with outside air whenever desired.

d. Filters supply and exhaust air.

This system is divided into three subsystems. A supply system, an exhaust system, and a heating-cooling system, as shown in Fig. 5-14. The supply and purge systems are split in two 50-percent capacity installations. The heating-cooling system has three 50-percent capacity fan-coil units, each with two coils, because they function as part of the emergency equipment. One coil uses demineralized water from the closed-loop for cooling, or recirculates water through a heat exchanger, for heating. The other coil uses final heat sink water (ocean, bay, river, lake, etc.) only. Each of the fan-coil motors is powered from the emergency power bus. The system is provided with normal temperature controls. The components must be capable of withstanding, or be protected from, pressure differentials that may occur during the rapid pressure rise after an accident; they must also withstand the seismic shocks for which the power station is designed. The fan-coil units must be capable of handling normal-density air and high-density air-vapor mixture after an accident.

The supply system has dust filters to limit incoming dusts. The exhaust system has high-efficiency particle filters and activated charcoal filters. If the emergency cooling water is sea water, all good sea-water design practices must be observed to ensure that the system is operable and retains its integrity during standby periods. All components of the supply and exhaust systems, except ductwork and isolation valves, are located outside the containment structure.

Criterion No. 38 of the General Design Criteria for Nuclear Power Plant Construction Permits requires redundancy in the cooling system; past practice has been to provide a second 100-percent capacity system that cools by direct water spray in the containment building.

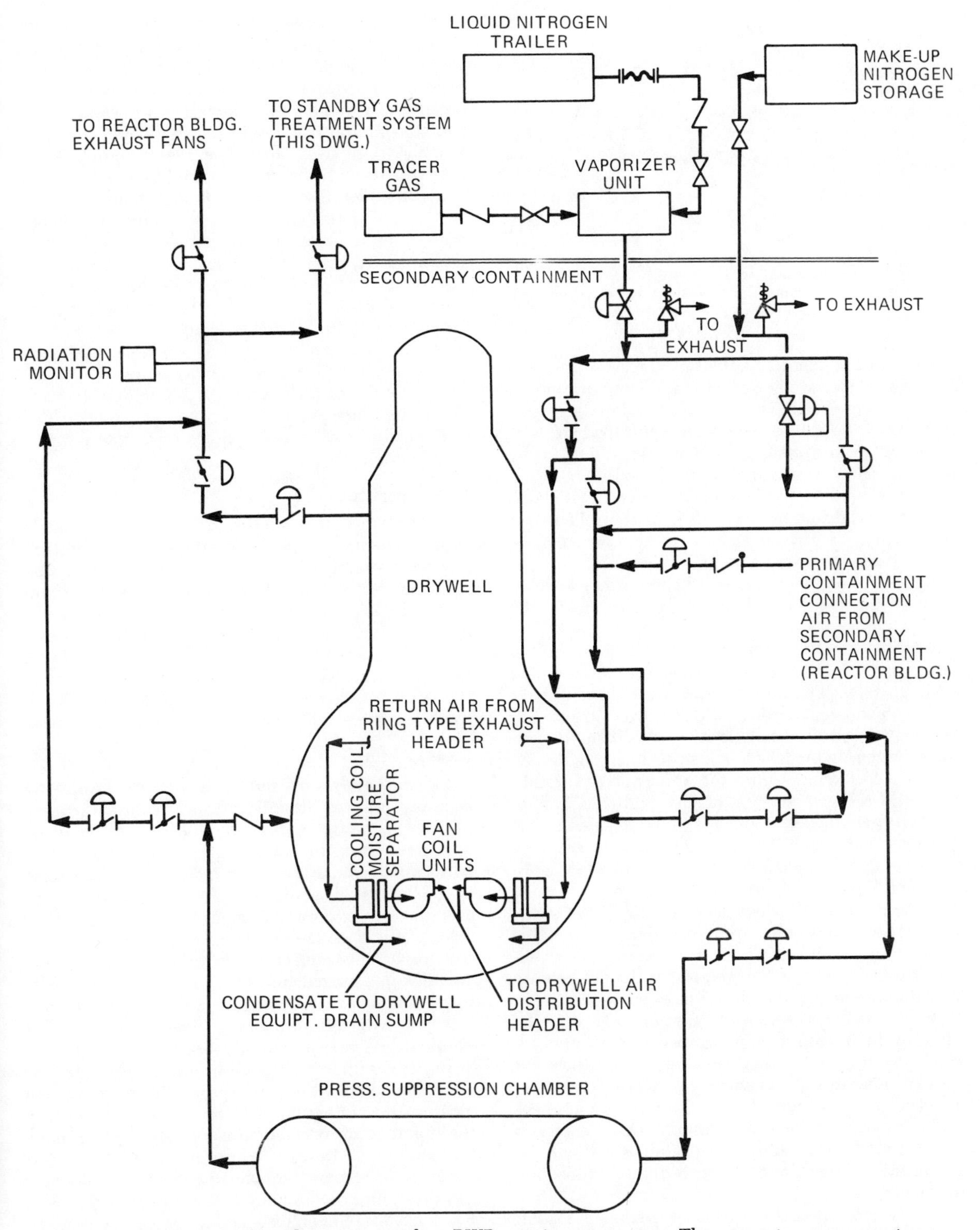

Fig. 5-13. Nitrogen supply and purge system for a BWR containment structure. These are not emergency systems. The N_2 must be purged with air before a man can enter.

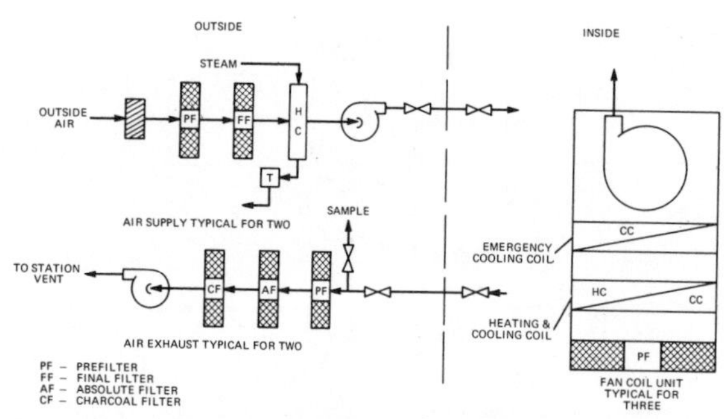

Fig. 5-14. Heating and cooling system for a BWR containment structure. Note the two isolation valves on each containment penetration.

In a post-accident situation in which the building has been successfully sealed, four long-term items must be considered. The first two are building pressure and temperature which are controlled by containment cooling systems (air recirculation) or by building spray systems. The third consideration is control of radioactive iodine-131 (I_2^{131}) released from damaged fuel elements. I_2^{131} is controlled by a combination of activated charcoal filters in the air-recirculation loops and a sodium solution in the containment spray. The iodine and the sodium react to form a water-soluble sodium iodide and remove the I_2^{131} vapor from the air.

The fourth consideration was introduced in the latter part of 1968, and was handled for the first time in 1969. This last consideration is control of gaseous hydrogen. Hydrogen is dissolved in primary coolant to control radiolytic oxygen. When the primary system is breached in the accident, hydrogen comes out of solution into the atmosphere. In addition, the residual activity of the fuel mass radiolytically decomposes water, producing gaseous hydrogen. The process may continue during the core decay period for as long as ten to twenty weeks until the energy level in the core becomes too low to sustain the reaction. The hydrogen content of the structure must be prevented from reaching a flammable or explosive mixture. Three alternatives are available for hydrogen control: the first is dilution: Pump outside air into the building while exhausting captured air under controlled conditions. This is the cheapest and easiest solution because all the required systems and equipment are already there for various other reasons. Obviously, however, this is dependent on the activity level in building atmosphere and so may have some restrictions. (AEC decisions at the end of 1971 indicate that dilution will only be permitted as a backup to either of the two following alternates.)

The second alternative is a recirculation system with a flame recombiner. This recombiner requires a gas burner, earthquake-proof gas supply and piping, oxygen supply and piping, and it must either operate (burn) all the time or have an acceptable ignition system. The third alternative is a catalytic recombiner using a paladium catalyst. This system requires an oxygen supply and has to guarantee that no liquid enters the recombiner; although water vapor is acceptable. Initial reaction to the two kinds of recombination, without an analysis in depth, indicates that catalytic recombination is better because it is simpler; its chief disadvantage being the need for a heater to vaporize any droplets of liquid. The heater can be either electric (a load on the emergency power supply) or steam. The decision would favor electric heat because the power equipment is already there and analysis may show that by the time this service is required, some emergency power, initially used for decay-heat removal, is available. The flame recombiner requires a great many more components than does the catalytic unit and hence appears more prone to failures.

Control of iodine and hydrogen are now required by Criterion 41.

5.13 Inspection

Inspection during construction is limited basically to normal dimensional verifications and nondestructive weld examination. All circumferential and axial welds in the steel structure will be 100 percent radiographed. Any weld that cannot be radiographed is inspected by either ultrasonics, magnetic particles, or by dye penetrant. Radiography is executed and interpreted according to Section VIII, Division 1, of the ASME Boiler and Pressure Vessel Code, while the other three examinations conform to Section III.

5.14 Testing

Design Criteria 52 and 53 require an acceptance leak-rate test at design pressure and subsequent periodic leak-rate testing over the life of the plant. The actual pressure test requires a minimum of seventy-two hours so several gross tests are made first, in the interest of saving time. Hatch seals are inflated and the space between them is pressurized. Both sides of the door are checked for leakage, with soap suds or halogen detector. All penetrations are individually pressurized and checked for leakage, with halogen detectors. Any suspect welds are individually checked. After all preliminary testing is complete, a building leak-rate acceptance test is performed.

Initially with low pressure concrete structures, the technique used was to pressurize the building and measure the air required to be pumped in to keep the building at constant pressure. There were many inaccuracies in this procedure and it was soon discarded. Next came the pressure decay method. In this procedure the building is pressurized and sealed. Readings of temperature and pressure were taken over a period of time and the leak-rate calculated. This method also had its inaccuracies and was soon modified to the reference method shown in Fig. 5-15.

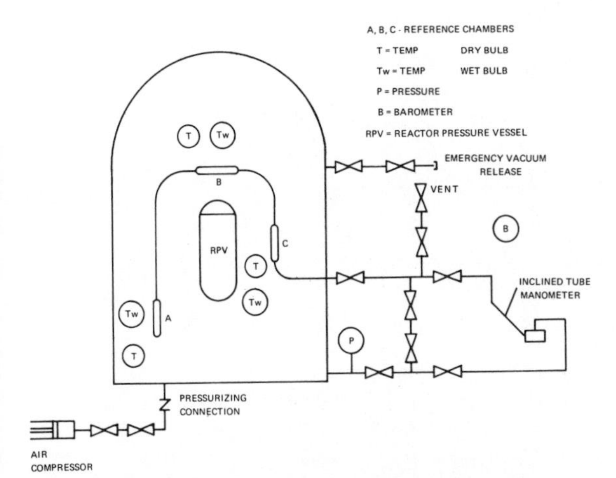

Fig. 5-15. Connections and instrumentation for the reference method of leakage testing.

5.14.1 *Reference Method*—Allowable building leakage varies according to plant location. The range runs from 2 percent to 0.1 percent per day of the building net volume. There are more installations at the low end of the range than there are at the high end. If the design leak rate is 0.2 percent per day in a 2 x 10^6 ft^3 building at 45 psig, then allowable free air leakage will be:

$$\frac{2 \times 10^6 \times 2 \times 10^{-3}}{1.44 \times 10^3} = 2.78 \text{ cfm at 45 psig}$$

$$2.78 \times \frac{59.7}{14.7} = 11.22 \text{ cfm free air}$$

As can be seen from the calculation, a high order of accuracy is required in order to measure the leakage with any degree of confidence. Therefore, during the test we measure inside and outside temperature, humidity, pressure in the building, and differential pressure between the building and the reference system. In the calculations, compensation is made for variation of temperature, pressure, and humidity. Because the absolute value changes are so small, the test runs for 72 hours with an option to continue for another 24 to 48 hours, if the results of the first 24 hours disagree with the results of the last 48 hours. In practice, readings are taken each hour and a leak rate calculated after each reading. The tests usually start at midnight; however, any mutually acceptable time is satisfactory. The basic requirement is that the test start and end at a time of day which is most likely to be a stable temperature situation. Cumulative percent loss as the test progresses is calculated as follows:

where

P = building absolute pressure, inches of water—the sum of building gage pressure and barometric pressure

P_w = H_2O vapor pressure in building, ins. of water

ΔP = difference between containment building-pressure and the reference system, ins. of water

T = building temperature, °R (°F + 460).

% Loss =

$$\frac{1}{P_1 - P_{w1}}\left[\frac{\Delta P_2\, T_1}{T_2} - \Delta P_1 - \left(P_{w1} - \frac{P_{w2}\, T_1}{T_2}\right)\right] \times 100$$

Subscript 1 denotes the value at the start of the test and subscript 2 denotes the value at hour, t, after start. Loss per day is cumulative percent loss divided by t hours multiplied by 24.

Building temperature and the dew point for determining the humidity pressure shall be weighted averages according to the volumes represented by each portion of the reference system. Since the pressure differential between the building and the reference system can go up or down according to temperature variation, the test must start with a pressure differential reading of 1.5 or 2.0 inches of water so that there is room for the differential pressure to change in either direction.

The manometer should be an inclined-tube-type that can be read to 0.01 inches and estimated to 0.001 inches. The barometer must be read to 0.01 inches of mercury. Building-pressure gages should read to 0.1 psig and estimated to 0.025 psig. Thermometers should be read to 0.5 degrees F.

5.14.2 *Running the Test*—A reference system is prepared comprising three or four volume chambers and the connecting tubing. The chambers are usually made of 2-inch thin wall copper pipe or tubing with brazed end caps. Copper is chosen to make the chambers responsive to ambient temperature changes in the building during execution of the test. Chamber length varies between ten and twenty feet. The reference chambers are connected with ¼-inch copper tubing. The assembled reference system is installed in the containment structure with each reference chamber in the center of the volume it is monitoring. Air temperature next to each reference chamber is measured. The reference system is then pressurized to about 10 or 15 psig with a mixture of air, and about 10- to 15-percent by weight of Freon 22 for leak detection. Using a halogen leak detector at its most sensitive setting, the reference system is inspected and repaired until there are no discernible leaks. The reference system is then pressurized to 15 or 20 psig and sealed. Temperatures at each reference chamber are taken over a twenty-four-hour period and a weighted average temperature calculated, considering the relative volumes of each chamber. Initial and final absolute pressures are compared, then compensated for temperature change. This twenty-four-hour check is repeated until zero leakage is shown in the reference system and then the system is blown down.

Humidity instruments and thermometers are installed alongside the reference chambers with their readouts outside the structure. Fans are provided to circulate the air all through the structure. The building is sealed and pressurization of the building and reference system is started. Start the circulating fans as soon as pressurization starts and run them all through the test. In order to save time, stop pressurization at about 25 percent of test pressure and examine all door seals, valve stems, valve seats, piping penetrations, etc., for leakage. Repair any leaks. Hold at this pressure for a half hour, or an hour, to be sure there are no gross leaks. It is good practice to pause at each 25-percent interval and check for leaks, so that when test pressure is reached there is reasonable assurance of a successful test. After reaching test-pressure, again check every possible source of leak. At this point particular attention should be given to inflatable seals and both doors in the air locks should be checked. During the pressurizing operation the inclined-tube, manometer equalization valves are open and the reference system is at the same pressure as the structure. At each pressure stage, gage connections, valve stems, etc., should be rigorously inspected to insure the absence of leaks in the measuring equipment. We prefer a removable section in the equalization line between the valves so that the section can be removed and the cross connection capped off.

When test pressure is reached, and after all known leaks have been corrected, seal off the reference system from the building. The differential pressure between the building and the reference system is zero; artificially bias the differential pressure reading by releasing some of the air from the reference system side of the manometer so the differential is about 1½ inches H_2O. Data should be taken every hour starting three or four hours before the official start of the test. During the course of the test, six-hour losses should be calculated for comparison with subsequent six-hour losses so that any failure causing an abnormal pressure drop can be detected before too much time is wasted.

The pressure loss equation includes a correction for humidity pressure. It has been included so that if the test is to be run on a BWR, it doesn't make any difference whether the pressure suppression chamber is wet or dry.

5.15 Ice Condenser Concept

In 1967 the Westinghouse Electric Corporation introduced its patented ice condenser containment concept. Basically, the system divides the containment structure into two compartments by a horizontal partition. The two compartments are connected by a vertical, annular, cold-storage ice compartment as shown in Fig. 5-16.

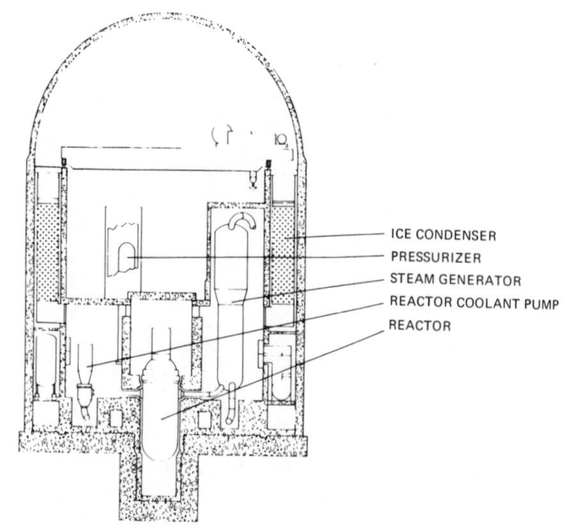

Courtesy, Westinghouse Electric Corp.

Fig. 5-16. Diagram of the patented Westinghouse Ice Containment System.

All energy sources are in the lower compartment. When an accident occurs, the steam is directed to the upper compartment through the ice compartment. The ice is 17 degrees F subcooled. Use is made of the sensible heat and the latent heat of the ice to absorb the heat energy released in the accident. In the normal building spray system or air recirculation system, water temperature increases 1 degree F for each Btu absorbed, per pound of water. In the ice-condenser system each pound of ice absorbs 144 Btu in the melting process. The system requires a refrigeration system and an ethylene-glycol circulation system to cool the ice machine and maintain refrigeration in the cold-storage compartment. The advantages claimed for the system are a lower design pressure—10 to 15 psig—and a smaller containment structure. The first two stations to use the concept, Donald C. Cook Nuclear Plant of Indiana and Michigan Electric Co. and TVA's Brown's Ferry Station, are now in construction. Evidence thus far indicates that each installation must be considered on its own merits. Each utility has individual equipment arrangement requirements that affect building size and these must be factored into the analysis to determine the economics of the situation.

As originally conceived, the system used borated ice so the melted ice contained a neutron poison when the liquid was used for recirculation. Further engineering developments added caustic to the ice so that dissolved sodium is immediately available to remove radioactive iodine from the atmosphere.

CHAPTER 6

Seismic Considerations

6.1 General

Nuclear power reactors and their supporting systems must include seismic influences in their design criteria for two basic reasons:

a. The public must be protected from the accidental release of radioactivity.

b. When a reactor is shut down during or after an earthquake shock, its safe and orderly shutdown must be guaranteed. It must also be kept safe after shutdown.

The public can be protected from activity release by adequate design measures on the *complete* pressure envelope of any system containing radioactive fluids. This is a static requirement and is relatively simple to implement. The second requirement, to provide a safe and orderly shutdown and maintain the reactor in a safe condition is a very complex requirement. Valves must open and close, motors must start, power must be supplied, and piping and instrument systems must operate. All the necessary components must absorb the shocks and either continue operating or start up. Some pieces of rotating equipment may be permitted to stall or jam as long as their pressure envelope retains it integrity. Emergency power supplies must start and deliver power when primary supplies fail. It is a matter of record that the San Onofre Nuclear Station operated normally through the Los Angeles earthquake in early 1971, because it was designed to accept the shock.

6.2 Terminology

In designing a nuclear installation we talk about two earthquakes—the "operating basis" earthquake and the "design basis" earthquake. The operating basis earthquake is classified as an upset condition at full power; the design basis earthquake is classified as a faulted condition. The operating basis earthquake (OBE) is selected as the most probable earthquake liable to occur after a thorough geological study of the area. The design basis earthquake (DBE) is usually taken at twice the intensity of the OBE. The loading is a combination of the horizontal load as determined by the seismicological study plus a vertical component of 75 percent of the horizontal load. All Seismic Class I systems must accept the OBE and continue normal operation.

As an "upset condition at full power," all OBE-induced loads and movements are included within normal design stresses in addition to any other loads that are present, such as thermal and pressure loads.

The DBE is different. The reactor coolant boundary must retain its integrity and the reactor must be brought to an orderly shutdown and maintained in a safe condition. As a consequence, the following is required:

a. Reactor control rod drives and back-up shutdown system must remain operational and shut down the reactor.

b. Control rod channels in the core must retain their geometry within the tolerances that prohibit control rod hang-up.

c. Fuel elements must retain their geometry sufficiently so as to avoid interfering with emergency core cooling.

d. Primary pumps (or BWR recirculation pumps) may jam and cease operation, but the pressure envelope must retain its integrity.

e. Emergency cooling systems must start and cool the core.

f. Systems not essential to safe shutdown of the reactor may fail, provided the activity release is not excessive to endanger the public.

g. Nonessential portions of safety systems must be isolated from essential portions.

In the DBE, stresses may exceed the yield point, permitting deformations to occur. The codes recognize this condition and prescribe the methods of calculation and permissible stress levels for the situation.

At the start of design of an installation, everything—buildings, structure, tanks, vessels, piping, wiring, cable trays, pipe supports, air ducts, lights, etc., are put into one of three Seismic Classes:

Seismic Class I: Those structures, components, systems, and instrumentation whose failure might cause or increase the severity of a loss of coolant accident or result in an uncontrolled release of excessive amounts of radioactivity and those structures, components, or systems vital to the safe shutdown of the reactor.

Seismic Class II: Those structures, components, systems, and instrumentation which are important to reactor operation but not essential to safe shutdown and isolation of the reactor, and whose failure could not result in the release of substantial amounts of radioactivity.

Seismic Class III: All other structures and equipment whose failure could result in interruption to station operation.

6.2.1 *Seismic Class I*—In defining a Seismic Class I system, all parts of the system must be investigated. Thus a Seismic Class I pump is useless if a motor feeder or pipe support has failed, the starter has ruptured, or a pipe fallen on the feeder cable tray. All "secondary" effects must be investigated. An unimportant masonry partition can be promoted to Seismic Class I if, when the partition falls, it breaks a Seismic Class I instrument lead. A portion of a common air duct becomes vital if it passes over a vulnerable Seismic Class I item. In Seismic Class I design, the ground acceleration, spectrum, and damping factors are defined for the specified earthquakes. Dynamic analyses are performed including all possible loads—dead load, pressure load, thermal load, seismic load, etc.

From the preceding description of the requirements of the DBE and the definition of Seismic Class I, a list of Seismic Class I systems and structures can be prepared:

1. Reactor containment building
2. Spent fuel storage pool or facility
3. Emergency power generator building
4. Nuclear service water intake structure and pump house that service decay-heat and emergency cooling systems
5. Reactor building in a BWR facility
6. Portions of the auxiliary building, housing engineered safety features systems
7. Portions of the auxiliary or radwaste building, housing radioactive waste storage facilities
8. Control building
9. Housing for emergency electrical switchgear plus power distribution routes
10. Primary coolant system pressure envelope and all of its supports
11. Reactor shutdown systems including the control rod drives and the back-up shutdown systems
12. Emergency core-cooling systems (See Chapters 2, 3, and 4 for a definition of these systems for each type of reactor.)
13. Containment spray systems
14. Containment air cooling system (for PWR only)
15. Closed-component cooling system (for a BWR only if service water is not suitable for bearing or motor cooling)
16. Nuclear service water system
17. Emergency power generator fuel storage and supply system
18. Containment I_2^{131} control system
19. Containment H_2 control system
20. Instrument air system
21. Emergency AC power system
22. Emergency DC power system
23. Containment isolation system.

6.2.2 *Seismic Class II and III*—Seismic Class II and III systems are designed in accordance with the requirements of the plant plus any national, state, or local codes that may be applicable to the installation.

It should be noted that not all of a system need be Seismic Class I. As an illustration, all of the Chemical and Volume Control System of a PWR is important for reactor operation but not for shutdown. However, the concentrated boron solutions stored in this system may be used for the back-up reactor shutdown system. Then, in the boron injection mode of operation, the equipment, valves, and pipelines used are Seismic Class I, while the remainder of the system is Seismic Class II. The proper procedure is to analyze all modes of operation of a system to determine its seismic classification.

6.3 Load Values

Seismic criteria for a given site are the result of a detailed study of the general area and of the site in particular. The records are searched for any earthquakes that may have affected the region in any degree. The geologic construction of the site is determined by site borings and a general examination of the area establishes the history of the earth and any possible movements that may have occurred. Finally, a maximum probable earthquake shock is estimated and ground movements and accelerations calculated. This study, the shock value and ground accelerations selected, together with the justification for the selection, are presented to the Atomic Energy Commission in the Preliminary Safety Analysis Report, for approval by an independent organization.

In using these criteria, the design engineers take the earth acceleration, applied at ground level, and develop a table of loading referenced to elevation from ground level. The table takes into account the dynamic response of the structures plus any damping or amplification values developed. As an example, the ground level load value of the DBE may be 0.1g horizontal; however, 120 feet up, the horizontal DBE load on the reactor building crane support may be 0.33 g.

Each Seismic Class I equipment specification, then, must contain the seismic load values for the individual piece of equipment as it will be installed.

Typical seismic horizontal load values range from 0.1 to 0.2 g for the DBE, with vertical load components running 70 to 75 percent of the horizontal loads.

6.4 Applications

For most applications, conformance to seismic criteria consists of a dynamic analysis of the equipment and its supports, followed by proper installation. Piping, however, is different. Piping is subject to thermal growth which may result from temperature changes as high as 1000 degrees F. Hence piping must be supported by a device that permits the required thermal expansion during normal operation, but behaves as a rigid structural member would under shock, vibration, or rapid acceleration.

This is usually accomplished by a device known as a hydraulic shock arrestor. The hydraulic shock arrestor is basically a double-acting hydraulic cylinder with the two hydraulic chambers connected through a restriction orifice. As a piping system changes temperature, the piston moves very slowly and displaces the hydraulic fluid from one hydraulic chamber to the other, through the orifice, with little or no resistance. In an earthquake, when pipe velocities are high, the orifice resists flow so the cylinder acts as a rigid member and transmits earthquake loads from the piping directly into the structure.

A subdivision of piping support is the support of primary pumps. Note, in Fig. 2-3, the short pipe runs of a primary loop. The pipe is at least 27″ I.D. with a 1- to 1½-inch-thick wall, and has little or no flexibility. If the pump casing were to be used as an anchor point for the piping, it could not accept the loads imposed on the nozzles without extensive, expensive modifications. It is easier to treat the pump as a piping specialty and handle it as though it were a valve. The pump is spring mounted and its weight is carried on its own hangars. It also has horizontal and vertical stabilizers with shock arrestors, emphasizing the fact that the pump is a piece of dynamic equipment.

Earthquake design is of more continuing importance to the structure designer and the component designer than to the system designer. The system designer has to identify the components, systems and structures that require seismic consideration and then insure that the proper measures have been taken to permit them to operate during and after an earthquake.

CHAPTER 7

Primary System

7.1 General

The primary system, composed of all the pressure vessels, pumps, heat exchangers, valves, and piping needed to remove heat energy from the fissioning reactor core, is very carefully designed, precision fabricated, and even more carefully installed. The system is the heart of a nuclear power plant because it is this assembly of equipment that guarantees the integrity of the fuel and the containment of plant radioactivity. With the issuance of the 1971 ASME Code, all equipment; pumps, valves, and piping required in the design, fabrication, and installation of the primary system pressure envelope are Class I of Section III of the Code. All phases of its fabrication and installation require a code stamp and a code report.

A typical system for a BWR is shown diagrammatically in Fig. 7-1. (It is shown pictorially in Fig. 3-6.) Note that the system for a BWR is "open ended" and its boundaries terminate at the outer isolation valves of the main steam and feedwater lines. This arbitrary decision is implemented by making the containment isolation valves part of the containment system so that in an emergency the reactor is completely isolated from the outside.

A typical PWR primary system is shown in Fig. 2-2. A reactor plant can have from one to four loops, depending on the reactor rating. As an illustration, an 1100 MWe plant by the Westinghouse Electric Co. will have four loops, each with one steam generator and one coolant pump. The same size plant from Combustion Engineering Inc., or Babcock and Wilcox, will have two loops; each with one steam generator and two coolant pumps.

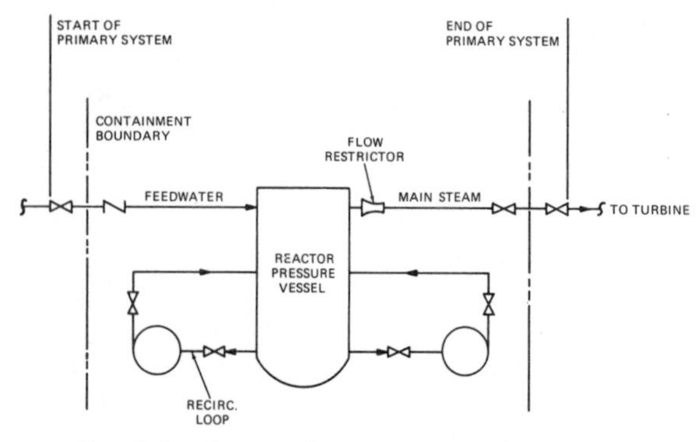

Fig. 7-1. Open end primary system for a BWR.

Courtesy, Power Engineering Magazine

Fig. 7-2. Reactor pressure vessel at Palisades Nuclear Power Station being moved across the site.

Reactor primary equipment is huge. Figures 7-2 and 7-3 show a reactor vessel rated at 2200 MWt (700 MWe) on a specially designed shipping skid outside of Consumer Power Company's Palisades Nuclear Power Station, and then up-ended inside the plant, ready to be installed in its foundation. The vessel's enormous size can be estimated by comparing it with the men in the photograph. Steam generators are still larger and heavier than the pressure vessels. Figures 7-4 and 7-5 show a steam generator being welded and moved into the containment structure. Notice the specially designed pivot at the left in Fig. 7-5, used to assist in installation. The main purposes of the special shipping and installation fixtures are for protection from damage and control of temporary loads imposed during installation.

7.2 Reactor Pressure Vessels

The reactor pressure vessel contains the fissioning nuclear core and is actually the energy-producing vessel in the power plant. A 2714 MWt BWR vessel is 72 ft 0 inches long and has an inside diameter of 238 inches. It is designed for 1250 psig with a wall thickness of 6¼ inches of base metal plus ⅛ inch of stainless steel cladding. The top head of the BWR vessel is not clad because the steam separators and

Courtesy, Power Engineering Magazine

Fig. 7-3. Pressure vessel of Fig. 7-2 up-ended ready for installation on its supports. The vessel has remained fastened to the shipping skid until it is vertical.

dryers (shown in Fig. 3-2) scrub the liquid out of the steam vapor before it contacts the head.

A PWR pressure vessel, rated 475 MWt (151 MWe) higher than the BWR vessel just described, has an inside diameter of 173 inches and a length of 43 ft 10 inches. Its wall has 8⅝ inches of base metal plus 7/32 inch of stainless steel cladding. The PWR vessel weighs 858,000 pounds, while the BWR vessel weighs 1,550,000 pounds. The differences in size of a BWR vessel and a PWR vessel are due to the larger core in the BWR and the steam drying equipment installed in the top portion of the vessel.

Initially, the vessels were made of SA-302[1], Grade B steel, an 80,000 psi, tensile manganese-molybdenum steel. In the normalized and tempered condition it could be used in thicknesses to 8 inches. By quenching and tempering, the material could be used in thicknesses up to 12 inches. Recently, the material was improved by the addition of nickel and the use of fine-grain melting practice. This improved material is now the accepted base metal material for thick-walled reactor pressure vessels and is designated SA-533, Grade B, Class I. All internal surfaces contacting primary coolant liquid are clad with stainless steel. The cladding is applied by an automatic welding process.

The welding process is a multilayer, multipass procedure that uses different filler rods in different layers. A typical first layer filler rod is Type 309 stainless steel which is rich in both chromium and nickel and creates an excellent transition zone from the base metal to stainless steel. The finish layers can be Type 308 stainless steel. After completion of the cladding, the top 0.050 to 0.100 inch of the cladding is sampled and analyzed to prove that the surface metal chemistry is what it should be.

However, this cladding layer has one shortcoming; the stainless steel is in a "sensitized" condition. Stainless steel is said to be "sensitized" when the carbon is not in solution but has precipitated out as a chromium carbide. Material in this condition is much more subject to corrosion than is regular stainless steel. Sensitization is caused by the slow cooling of stainless steel from about 1550 degrees F down to 1050 degrees F. The weld overlay process requires preheating of the base metal slab and transforms it into a large heat sink such that it is impossible to cool the cladding rapidly through the sensitization range. But as long as the material is carefully handled and processed, the sensitivity does not impose any unsurmountable problems.

7.2.1 *Nil Ductility Transition Temperature*—Carbon and low alloy steels exhibit a property known as "nil ductility transition" (NDT). This phenomenon is characterized by a sudden loss of ductility and the transformation of steel into a brittle metal similar to cast iron. Normal carbon steels with controlled chemistry reach the NDT point at about 0 degrees to −20 degrees F, but during exposure to neutron irradiation the NDT temperature rises. Hence, the AEC has recognized the potential hazard in Design Criterion 31. Since all reactor pressure vessels are designed to work only in the elastic range, the initial NDT temperature is specified to be between 20 degrees F and 40 degrees F, making the allowable hydrostatic test temperature between 80 degrees F and 100 degrees F. The change in NDT temperature is directly related to the integrated neutron dosage—the greater the total dosage, the greater the change. The NDT point has been known to change as much as 300 degrees F.

In specifying a reactor pressure vessel, the best way to handle the NDT change is indirectly. The basic thing to be avoided is the necessity of having to pressurize the vessel at a temperature lower than the NDT point. The specification therefore relates pressurization temperatures to circulating pump net

[1] Material specifications starting with the letter "S"are ASME specifications from Section II of the ASME Code. They are generally the same as ASTM specifications. The ASME number adds an "S" to the front of the ASTM number.

Courtesy, Combustion Engineering, Inc.

Fig. 7-4. Automatic welding of the top head on a steam generator in the shop.

positive suction head. Since an average coolant pump needs 5000 to 6000 hp, this same energy is used to

Courtesy, Power Engineering Magazine

Fig. 7-5. Steam generator moving through the construction opening at Palisades. A lifting pivot is resting on the crawler at the left.

warm the system initially. If the reactor pressure vessel can be pressurized at startup sufficiently *at all times* over the life of the plant (usually 40 years), to permit operation of the coolant pump then a 300 degree F NDT change can be tolerated because the service temperature of the vessels is approximately 600 degrees F. What is required are programs, recognizing NDT shift, that control temperature and pressure so that at no time during either warmup or cooldown, are excessive loads imposed on an irradiated vessel wall. It is appropriate to note at this point that a BWR can operate at up to 20 percent of rated power without using its recirculation pumps. Therefore, the preceding discussion of NDT is of less consequence for a BWR than for a PWR since the BWR vessel pressure is developed by boiling, directly in the vessel.

NDT temperature shifts are predicted from samples irradiated in materials-testing reactors and from tests performed on the PM-1A pressure vessel after the reactor was shut down and dismantled. Nevertheless, surveillance specimens are placed in each reactor vessel as a control during the life of the vessel. The surveillance program follows ASTM E185, "Standard Recommended Practice for Surveillance Tests on Structural Materials in Nuclear Reactors." In practice, NDT shifts on irradiated samples are compared with predicted shifts to maintain control over the vessel material.

7.2.2 *Sealing*—All vessels, as shown in Figs. 2-4 and 3-2, have a removable flanged head, fastened with studs and nuts, for initial installation access and routine refueling. Precise joint loading is achieved by elongating the stud under carefully controlled conditions and then running the nut down to solid contact to maintain the preload on the stud. The joint is sealed by two independent concentric O-ring seals and the annular space between the two seals is continuously monitored for leakage. The O-rings are metal and are normally made of Inconel-600 or 750-X plated with silver. The use of O-ring seals may be taken as a general indication of the order of accuracy required in parts of reactor pressure vessel fabrication. O-ring diameters are about 6 to 10 inches larger than the vessel internal diameters previously given; the O-ring groove diameter requires a dimensional accuracy of −0.000 inch, +0.005 inch. Thus, this machining operation has an accuracy of 0.005 inch in 238 inches.

7.2.3 *Fabrication*—Fabrication is a series of carefully controlled operations with quality control checks after each step. The first of the controls starts with chemical analyses and destructive testing immediately after pouring the base metal. Each step in the procedure, from acceptance of raw materials at the shop to final shipment, is documented so that the complete history of the vessel is known. At the present time there are three fabricators in the United States who manufacture these vessels: Combustion Engineering, Inc., Babcock and Wilcox Company, and the Chicago Bridge and Iron Company.

It is felt that the present generation of 1100 MWe reactor pressure vessels represents the largest vessels that can be shipped assembled, with any reasonable degree of economy. Present-day reactor pressure vessels are usually too large to ship overland in one piece. The Chicago Bridge and Iron Company has already assembled a BWR pressure vessel, 207 inches I.D. × 63 ft 2 inches overall length, on the site at Monticello, Minnesota. The plant rating at this site is 472 MWe. The decision was made to site fabricate the vessel because of transportation clearances rather than vessel weight. In this case, barge freight was available as far as Minneapolis. No satisfactory road or rail route could be found from Minneapolis to the site at Monticello. Additional information is available in the Final Safety Analysis Report for the Monticello Nuclear Generating Plant of the Northern States Power Company. In late 1969, a study made for a 1700-1800 MWt plant 50 miles from Mexico City reported an additional $2,000,000 cost for field fabrication exclusive of shipping cost differentials.

With the exception of the small rod drive or in-core instrumentation nozzles located in the bottom head, all other nozzles are located above the core. The main reason for this location is to keep the core covered with coolant if a pipeline should fail. Reactor pressure vessels are supported from support pads placed beneath the main coolant or steam nozzles, or support skirts at the bottom, as shown in Fig. 3-2. Supports just below the main nozzles minimize thermal motion transmitted to outside pipelines.

7.3 Primary Pumps

This description treats a BWR recirculation pump as a primary pump. The modern primary coolant pump is a vertical, single-stage centrifugal pump with a controlled leakage shaft seal. Volume capacities range from 25,000 gpm up to 90,000 gpm per unit depending on the reactor type and size.

Fig. 7-6 illustrates a primary pump on the test stand. The visible portions of the unit are the drive motor and the coupling barrel. The pump casing and nozzle are hidden below the floor grating. Dynamic heads range from 300 ft to 800 ft. The pressurized water reactor pumps are in the higher end of the volume range, but in the lower end of the head range. The BWR recirculation pumps are the reverse. Previous to the 1969 equipment changes, the General

Courtesy, Westinghouse Electric Corp.

Fig. 7-6. A Westinghouse primary pump on the test stand. Only the motor, oil coolers, coupling barrel, and top casing flange are visible. The casing and piping are below the floor. (See Fig. 7-7.)

Electric BWR recirculating pump had a design head requirement of 450 ft to 500 ft. When General Electric substituted a throttle valve for the variable-speed pump an additional 300 ft of head was needed for proper operation of the valve.

The pumps require 4500 to 7000 hp with reactor coolant at operating temperature and pressure. Cold operation with reactor coolant at room temperature requires an additional 1500 to 2000 hp. The motors are usually rated for a maximum of one hour's operation at the cold conditions and unlimited, continuous operation at the hot conditions.

Initially, the motors were air-cooled but facility designers are now turning to "water-cooled" motors to keep the large cooling load out of the ambient air. "Water cooling" is an oversimplification. There is

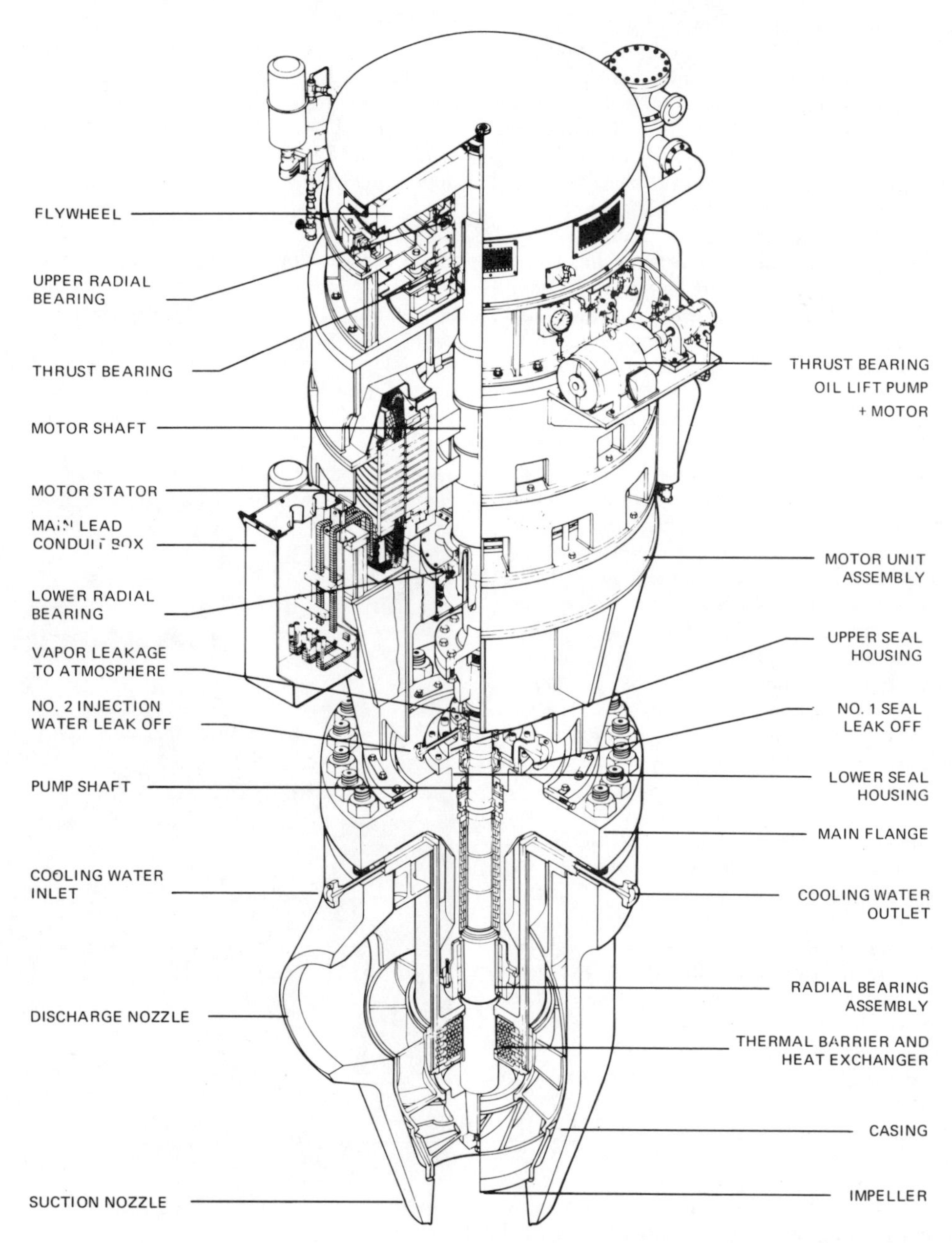

Courtesy, Westinghouse Electric Corp.

Fig. 7-7. Cross section of a primary pump. Note the flywheel at the top of the assembly and the externally cooled thermal barrier protecting the radial bearing from primary coolant temperature.

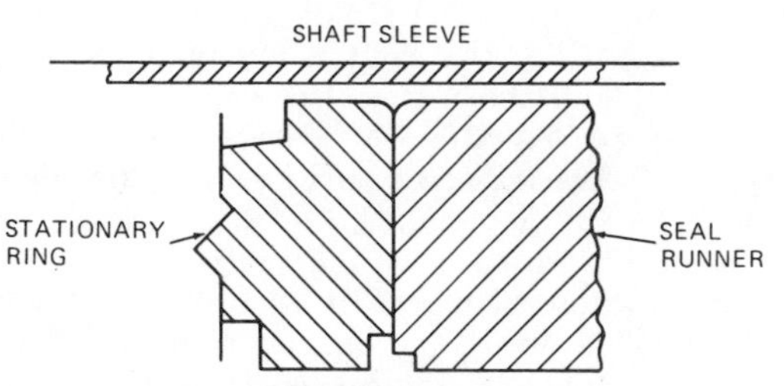

Fig. 7-8. Typical film seal opposing faces. The faces are not in contact during operation.

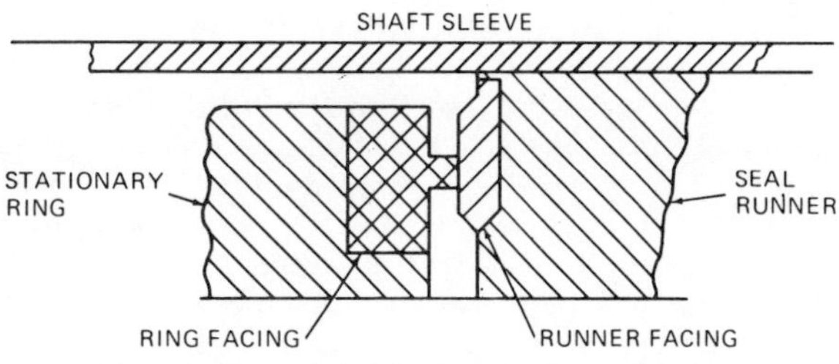

Fig. 7-9. Contact seal. The narrow stationary surface is usually graphite and runs against a hard carbide.

actually a closed loop, driven by the motor cooling fan, in which the air circulates through the motor and a finned-tube cooler. Cooling water from a closed cooling loop flows through the motor cooler.

Figure 7-7 is a cross section of a Westinghouse manufactured primary pump. Note the fly wheel on the top end of the motor. Its purpose is to provide sufficient energy, after a power failure, to cool the reactor core until the emergency cooling systems, powered from a diesel generator and a battery bank, take over the cooling load. Flywheels are now being ultrasonically inspected to prove absence of flaws in the material even though they are not part of the primary pressure envelope. This inspection is performed because of the engineered safety feature aspect of the flywheel. The motor thrust bearing is a Kingsbury-type bearing. The radial bearing in the pump is proprietary graphite, lubricated and cooled by cool water injected via the cooling water inlet. Below the pump bearing is the thermal barrier which provides a cool (150 degree F), working, ambient temperature for the bearing. The barrier is essentially a cooling coil assembly. Below the thermal barrier is a mixed-flow-type impeller. The visible portions of Fig. 7-6 can be correlated with the part identifications in Fig. 7-7.

All pumps should be specified with renewable impeller and casing wearing rings. The pumps should also have a spacer-type coupling to permit replacement of the shaft seals without disturbing the motor or the pump casing. The seals should be incorporated into a single seal cartridge to facilitate replacement. The pump in Fig. 7-7 does not need isolation valves to change the seals. Between the thermal barrier and the bearing assembly is a back seat, and when changing a seal, the shaft-impeller assembly is lowered until it has seated on the back seat sealing the coolant into the pump. The system must be depressurized, but not drained, to replace the shaft seals. This back seat should be specified in a pump procurement document to insure that it is furnished.

7.3.1 *Pump Seals*—All pumps use multiple controlled leakage shaft seals. Each of the manufacturers treats his seal designs as proprietary information and will use various combinations of film seals, contact seals, and throttle bushings. The illustrations that follow are illustrative of principles only and do not come from any manufacturer.

A film seal has two relatively broad surfaces, shown in Fig. 7-8, opposing each other and separated by a thin film of liquid about 0.0003-inch to 0.0005-inch thick. The seal surfaces never actually contact each other. This type of seal requires a controlled leakage flow of the order of a couple of gallons per minute for proper operation. The two opposing surfaces are usually made or faced with some type of ceramic such as aluminum oxide or boron carbide. The seal also requires a certain minimum pressure across the faces to operate properly. This pressure is usually selected to be equal or slightly less than the net positive suction head. It must be remembered, however, that the engineer *cannot assume* that this is true and the minimum operating pressure for the seal must be checked for every application. The engineer should consider the use of a pressure switch to prevent pump startup unless the minimum seal pressure and/or NPSH is available.

A contact seal, as shown in Fig. 7-9, has a relatively narrow, stationary, wearing surface contacting a broad, hard, running surface. Typical contact materials in this seal are graphite for the stationary wearing surface and ceramic for the running surface. This is a true zero leakage seal, as no flow is required through the seal for correct operation, but it fails suddenly, since there is normally no leakage through it. The graphite seat is replaced on a regular preventative maintenance schedule. Flow direction for both the film seal and the contact seal is in toward the shaft.

A third seal, the throttle bushing, Fig. 7-10, was

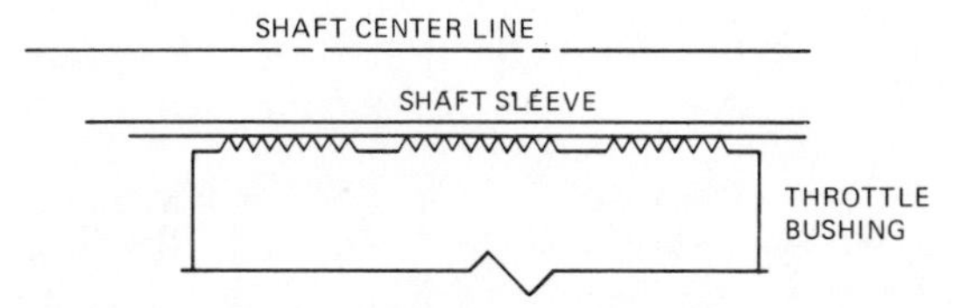

Fig. 7-10. Typical throttle bushing seal arrangement.

the first seal built for a large nuclear pump in the early 1960s. This design is a variation of the familiar labyrinth seal on a steam turbine and works on the same principle. Each close-fitting ring around the shaft sleeve acts as a restriction orifice and a series of orifices are built-up to take the total pressure drop required. This seal, of course, requires continuous flow through it, to be effective. It is most frequently used today as a third seal, backing up two of the other type seals. One manufacturer uses it to by-pass water around a contact seal to reduce its normal pressure duty, even though the contact seal is built and tested for the full pressure requirement.

Elastomeric O-rings are used for static joints in the various parts of the seal assembly. Neoprenes have been used in the past, but ethylene-propylene terpolymer has a higher radiation resistance and can be molded into suitable shapes. Casing gaskets are austenitic stainless steel spiral, metallic gaskets with a low-chloride asbestos filler (less than 200 ppm).

The pump bearing is a self-aligning graphite bearing with a carbon binder. It is water-cooled and lubricated, and runs against a Stellite facing on the shaft. Graphite is selected for the bearing material for both its radiation resistance and its self-lubricating properties.

The net positive suction head (NPSH) of these pumps varies with size and is in the range of 75 ft to 150 ft of head. The pumps in today's nuclear plants require about 100 ft of NPSH. This NPSH is not important during normal plant operation at temperature and pressure. It is, however, a limiting factor in the start-up program of a pressurized water reactor where pump energy is used to warm the system before going critical. The reactor pressure vessel must be so designed that the system can safely be pressurized cold at the end of forty years, to provide sufficient NPSH. A boiling water reactor can be started on natural circulation and the pump started after the system is warm.

The pumps were designed and built initially under the rules of Class A, of Section III of the ASME Code, to insure design criteria, material control, and inspection. In the fall of 1968, the draft "Code for Pumps and Valves for Nuclear Power" was issued; the Summer, 1969 addendum to Section III of the ASME Code made the use of Class I in the Pump and Valve Code mandatory for use in systems classified as Class A of Section III. The pump has now been included directly, in Section III of the ASME Code.

All of the large primary pumps are coming into service in the early 1970s. Hence, it is prudent to require hot tests at system temperature and pressure on the pump and seal before shipment from the factory, to insure that all is in order and produce a certified performance curve. Additionally, the author prefers to accumulate as much as 100 hours' operating time, with at least half of it continuous, and at least a half-dozen stops and starts before shipment.

The casing and impeller are of cast stainless steel, SA-351, Type CF8; the shaft is a forging, SA-336 Grade F8M; and the motor flywheel is of carbon steel, SA-515 or SA-516, 100 percent ultrasonically inspected. The wear rings should be Type 304 stainless steel, SA-240 if plate, or SA-336 if forged, and they should have a hardness difference greater than 70 Brinnel Hardness Numbers. Casing studs and nuts must be made of ASME specification materials; Type 316 stainless steel, suitably treated to avoid galling, is preferred.

Primary coolant pumps are made in the United States by three manufacturers: Westinghouse Electric Corp., Byron-Jackson Division of Borg-Warner Corp., and the Bingham Pump Company. Other manufacturers have been reluctant to enter the field because of the multimillion-dollar investment required for additional facilities, including a hot test loop.

7.4 Steam Generators

The steam generators described here are used only in pressurized water reactors. Each of the three PWR manufacturers designs and manufactures his own

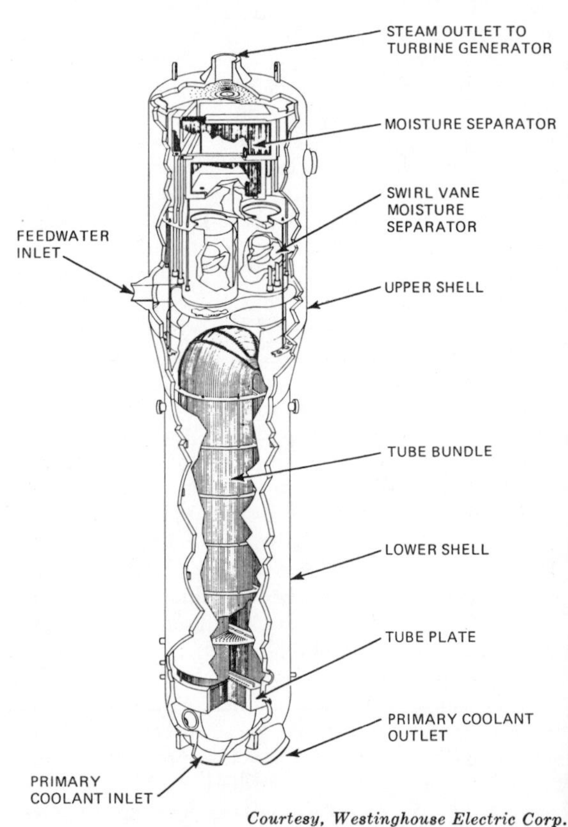

Courtesy, Westinghouse Electric Corp.

Fig. 7-11. U-tube steam generator for a saturated steam PWR. Steam drying equipment is the type used by Westinghouse Electric Corp.

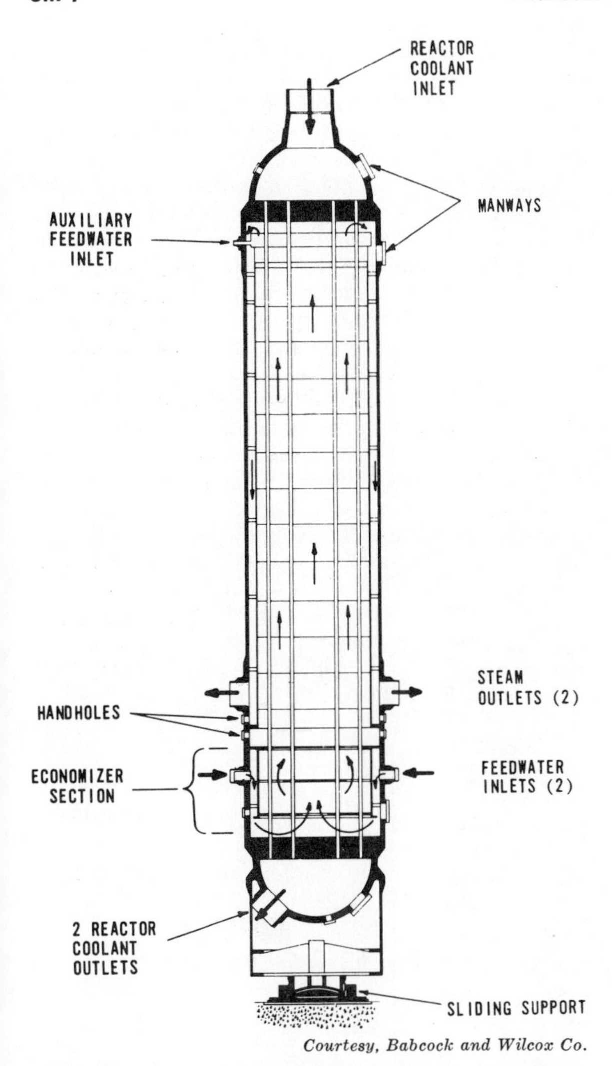

Courtesy, Babcock and Wilcox Co.

Fig. 7-12. Once-through steam generator of Babcock and Wilcox Co. Note the feedwater inlet just above the bottom tubesheet.

steam generator. As noted in Chaper 2, Westinghouse Electric Corp. and Combustion Engineering Inc. use a U-tube generator as shown in Fig. 7-11, and Babcock and Wilcox Co. uses a "once-through" generator shown in Fig. 7-12. With a generating station of approximately the same power level—1000 to 1100 MWe (3100 to 3400 MWt)—steam generators vary in size. Westinghouse will furnish the rating in four steam generator units while the other manufacturers achieve the same capacity in two units. Generator design requires Class 1 of Section III of the ASME Code on the tube side (primary coolant), and Class 3 on the shell side (steam side); some manufacturers will furnish a Class 1 shell side if requested. The tube side has a carbon steel bonnet, clad internally with stainless steel. The shell and baffles are carbon steel, and the tubes are Inconel. The tube plate (tubesheet) is carbon steel, faced on the tube side with weld overlaid Inconel. The tube to tubesheet joint is welded, using special machinery developed for the purpose. The carbon steel plate is either SA-533 Grade B, or SA-516 Gr. 70, depending on thickness and application. Carbon steel forgings are SA-508.

Figure 7-11 is a typical U-tube unit. The lower shell is the evaporator containing the tube bundle, and the upper shell is the steam drum containing the steam separating and drying equipment. Feedwater enters the generator through the feedwater inlet and is distributed around the circumference of the generator by a pipe manifold. As it leaves the manifold, the feedwater mixes with the hot separated water from the dryers and flows down the annulus, between the shell and the tube bundle shroud. At the bottom, the preheated water flows radially into the tube bundle where it is heated and evaporated. The rising wet steam is dried in two stages and leaves the generator for the turbine generator set. Steam leaving this unit is not radioactive as is the steam in a boiling water reactor. The steam has a slight moisture content, 0.20 to 0.25 percent.

The flow pattern described has two advantages: 1. The cool feedwater protects the shell from exposure to live steam. 2. By the time the feedwater reaches the tube plate it has been sufficiently heated to reduce the thermal stresses that occur across the tube plate.

The Babcock and Wilcox Co. has two steam generator designs. Up to mid-1969 they offered a design in which feedwater enters the annulus between the shell and the shroud, about halfway up the generator, flows down the annulus to the tubesheet and then enters the tube bundle radially. As the water rises up through the bundle it is evaporated and superheated. The steam turns down into the upper portion of the annulus, flows down the annulus a short way, and then exits. The preheating effect in the annulus tends to reduce the effective temperature difference of the generator and, hence, reduce the capacity. This design is shown in Fig. 7-13.

In mid-1969 they started to offer the revised steam generator design, shown in Fig. 7-12. Feedwater is introduced into the tube bundle just above the tubesheet. This increases the effective temperature difference and also provides true counter current flow. The steam leaves the tube bundle and goes down the annulus between the shroud and the shell, exiting near the bottom. This flow pattern tends to equalize the expansion of the shell and tubes and reduce stresses in the tubesheets. The design, however, requires an additional modification: the shell side of the tubesheet is overlaid with a ½-inch layer of nickel-chromium-iron alloy to help control the additional thermal stresses. Feedwater temperature requires a lower temperature limit control system so that thermal stresses across the lower tubesheet are

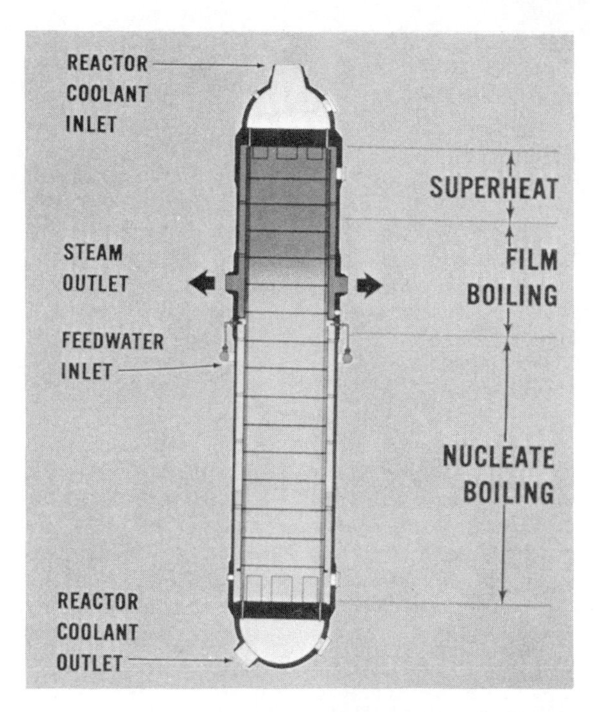

Courtesy, Babcock and Wilcox Co.

Fig. 7-13. Once-through steam generator offered prior to 1969. Note the feedwater inlet midway up the shell.

controlled. In extreme situations it may be necessary to raise the feedwater temperature to limit thermal stresses across the tubesheet.

Thermal designs are proprietary and no information about them is made public.

Initially, one of the manufacturers shipped his steam generators in two pieces to reduce shipping weights and clearances. The required field weld was not consistent, however, so now all generators are shipped fully assembled.

7.5 Pressurizer

The pressurizer supplies and controls the system pressure in a PWR, as described in Chapter 2. The unit is a vertical cylindrical vessel, as shown in Fig. 7-14, and the sizes do *not* change with a change in reactor rating. The relief valves and spray nozzle are at the top in the vapor zone; the heaters are down in the liquid zone. Pressurizers have a volume of 1500 to 1700 ft^3, about half of which is filled with liquid. Mechanical code safety-valves and power-operated relief valves are connected to vapor space nozzles and the surge line is connected to the bottom of the vessel.

Specific vessel configurations vary from manufacturer to manufacturer but a typical pressurizer has an inside diameter of about 7 ft 6 inches and an overall length of about 45 ft. Vessels are supported either by lugs about midway up the shell, or by skirts down at the bottom. Pressurizer base metal is carbon steel, SA-516 Grade 70, 6 to 6½ inches thick, clad internally with a minimum of ⅛ inch of austenitic stainless steel.

Heat energy is supplied by multiple banks of replaceable electric heaters; typical total connected heater load is 1500 to 1600 KW. The heaters are arranged electrically so that on startup all the heater banks are used in order to minimize pressurization time. As the unit comes up to pressure, heaters are turned off until finally, only one small group of heaters operates continuously to replace insulation heat losses. A second, small group of heaters responds to the demands of the pressure control system as required. Heating wire is hermetically sealed into an SA-213, TP 316L tube with magnesium oxide insulation.

Design pressure is the same as the rest of the primary system, but design temperature is 50 to 100

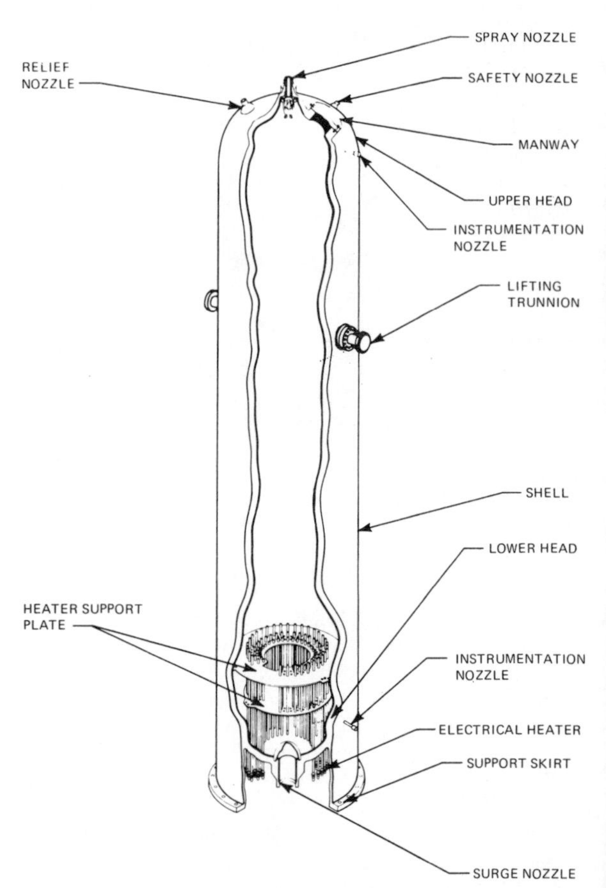

Courtesy, Westinghouse Electric Corp.

Fig. 7-14. Cross section of a PWR pressurizer. This unit is skirt supported with bottom-entry heaters. Other variations are side-entry heaters and shell mounted support brackets.

degrees F higher than the saturation temperature of the steam corresponding to design pressure.

Each of the reactor manufacturers fabricates his own pressurizer.

7.6 Primary System Valves

Primary system valves are refined versions of familiar valve types. The greatest change has been in the development of operators for large-size valves to permit rapid closing yet prevent water hammer and to avoid the mechanical shock of the valve plug impacting on the seat during an emergency closure.

At the start of the nuclear industry there were no valve standards. Each manufacturer rated his valve design in his own manner and confirmed it by test. Body and seat leakage test standards were and still are available as a standard practice of the Manufacturers Standardization Society (MSS). External dimension standards were and still are available in ANSI Standards and MSS Standard Practices, for various types of valves. There are no standards or codes to this day that control valve internal designs. Section III of the ASME Code applies only to the pressure containing boundary of valves 4 inches and larger. The code may be applied to valves smaller than 4 inches by direct statement in the procurement specification. Before the issuance of valve codes, ANSI B31.1.0 and the Nuclear Code Cases ("N" series) and Section III of the ASME Code were used. Valve quality was controlled by a combination of specific requirements and parts of Section III of the ASME Code. Today, the pressure boundary must conform to Class 1 of the new edition (1971) of Section III, and the design engineer must specify his seat tests. Class 1 of the code controls the design, materials, and quality of the pressure boundary; the system design engineer must add requirements for the valve internals by direct statements in the procurement specification.

Valves are not permitted to leak to atmosphere at any time—hence, two alternatives are possible: 1. The valve shall never leak along the stem, or 2. The valve shall be arranged so that the leakage is captured and directed into the radioactive waste system. Small valves, up to ½ inch or ¾ inch, with bellows seals may be competitive economically. Above this size, bellows are too expensive and a double packing arrangement with a lantern ring, as shown in Fig. 7-15, is used. The port opposite the lantern ring is piped to waste and all leakage passing the first set of packing is collected.

Valves are of stainless steel, forged or cast, depending on the size. Whenever the valve is essentially a commercial product, the valve body is Type 316 stainless steel, because these are available normally. If the valve is special, it may be either Type 304 or Type 316 stainless steel; the deciding factor is cost. Type 316 stainless steel is more expensive than Type 304, but it is also stronger. Thus, if the choice is between a 1500 lb standard valve in Type 316, or a 2500 lb standard valve in Type 304, the 1500 lb valve may be cheaper.

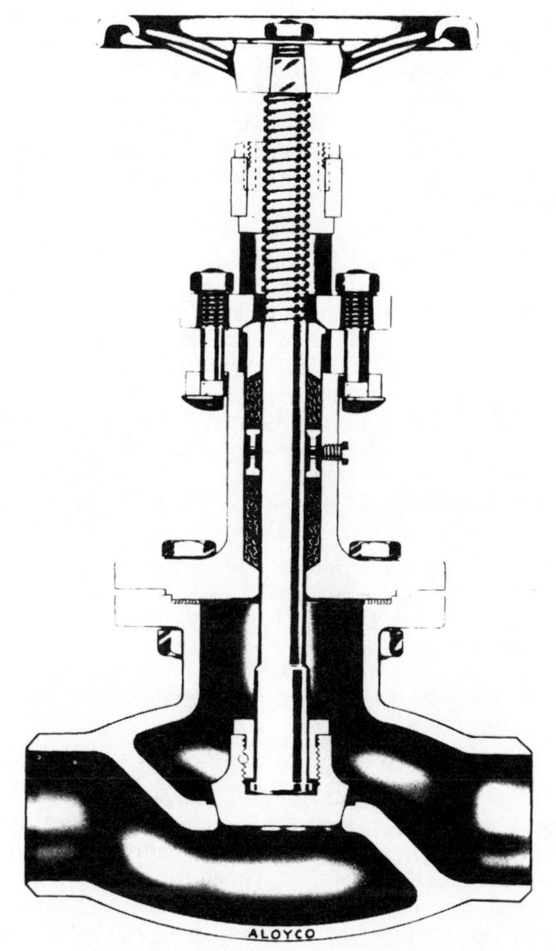

Courtesy, Aloyco

Fig. 7-15. Hand operated globe valve with double packing, lantern ring, and packing leak-off.

The main steam isolation valves in the BWR, which do not contact liquid, are carbon steel, ASME SA-352 Type LCB, or SA-216 Type WCB, rather than stainless steel. BWRs are supplied with two recirculation pump gate pattern isolation valves, plus a ball type throttling valve with a small by-pass on the discharge side of the pump. The gate valve is a venturi-type valve with a 22-inch port and 24-inch body connections. The valves in the recirculation loops are of stainless steel because they are in continuous contact with primary water.

7.6.1 *Valve Design Requirements*—Valve seats should be hard-faced with an erosion-resistant material such as one of the cobalt-tungsten alloys. Seat leakage requirements should be 2 cc-per-hour, per inch of seat diameter at design differential pressure or less, for valves to isolate components under pressure, auxiliary systems and containment isolation valves. Other valves can comply with the MSS specification of 10 cc per hour per inch of valve size. The author prefers a seat test of 3 minutes regardless of valve size.

In 1969, experiments were conducted using a Type 410 or 416 nitrided seat ring. Initial experience indicated that the seat might be too brittle, but more work is required before a definite conclusion can be drawn. It would be advantageous for the design engineer in the 1972 to 1975 period to investigate the final conclusions for the use of this material.

Valve stems can be an austenitic steel, an Inconel, or a hardened 17-4 PH chrome-nickel steel forging. The exact designation for the 17-4 PH is ASTM A-564, Grade 630, heat-treated to meet the following specifications:

Ultimate tensile strength, min	150,000 psi
Yield strength, min	135,000 psi
Elongation in 2 inches, min	17%
Reduction in Area, min	58%
Hardness	
Brinnel	332
Rockwell	C34
Charpy V-notch	20 ft lbs at 10°F.

Valve body designs should provide replaceable seat rings and backseats for the stems. All valves, not in throttling service, should be backseated when they are open, to decrease dependence on the stuffing boxes. Primary-system valves should be equipped with operators either because of their emergency duties or their inaccessibility. Gate valves should have either a split wedge or a "flexible" wedge to permit sealing in both faces. In specifying tests for gate-valve patterns, the seat should be tested across the complete wedge from each side, and from inside the wedge out to each wedge seat. This second test is particularly important in the case of isolation valves on a BWR primary system.

7.6.2 *Body Gaskets*—Bonnet-to-body gaskets are of two types—solid metal and spiral-wound stainless steel with a soft filler. The spiral-wound gasket is Type 316 stainless steel, with an asbestos filler; the asbestos should not have more than 200 ppm of chlorides. When using this gasket type, gasket compression should be controlled by compression gauging. In Fig. 7-16, gasket compression is fixed by seating the valve bonnet on the body.

Solid metal gaskets are used on the higher pressure applications and are usually captured in a tongue-and-groove design. Specific materials should be chosen according to the application—some materials are aluminum, Monel, plated and plain soft iron, and stainless steel. Gasket loading is accomplished by controlled bolt loading rather than compression

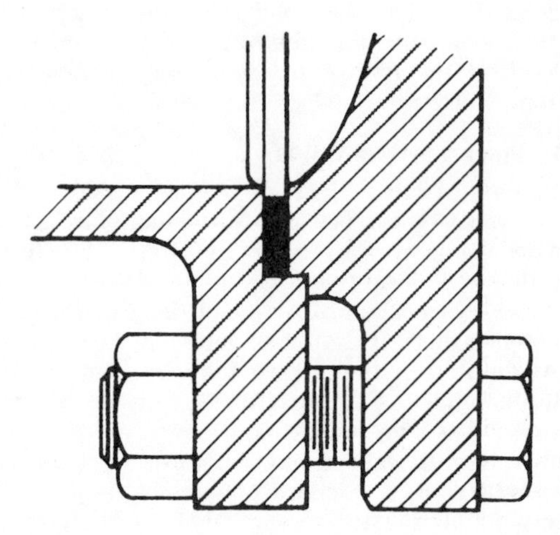

Courtesy, Aloyco

Fig. 7-16. Gauging gasket compression by seating the bonnet on the valve body.

gauging. Additionally, some designs include a canopy seal ring although it is not too popular. On large valves, double bonnet seals may be used with an annular leak-off in the same manner as on a reactor pressure vessel.

7.6.3 *Stem Packing*—Valve stem packing has evolved from an unsatisfactory, grease lubricated packing to a graphite-impregnated asbestos with corrosion inhibitor for storage, that can accept a dosage of 10^7 *R*. "*R*" is the designation for a roentgen, a unit of energy. A description of this unit, its effects on material, and the resistance of several materials to radiation damage is given in Chapter 12. The use of grease is to be avoided entirely because the grease deteriorates under ionizing radiation and forms tars and gums. If the application is such that the graphite-impregnated packing has too much friction, then TFE resins may be used under very carefully controlled conditions. TFE resin starts to degrade at a total dosage of 5×10^5 *R*; it can safely be permitted to operate up to a dosage of 1×10^5 *R* before replacement. The objection to this, however, is that the total dosage in an operating year may exceed the tolerance limits for TFE.

7.6.4 *Operators*—Operators may be any of the conventional types—electric, pneumatic, or hydraulic. In applying an electric operator, insulation for motor and wiring must be able to accept the radiation. Class B or H insulation is acceptable for motors. Lubrication must be acceptable for the anticipated radiation levels; most of the petroleum companies have suitable lubricants and are willing to cooperate. The electric gear operator should have, as a minimum, torque and end position limit switches to protect the equipment. Any other signal devices are additional. Electric operators are usually applied to large valves that fail "as is." Throttling valve operators should have continuous position indicators; ON-OFF operators should have limit-switch indication. All position indication should originate at the valve, not at the control.

Pneumatic operators are used as in conventional applications. The only vulnerable part in a pneumatic diaphragm operator is the diaphragm itself. Natural rubber and ethylene-propylene terpolymer are the best diaphragm materials with a damage threshold of 10^8 *R*. The two next best diaphragm materials are neoprene and Hypalon, with a damage threshold of 10^7 *R*. Valves with pneumatic operators that must isolate the primary system in an emergency are provided with air reservoirs and supply systems similar to Fig. 5-10B. For higher operating forces, cylinder operators are used with higher air pressures. Operators are all double acting with the addition of springs, to drive the valve to the safe position if pneumatic power should fail. Thus, a single failure will not prevent the valve from closing.

Hydraulic operators are not used because of the inconvenience of supplying an acceptable hydraulic fluid and they offer no advantage over the pneumatic cylinder. However, demineralized water has been used successfully as a hydraulic fluid.

7.7 Piping

The material in contact with primary coolant liquid is always an austenitic stainless steel; and the alloy is usually Type 304. In the BWR, carbon steel piping is used for the primary *steam.* Economics plays an important part in pipe selection as it does in every engineering decision. Since stainless steel is required in contact with the liquid, an investigation should be made of using the more economical carbon steel pipe internally clad with stainless steel. Experience in the market of the late 1960s has indicated that the pipe sizes up to 12 inches are just as economical in solid stainless steel as in clad carbon steel, for wall thicknesses up to schedule 140. In the size range of 20 inches and larger, clad pipe is more economical than solid stainless for PWR design conditions (2500 psig and 650 degrees F); for BWR design conditions (1250 psig, 575 degrees F), solid stainless steel may be the more economical choice. The present stainless-steel pipe standard, ANSI B36.19, goes only as large as 12 inches O.D.; sizes above that require complete specification and may draw on ANSI B36.10 for size information. Acceptable solid stainless steel pipe specifications are ASME SA-312, SA-358, SA-376, SA-409, SA-430, SA-451, and SA-452. Clad piping requires a complete specification. Carbon steel BWR piping is SA-155, Type KCF-70.

Large size, solid stainless steel pipe should be fabricated of ASME SA-240 plate, rolled and welded. The plate must be inspected according to the Code and then the weld should be examined by two nondestructive methods. One method should be recordable and volumetric—either radiography or ultrasonic. The second method should be surface sensitive, either dye penetrant or eddy current. Eddy current examination is cheaper to perform on a production-line basis than is dye penetrant, and if executed before the volumetric examination, can save money by avoiding needless volumetric examinations. Acceptance standards should be those of Class 1, of Section III of the ASME Code. Welding procedures, operators, and machines must be qualified according to Section IX of the ASME Code. Dimensional tolerances such as wall thickness, O.D., ovality, etc., must be given.

Minimum wall thickness should be calculated according to Section III requirements. Then the engineer must design his weld joint taking into account all the tolerances of the pipe and the fitting or other connecting items, to arrive at a nominal wall-thickness. Special sizing at pipe ends with reduced eccentricity and ovality tolerances at weld joints make it possible to use a lighter weight, nominal pipe wall. The engineer must compare the cost of preparing a weld joint from pipe, with and without special sizing, before making a final decision.

If clad pipe is used the procedure is basically the same. Base metal carbon steel should be ASME SA-516, Gr. 70, although ASTM A-106 has been used in the past. This is the same steel used in the BWR pipe. The stainless-steel cladding should have a surface finish comparable to seamless pipe; thus, weld overlay, if used, requires surface-finishing operations. Field welding of clad pipe should be held to a minimum because of the dissimilar metals involved but two methods can be used where necessary. In the first method, the stainless cladding is cut back as in Fig. 7-17, the ends prepared and a carbon steel weld made. When the carbon steel weld is completed, it is radiographed, after which the back is ground flat. Stainless steel is then overlaid into the groove to complete the weld. The overlay should be inspected volumetrically after grinding to insure the integrity of the cladding bond.

The second method is to safe-end the pipe in the shop as shown in Fig. 7-18, and then make a single weld in the field. The pipe is safe-ended in the shop by buttering on stainless steel. The weld buttering is examined volumetrically and then the weld end is prepared. In the field, a single stainless steel weld is made and examined. The more difficult bimetallic weld is done in the shop under better conditions.

Both weld methods have been used successfully; the author prefers the buttering technique that gives the simpler field weld.

7.8 Insulation

Insulation for the primary system has many requirements in addition to the insulating properties. The material must be very low in halides, accept the

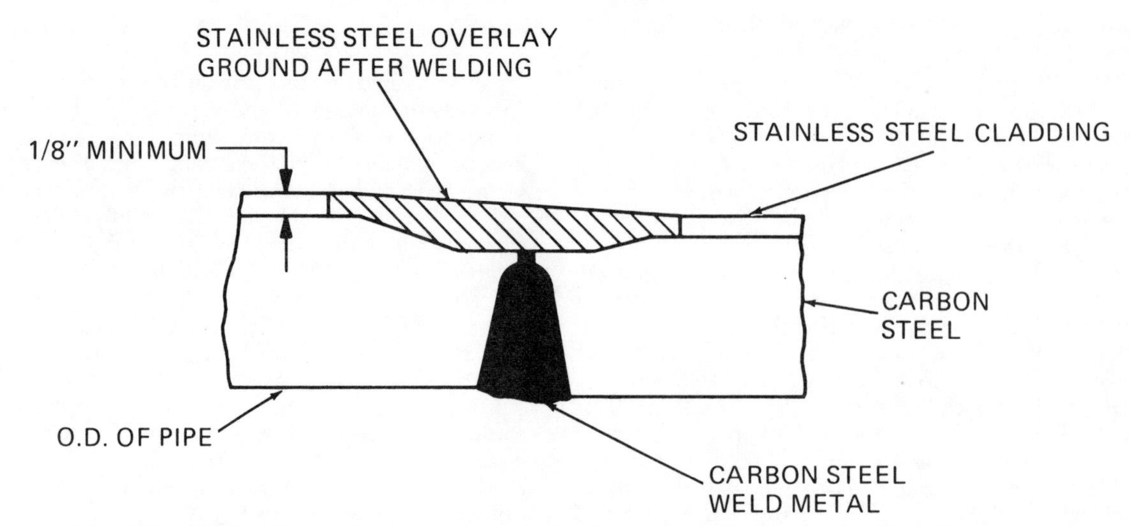

Fig. 7-17. Field weld of a primary pipe joint. In this design the carbon-steel weld is completed and inspected before placing the stainless steel overlay.

operating basis earthquake, and be completely removable and replaceable for inservice inspection. It must also accept the total 40-year neutron dosage of the plant.

Asbestos and calcium silicate have been used successfully, but they have their disadvantages. The normal mineral insulations have a tendency to dust and powder, making it harder to maintain proper cleanliness. In the event of a leak or spill they act like a sponge and present a radiation and disposal hazard. When removing them for equipment inspection the insulation is susceptible to damage and breakage. Providing metal shields to make them strong enough to withstand the necessary earthquake shock can incur appreciable additional costs.

Today, extensive use is made of all metal reflective insulation made of stainless steel or a combination of aluminum and stainless steel. This insulation, rigid and easily removed and reinstalled, has been used on many parts of the primary system where accessibility is required for inservice inspection. In specifying this material the engineer must give hot and cold surface temperatures, allowable heat-loss per unit area, temperature and velocity of air moving over the outer surface, and maximum allowable insulation thickness.

This insulation is all shop prefabricated and then installed in the field.

7.9 Design Specifications

The key concept in a primary system specification, whether for a component or installation, is meticulousness. Nothing is to be overlooked or assumed. It is far better to write a rigid specification and receive a limited number of high proposals than to issue a "slipshod" specification, receive low bids, and subsequently become embroiled in negotiated "extras." Never ask the contractor what he will do; tell him what must be done. The specification should include a general description of the work to be done,

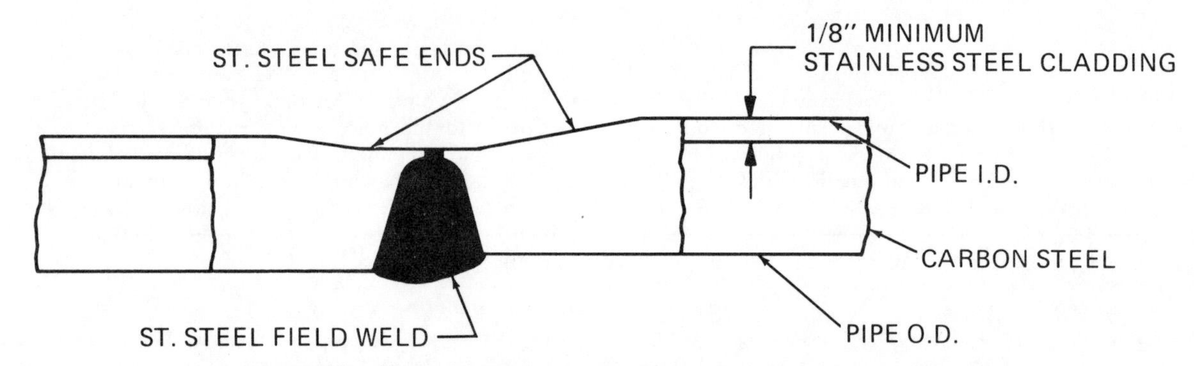

Fig. 7-18. Field weld of a primary pipe joint with buttered safe ends. The safe ends are prepared and inspected in the shop.

a description of the operating cycles of the equipment over the life of the plant, and a specific description of the work to be included in the bid.

The "work to be included" encompasses such things as the hardware, spare parts, engineering, design reports, drawings, quality assurance reports, and operating and maintenance instructions. In the present NSSS concept all equipment, piping, and valving (except supports and insulation) are furnished by the NSSS supplier.

7.9.1 *Documentation*—Requirements for quality assurance reports should be elaborated upon to prevent future misunderstandings. The reports must be kept at the site for the life of the plant. The specification must make a distinction between reports needed for licensing purposes and those needed for the base-line inspection for future inservice inspections. When the same report can serve both purposes, the specification should state this, and require duplicate copies, one for each file.

The following list of reports and documents, which may not be complete, is illustrative of the material required:

a. Raw material chemical analyses
b. Raw material destructive tests (tensile, impact)
c. Weld overlay, top layer analyses
d. Welding procedure qualifications
e. Welding machine qualifications
f. Welding operator qualifications
g. Nondestructive examination reports
h. Dimensional acceptance reports
i. Cleaning procedures
j. Deviation requests
k. Material review board reports
l. Defect repair reports
m. Heat treatment procedures and temperature charts
n. Proof tests (hydrostatic, thermal, flow, leak, etc.)
o. Packaging reports
p. Receiving inspection reports
q. Piping cold spring and hanger preloading.

7.9.3 *Materials*—All materials used in the pressure boundary must be taken from Section III, or the Code case interpretations. Specifications should call out exact materials of construction wherever possible. The chapter has listed acceptable materials for each of the parts of the pressure boundary. The specification should require that material samples of the reactor pressure vessel be supplied for future reference. One sample should be of the base material plate before start of fabrication, and a second sample should be supplied that has accompanied the pressure vessel through all welding and heat treating processes. NDT temperatures should be specified for all ferritic materials, in conformance with Criterion 31. The temperature selected should be based on a normally heated room—about 70 to 80 degrees F.

7.9.3 *Operating Cycle*—Design specifications must give complete operating cycle data for all mechanical components and piping systems, over the full life of the plant. Temperatures and pressures at the beginning and end of each cycle must be given, as well as the time period in which the change takes place; a normal start-up may take eight or ten hours, but an emergency shutdown may take only two or three minutes. The purpose of the data is to permit the mechanical designer to perform a proper fatigue analysis based on the predicted operation of the plant.

The following different situations must be considered in describing the operating cycles:

a. Shop hydrostatic tests
b. Field hydrostatic tests
c. Startups and shutdowns during plant testing and debugging
d. Normal full-power startups
e. Normal full-power shutdowns
f. Normal hydrostatic testing during refueling
g. Normal load following power-level changes
h. Emergency shutdowns during testing and normal operation.

The number of cycles for each of the above situations must be given over the complete life of the component.

7.9.4 *Design Data*—All necessary design data must be given in the specification. For a pump it would be a total dynamic head; mass flow rate; available cold, net-positive suction head; required cold operating time; operating suction pressure, hot and cold; maximum permissible shut-off head, if any; design temperature and pressure; external nozzle loads; earthquake acceleration spectrum and damping factors; etc. Other components have similar criteria. Earthquake design criteria must also include the operating and design earthquake "g" loading to be used with the acceleration spectrum and damping factors.

7.9.5 *Fabrication*—The specification must define fabrication requirements and restrictions and should require written fabrication, heat treating, welding and cleaning procedures. Requirements should be inserted prohibiting the use of contaminating substances or tools on stainless steels. Typical offenders are: previously used grinding wheels or burrs, carbon-steel wire brushes, and marking inks or machining coolants containing heavy metals, halides or sulfur.

Major defects in either raw material (plate, casting, rod) or in welds should require a complete description of the defect, plus drawings or weld maps locating the defect, the proposed procedure for removing the defect, the examination to confirm removal of the defect, the procedure for repairing the defect, and the examination to confirm satisfactory repair of the defect.

Heat-treating procedures should define heating rates, soaking time, cooling and/or quenching procedures, uniformity of furnace temperature, and require continuous furnace temperature recording charts.

Cleaning procedures should define the steps in the process, reagents to be used, water analyses, methods of verifying satisfactory cleaning, temperatures, etc.

7.9.6 *Quality Assurance*—The specification, by law, must require a written quality-assurance program that will start with the conceptual design of the component and continue through fabrication, testing, and installation of the equipment. The requirements of the program are given in the Code of Federal Regulations, Part 10, Chapter 50, Appendix B, which is reproduced in this book as Appendix II. It must be remembered that Appendix B requires only a formal program delineating the manner in which actions will be performed. The parameters which will be measured to prove quality and conformance to specification, must be given in the specification. The specification should augment and amplify the inspection requirements of the various codes. A typical illustration of this might be two radiographic films in the same cassette, exposed and developed for single-film viewing. Drawings should show dimensions and tolerances.

All flow tests and hydrostatic tests required for acceptance, should be given in this section. And, of course, for every test or examination required, acceptance criteria must be given.

The codes give requirements for nondestructive examinations and hydrostatic proof tests only. The specification must add additional requirements such as the following:

a. Each drawing showing a weld shall also indicate the required examinations and nondestructive tests.
b. Hold points in fabrication beyond which the component may not proceed without completion of all prior examinations. An example of this might be approval of all shell-section longitudinal welds of the reactor pressure vessel, before assembling the sections into a complete shell.
c. Examinations or inspections that the purchaser requires be witnessed by his own inspection staff. Examples of this are cleaning operations, hydrostatic testing, and pump performance testing.
d. Conditions that must be controlled during inspection operations such as: working environment, inspection tools and standards, and personnel qualifications.
e. Procedures for submitting documents and data for approval.

The 1971 edition of Section III of the ASME Code requires that all holders of a Class 1 Nuclear Stamp have an approved Quality Assurance Program that conforms to Code requirements. This fabricator should be required to demonstrate that his program conforms to all requirements of Appendix B (Appendix II in this book).

7.9.7 *Cleaning and Cleanliness*—Extreme cleanliness is required so that there will be no interference in the operation of control rods or other equipment that have very close tolerances. Additionally, irradiated dirt particles present a disposal problem and an unnecessary load on the purification system. The systems must be thoroughly cleaned before plant startup. The American Nuclear Society under the auspices of the ANSI has written definitions of cleanliness for use in nuclear plants and components, but they have not yet been published as official standards.

The specification should have the philosophy that the best way to assure that the assembled plant is clean enough for use is to insure that each component or piping sub-assembly is brought into the reactor building sealed, with the required degree of cleanliness. This means that shop-fabricated components are cleaned and sealed at the shop before shipment. Piping assemblies are cleaned and sealed either at the home shop or at the site pipe shop. Equipment should be unsealed only when it is being worked on and should be resealed by the work crew each night at the close of the workday. Additionally, the author requires that piping be sealed when work stops for 30 minutes or more, as at the lunch break, or when going for another sub-assembly.

Cleaning should be restricted to grinding, wire brushing, blasting, degreasing, and detergent cleaning—all under controlled conditions. Acid cleaning should not be used again after pickling. Grinding should be accomplished with new, unused grinding wheels or burrs. Wheels used on P-1 materials (carbon steels) should never be used on stainless steels, nickel-chromium-iron alloys, or nickel-iron-chromium alloy. Wheels should be aluminum oxide or silicon carbide bonded with resin. Wire brushes should be new stainless steel brushes.

Blasting may be clean, dry, air blasting or clean, halide-free water honing. Blasting material should be new, clean, iron-free alumina or silica. If liquid honing is used, then all the areas honed should be thoroughly rinsed with demineralized water before the honing water dries.

Degreasing must be segregated into two classes—parts without crevices and parts with crevices. Parts without crevices can be degreased with inhibited trichlorethylene or perchlorethylene. Complicated parts or parts with crevices should be degreased with alcohol, acetone, or other halogen free solvents. All degreasing solvents must be free of sulfur and heavy metals.

Equipment should be cleaned with a detergent after final assembly. This cleaning should be with a hot, flowing detergent using trisodium phosphate or a non-ionic detergent such as MIL-D-16791. Next it should be thoroughly rinsed with hot, flowing demineralized or distilled water and then dried with warm, filtered oil-free air. Immediately after completion of drying, the component or sub-assembly should be sealed to prevent entrance of dirt or moisture. It is best practice to perform hydrostatic tests, with halide-free water, just prior to the final detergent cleaning.

Cleaning operations for parts without crevices can use tap water, provided the part is rinsed or dipped into demineralized or distilled water before the tap water dries. Final cleaning, and cleaning of parts with crevices, should always use distilled or demineralized water. The demineralized or distilled water should not have more than 10 ppm total solids nor more than 0.05 ppm chlorides.

7.10 Inservice Inspection

With the publication of the "Draft Code for Inservice Inspection of Nuclear Reactor Coolant Systems" in the fall of 1968, the requirements for equipment inspection on a routine basis during scheduled plant shutdowns, crystallized. The code prescribes an inspection program performed during annual refueling shutdowns that must be completed over a ten-year period. The draft code was formally issued in the fall of 1969, as Section XI of the ASME Code and reissued as a mandatory section in the 1971 edition of the code.

The problems to be solved because of the inspection requirements are of two types. The system and component design engineers must provide sufficient access to all parts of the equipment to permit inspection. The inspection equipment suppliers must design adequate remote inspection equipment, using existing techniques or inventing new techniques to suit plant situations. Utilities and consulting-engineer firms began to recognize the access requirements of the new code as soon as it was published. Vendors stated that their equipment had the necessary accessibility, but refused to quote on the original base-line inspection. "Base-line inspection" is the term given to the inspection performed just before start-up to determine the initial condition of the plant.

The boundary of the inspection program is the primary coolant system—reactor pressure vessel, pumps, piping, steam generator, and pressurizer. For a BWR, the boundary ends at the isolation valve outside containment on the main steam and feedwater lines. For connecting systems, the boundary extends to either the first positive-acting isolation valve outside containment or to the second of two block valves inside of containment, if both valves are normally closed during reactor operation. The areas subject to examination and the extent of the examination are given in Appendix IV which is a reprint of Table IS-251, from the 1971 edition of Section XI of the ASME Code.

Access must be provided to all pressure-containing welds of the pressure boundary so they may be examined at least once during a complete inspection cycle. Welds must be examined volumetrically and it is here that the state of the art of volumetric examination must be updated. At the present time it is impossible to radiograph the active parts of a pressure vessel; thus, ultrasonic examination is required. Radiographing of the remainder of the system that is outside the primary biological and missile shield is a question of plant economics. With enough time, radiation levels will decay sufficiently to permit radiography. The evaluation requires a knowledge of film speeds, radiation decay rates, eligible manpower, and the cost of plant down-time. It is believed in the industry that by 1975 to 1977, adequate examination equipment will be available.

Visual examination is intended to be executed by the naked eye, or with mirrors, periscopes, boroscopes or fiber optical systems. Weld surface examinations will be executed with the most convenient of magnetic or dye penetrant examinations.

The plant design engineer's most important consideration is accessibility. This takes the form of inspection covers, storage areas, removable internal parts, platforms, stairways, ladders, etc.

The AEC is now requiring that an Inservice Inspection Plan for the plant be prepared during the design stage. The plan must include the complete ten-year inspection cycle and give the details of the actual items to be inspected (such as Valve No. 10A or Weld 1-19), the methods to be used, and when the inspection is to be performed. The results of each inspection are compared to the base-line inspection to determine plant condition. The need for plant repairs or additional examinations is based on the results of the comparison of the base-line and the inservice inspections.

CHAPTER 8

Residual Heat Removal Systems

8.1 General

After reactor shutdown, residual heat removal (RHR) systems remove the afterheat from the core present there as a result of fission product decay energy and residual gamma activity. In power plants the complete residual heat removal system is a combination of the turbine steam by-pass plus the feedwater system and additional residual heat removal equipment. Power decays along an exponential curve and drops to less than 10 percent in one or two seconds, but does not reach 1 percent of power level for some six to eight hours. An equation for calculating decay heat is given in Chapter 11. The initial part of the decay heat is dissipated by generating steam at full pressure, running a steam-driven feedwater pump and by-passing the excess steam to the condenser. Heat dissipation continues in this manner until the heat load drops to within the capability of the residual heat removal equipment. The reactor vendors differ in the sizing of their residual heat removal systems but it is usually in the range of 0.4 to 0.6 percent of full core thermal power. Transfer from steam generation to RHR operation is usually designed to occur at 300 degrees F and 300 psig.

The residual heat removal equipment is connected to the emergency core cooling systems and operates at the low end of the pressure range. (See Figs. 2-8 and 3-15 for typical flow diagrams.)

In a PWR, the system acts as the low pressure safety injection system, as shown in Fig. 8-2. The connections from the reactor building sump and the refueling water storage tank to the suction of the RHR pump are used only during emergency cooling situations. In the BWR, the system acts as a back-up to the low pressure core-spray system while in the low pressure core injection (LPCI) mode, and as a heat sink during containment cooling. In the LPCI mode, pump suction is taken from the pressure suppression pool with the condensate storage tank as a back-up water supply. In the containment cooling mode, the system recirculates pressure suppression pool water through its heat exchangers, to cool the pool water.

Connections on the suction side of the RHR pumps and discharge side of the heat exchanger connect this system to the refueling canal and the spent fuel cooling system. This permits the heat load of spent fuel elements to be physically followed as an element is removed from the reactor core and transferred to storage, and utilization of the RHR pumps to fill and drain the refueling canals in a refueling procedure. The connections also permit the use of the RHR cooling capacity in the spent fuel storage canals if a situation requires complete unloading of a reactor core for maintenance reasons.

8.2 Equipment Sizing

8.2.1 *Heat Exchangers*—This system is directly involved in the safety of the reactor core, hence, all equipment is duplicated. Two heat exchangers are provided, each sized for 100 percent of safety duty. During a normal shutdown for refueling, they are both used to cool the primary system as rapidly as permitted (usually 50 degrees per hour). Sizing calculations are performed for each of the operating situations, and for several conditions within each situation, to determine the final sizing.

No specific design guidance can be given. The core designer will specify the minimum coolant velocity he can accept during the various stages of cooldown, and the cooling requirements for emergency injection. The exchangers must be designed within those limitations of temperature and flow.

8.2.2 *Pumps*—Pumps are also duplicated with 100 percent of capacity available for each heat exchanger. The installations differ with the various manufacturers. The BWR has one 100 percent unit for each exchanger, plus a 100 percent spare. The PWRs are supplied with two and three 100 percent units. When the third unit is supplied, it is a standby for either of the other two pumps. Methods of connection vary with the reactor vendors. Figure 8-1 shows the arrangements used in PWRs while Fig. 3-15 shows the arrangement in a BWR.

The pumps are powered from emergency power buses which are supplied from the turbine generator set, an outside power source, and the plant emergency diesel generators. Each 100-percent pumping capacity is on a separate bus, so failure of one bus will only affect one pump.

8.3 Quality Level

System quality is slightly less than that of the primary system. Failure of this system cannot cause a nuclear incident; however, failure may increase the severity of an incident. Since the RHR system contains reactor coolant during operation and is an en-

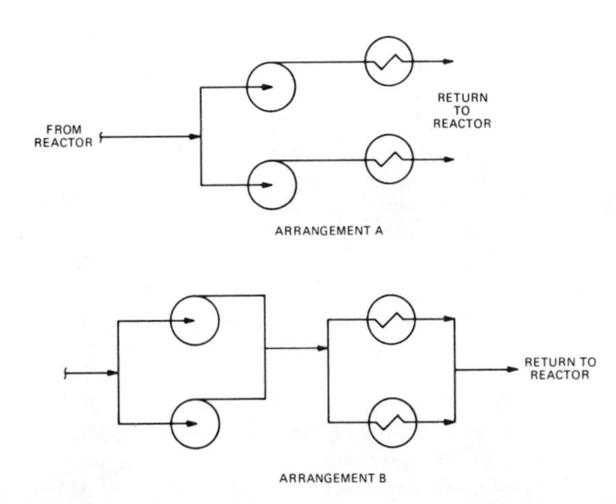

Fig. 8-1. Diagrammatic arrangements of pumps and heat exchangers in RHR systems. Arrangement B is no longer acceptable because it does not comply with the single-failure criterion.

gineered safety feature, the system should be built in conformance with Class 2 of Section III of the ASME Code. Cleaning and cleanliness requirements are the same as for primary equipment. All material in contact with the coolant must be stainless steel, so all precautions given in Chapter 7 for handling stainless steel must be observed.

8.4 Equipment Design

Equipment designs must accept both the operating basis earthquake and the design basis earthquake and continue to operate according to Design Criterion 2.

Figure 3-23 shows the RHR heat sink as the Nuclear Service Water System. The arrangement in that illustration is satisfactory so long as the service water is fresh water. If a seashore site is picked and salt water is used for cooling, then the author prefers an intermediate, buffered fresh water system interposed between the RHR heat exchanger and the service water, to avoid the corrosion problems engendered by stainless steel in prolonged contact with stagnant salt water.

8.4.1 *Heat Exchangers*—Heat exchangers today are U-tube design, two-pass tube side and single-pass shell side, although floating head designs have been used in the past. The primary coolant side is built to Section III, Class 2; the heat sink coolant side is Section III, Class 3 of the ASME Code. The units should be in accordance with TEMA Class R. Fixed tubesheet heat exchangers have never been used because the temperature difference between the shell and tubes at the start of the cycle is of the order of 175 degrees F to 200 degrees F. Exchangers may be either horizontal or vertical. It should be noted that in the hot standby mode the RHR exchanger of a BWR is used as a condenser.

Construction can have bolted bonnets or channels and a removable tube bundle. The tube-to-tubesheet joint is a light contact roll followed by a tube-to-tubesheet weld. Joints are then given a 100 percent dye penetrant inspection. Primary coolant side pressure welds should have 100 percent radiographic inspection; the heat sink side should have normal, Class 3 weld inspection. The tubesheet should either be the same material as the tubes or carbon steel clad with the same material as the tube. The actual choice depends on the procurement economics of the moment. The tube and tubesheet combination must permit a proper weld to be made. Gaskets should be either spiral metallic with an asbestos filler or metal-jacketed asbestos. Six-inch or eight-inch inspection holes should be provided in the channels, along the shell, and at the inlet and exit of the tube bundle. In PWRs, the primary coolant is in the tubes and component cooling water (the heat sink) is in the shell. The tubes should be seamless stainless steel, ASME Specification SA-249. The author prefers Type 304L tubes, although Type 304 and 316 have been used. The tubesheet can be stainless clad carbon steel, either plate or forging, SA-240 or 479, depending on the size and thickness. The cladding should not be less than 3/8-inch thick and be given a dye penetrant examination for surface defects, plus an ultrasonic examination for the metallurgical bond. The channel or bonnet should be internally clad carbon steel, examined in the same way as the tube sheets. The shell, tie rods, spacers, and baffles should be of carbon steel, ASME SA-515 or 516 with a 1/8-inch corrosion allowance on the shell.

In BWRs the heat sink varies and the reactor vendor uses either the service water directly, or an intermediate closed loop. Service water, at an ocean site, is sea water which the author prefers not to use in a decay heat cooler. If fresh water, or an intermediate loop is used, construction is the same as for a PWR except that the channels can be plain carbon steel. If sea water is used directly, it should be in the tubes which would be cupro-nickel; the tubesheet and channel should then be clad with cupro-nickel. Stainless steel should never be used with sea water as it is very prone to corrosion from stagnant or low-velocity sea water.

8.4.2 *Pumps*—The decay heat pumps are usually multistage, horizontal centrifugal pumps with mechanical seals. Dynamic head varies with the plant design but is of the order of 300 ft to 350 ft. Volume is a function of the plant power rating; with 3100 to 3400 MWt plants requiring about 4000 gallons per minute.

Particular attention must be given to the net positive suction head (NPSH) of this pump because it must recycle water from a PWR's reactor building sump or a BWR's pressure suppression pool in an emergency. Final treatment of the pump is a combination of the lowest practical NPSH for this pump, combined with careful pump location and piping.

The pump is always placed lower than a PWR sump and the piping is arranged with a check valve to keep the suction line full of water in standby periods. In a BWR the pump must be low enough to take suction from the pressure suppression pool.

In PWR plants the pump is all stainless steel construction. In BWR plants the casing is cast steel with stainless steel impeller, shaft, and shaft sleeves.

A double mechanical seal should be used with graphite bearing against a cobalt-tungsten alloy. Wear rings should be furnished on both impeller and casing—for a PWR, both rings should be stainless steel; for a BWR, the casing ring may be bronze or stainless and the impeller ring can be stainless. The pump design should be such that the impeller can be removed or the shaft seal replaced without disturbing the motor or piping. This can be accomplished by using a spacer-type coupling. The pumps are located upstream of the heat exchanger and require bearing cooling because at the beginning of operation, liquid temperatures in the pump are 350 to 300 degrees F. The pump seal should also be checked to determine if the seal design requires cooling since the maximum uncooled seal temperatures are about 150 degrees F. All cooling requirements are serviced by the emergency branch of the component cooling water system.

8.5 System Design

The piping design must separate the two parallel halves of the system so that no single failure of a pump, valve, or pipe can prevent the other half of the system from operating. Thus, each half must be physically protected and installed in a Class I seismic structure. All pipe routing must be examined for nearby non-critical items or structures that could fail and, in so doing, damage this piping system. Each pump and heat exchanger must be installed in a protective cell. This same separation and protection of system components must extend to controls and electrical components.

The system design boundaries start and end at the second isolation valve from the primary loop. Piping should have 100 percent radiographic inspection of all possible welded joints. It should be noted that this requirement is more stringent than the requirements of the code. All system operating valves must be remotely controlled to permit operation and control during an emergency; valves furnished to permit equipment removal may be manual. Seat leakage in

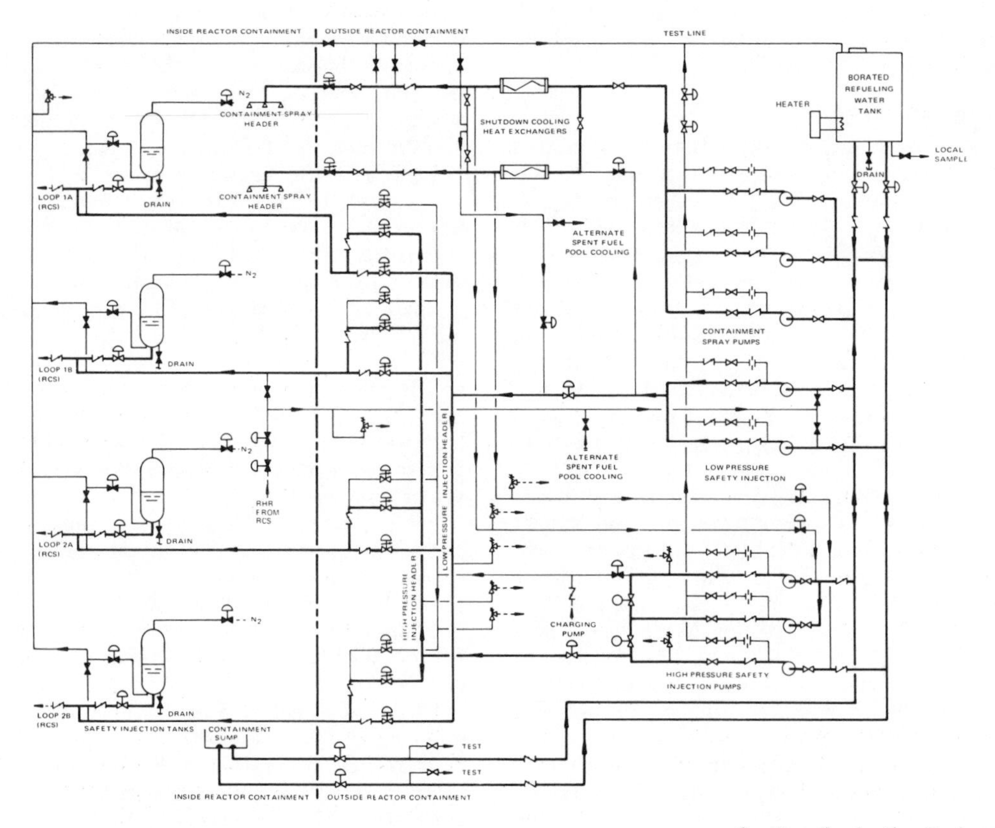

Courtesy, Combustion Engineering, Inc.

Fig. 8-2. The RHR system is part of the overall safety injection system. The heavy lines denote the RHR system.

conformance with MSS SP-61 is satisfactory, except for containment isolation valves which should be 2cc-per-hr per inch of seat diameter, at design differential pressure. Valve construction should be the same as for the valves in Chapter 7. Piping and valves for a BWR installation should be carbon steel; but for a PWR installation, all materials must be made of stainless steel.

The BWR-RHR system is part of an interconnected piping complex involving residual heat removal, low pressure core injection, reactor hot standby, and containment spraying. The PWR system is also interconnected with the safety injection system as shown in Fig. 8-2. All parts of the system needed during emergency cooling or injection operations must comply with Design Criteria 35, 36, and 37. These criteria require special recirculation lines that permit recirculation of borated water from the refueling water storage tank, or buffered water from the pressure suppression pool, up through the RHR components and piping, as close to the core as practicable. This procedure serves to demonstrate the emergency operation of the system. Additionally, facilities must be designed to permit an individual operating test of each pump and remotely operated valve. In a BWR the carbon steel pipe lines are filled with demineralized water in the standby condition so that a corrosion allowance of about 0.10 inch should be added to the carbon steel.

8.6 Reactor Core Isolation Cooling

The RHR systems just described for the BWR and PWR are emergency systems. The BWR has still another decay heat removal system, the Reactor Core Isolation Cooling System described in Chapter 3, used when the reactor is inadvertently isolated from the turbine. This, however, is not an emergency system. The system has only a turbine-driven pump, the piping, and valves. For design purposes the equipment is treated in exactly the same manner as the RHR system in all respects—materials, codes, seismic, etc., are all the same. The turbine is of standard materials, but its design must conform to Seismic Class I.

CHAPTER 9

Emergency Systems

9.0 Description

Emergency systems are provided for use in the unlikely event of an accident. The systems are provided in parallel and each of the parallel paths has 100 percent capacity. Capability is provided for shutting down the reactor, cooling the core, and cooling the containment building.

9.1 Reactor Control Shutdown Systems

Design Criterion 26 requires two, independent shutdown systems. In the BWR, the first system is the cruciform rods with their hydraulic drive packages. The second system is the standby liquid control system using borated water that is shown in Fig. 3-22. For the PWR, the first shutdown system is the combination of the control element assemblies and the dissolved boron. The second system is the high-concentration boron injection by the high-pressure charging pumps. Each of the systems is capable of shutting down the reactor and maintaining it subcritical as it cools down to room temperature.

The PWR high-concentration boron injection is a sub-system (Fig. 2-7). On emergency injection, the signal closes off the volume control tank outlet and opens a flow path from the concentrated boric acid tank, through the charging pumps, into the primary system.

All flow paths for boron injection, either the BWR sodium pentaborate or the PWR boric acid, must be maintained at a temperature high enough to avoid crystallization in the pipelines. This is usually accomplished in the PWR by correlating normal purification temperature with concentrated boron solution solubility. The piping from the concentrated boric acid tank to the charging pumps must be heat traced. The use of sodium pentaborate in the BWR, permits lower solution temperatures for the emergency system—between 75 and 85 degrees F.

All equipment and piping in the reactor shutdown control systems are stainless steel. Piping and equipment are Type 304; valves are Type 304 or Type 316. The valves may be either of the two stainless alloys; Type 316 is usually specified, because it fits into the manufacturer's normal material production and there would be no economic advantage in insisting on a Type 304 valve.

9.1.1 *Pumps*—Pumps inject against full reactor pressure so the dynamic head will be of the order of 1100 psi for a BWR, and 2300 psi for a PWR. A reciprocating pump is usually used, although in some of the larger reactors centrifugal units have been supplied. Design temperature should be 175 degrees F, providing a safe margin above operating temperature. For a reciprocating pump, casing design pressure should be the same as that of the primary system. For a centrifugal pump, casing design pressure should be shut-off head, plus static head, plus 100 psi. The static head is either the top of the storage tank, or the highest shut-off valve on the discharge side, whichever is greater. Note that this design pressure requires a statement in the pump procurement specification defining shut-off head. An adequate requirement is that shut-off head shall not exceed 25 percent of design head. This requirement, however, can result in an excessive design pressure. An alternate approach to shut-off head selection that can result in lower pressures is as follows:

1. Calculate the required dynamic head.
2. Review all possible sources of supply and assemble a qualified bidders' list.
3. Review, in depth, the applicable pump curves of the qualified bidders.
4. Using only current information from qualified bidders, select the lowest shut-off head that will permit the desired minimum number of bidders to submit proposals.

A centrifugal pump should have a Type 304 or 316 stainless-steel cast casing and impeller with replaceable casing and impeller wear rings. The castings should conform to ASME SA-351 Grade CF8 or CF8M. The wear rings should either be over 400 Brinnel Hardness Number (BHN), or if lower than 400 BHN, the two rings should have a material hardness difference of 70 BHN or more. The shaft should be forged Type 316 conforming to SA-336 Grade F8M. The shaft should be sealed with a double mechanical seal using Teflon or graphite bearing against stainless steel. Teflon is acceptable in this service because the fluid pumped is not radioactive. Castings should have a 100 percent recorded volumetric examination, either radiographic or ultrasonic. Pressure-envelope design should conform to Class 2, of Section III of the ASME Code.

A reciprocating pump should have a Type 304 forged body of SA-336 Grade F8 material. The plunger should be 17-4PH stainless steel corresponding to ASTM A-564, Grade 630. Shaft packing should be either Teflon or a graphite-lubricated low-chloride asbestos. Pump check valves should be hard-faced with a cobalt tungsten alloy such as Stellite. Valve seats and check valves must be replaceable. The pump unit must include a relief valve, rated at 100 percent flow, for pump protection. If a modulating variable-stroke pump is used (as in a PWR to control pressurizer liquid level in normal operation), it must be arranged to have 100 percent stroke on minimum control signal value. This will permit a "fail safe" emergency signal by simply de-energizing the control circuit in the emergency. This pump shall conform to Class 2, of Section III of the ASME Code. In addition, the body forging should be radiographed and the plungers given a surface examination (magnetic particle or dye penetrant) to Class 1 standards. The plungers must be examined after machining, and additionally, at the option of the manufacturer, may be examined as a raw forging. Check valve seats should also be given a surface examination.

The hydraulics of the pumps should conform to the applicable portions of the Hydraulic Institute Standards; the reciprocating pump is classified as a power pump. Each pump should have a certified performance curve obtained at operating temperature. The reciprocating pump should also have a 100 percent capacity, relief valve test as part of its performance testing.

9.2 Core Cooling

Design Criterion 35 requires emergency core cooling to cool the core adequately, and to limit fuel-cladding damage to negligible amounts. The criterion requires sufficient redundancy so that no single failure will prevent accomplishment of cooling. Criteria 36 and 37 require inspectability and testability. Prior to the issuance of the revised General Design Criteria, in 1971, the criteria in effect required two emergency core cooling systems. The criterion has been changed so that a single system with sufficient redundancy (not yet defined) could meet the criterion. For the foreseeable future, however, it is believed that two systems will continue to be used, since there is no doubt that they have sufficient redundancy.

The core designer will set the criteria for design such as flow rate and duration. The general approach used is to provide an external guaranteed water supply of limited duration. The size of the water supply is such that it can fill the reactor vessel several times before it is exhausted, after which water is recycled. The reactor containment structure is designed as a pressure tight and watertight structure whose lowest level acts as a tank to collect the water for recycling. The physical arrangement and instrumentation is such that once cooling water starts to overflow out of a ruptured primary system, a closed cycle with cooling capability is established and cooling can be continued as long as necessary. The portions of the building involved in the water cycle must have adequate finishes, either epoxy or vinyl, to avoid chemically contaminating the recycle water during routine, periodic, system-operating tests.

9.2.1 *BWR*—The BWR design provides two different parallel means for emergency core cooling. The first method has the high-pressure core spray system at the high-pressure end of the range, followed by the low-pressure core spray system at the low-pressure end of the pressure range. Both systems discharge through ring headers inside the reactor pressure vessel directly onto the core. The second method is a combination of steam blow-off to the pressure suppression pool at the high-pressure end of the range followed by RHR system operation in the low pressure core injection mode as soon as the pressure permits. Low pressure injection discharges directly into the reactor vessel. In all cases the water supply is taken from the pressure suppression pool while part of the RHR system controls the pool temperature.

High Pressure Core Spray Pump. This pump is a vertical, multistage variable speed, can or barrel-type pump with a mechanical seal. The casing is cast carbon steel; the impeller and shaft are alloy steel. Hydraulic thrust is eliminated in this design by arrangement of the stages in the machine. NPSH requirements for this pump at the inlet flange are quite low, because the first stage impeller suction is at the lower end of the pump. Typical pump design conditions specify 2200 ft of dynamic head; thus, if one stage develops 200 ft, eleven stages are needed. This results in a pump about 7 ft long; so 7 feet, less friction loss NPSH, is available within the pump casing itself. Hence, if the impeller requires 11 ft of NPSH, about 6 ft of it is available in the machine, and only 5 ft is required outside. The pump can operate over the whole pressure range of the reactor. The core designer will specify the flow requirements that the pump must deliver. Generally, the pump must deliver a low volume at high head and increase its delivery as the pressure falls. The designer must check the NPSH over the complete range of pump speeds to insure proper operation in all situations.

The pump is designed in accordance with Class 2 of Section III of the Code. A certified performance test should be performed showing performance at maximum design speed and performance over the required range of speeds and delivery pressures. Since the pump is required for emergency duty, the steel casing should have controlled chemistry and 100 percent radiography. The impeller should be radiographed; the shaft dye penetrant or magnetic particle inspected. The mechanical seal should be graphite against steel.

The reactor vendor supplies a diesel generator to power this pump in his scope of supply.

Low Pressure Core Spray Pump. Mechanical design and material of construction of this pump are the same as the high pressure core spray pump. However, it is a constant speed device powered only from the emergency buses; it does not have its own diesel generator. The pump comes on but does not begin to deliver water until the pressure differential between containment and the reactor falls below its shut-off head. Design head is approximately 285 ft; shut-off head is about 650 ft. Since pump delivery starts at high pressure and builds up along the pump performance curve as the reactor vessel pressure drops, the net positive suction head must be investigated over the delivery range. In performing this investigation, the temperature of the pressure suppression pool should be checked over the entire period the pump is expected to run, to insure that the correct vapor pressure is used in each phase of pump operation.

9.2.2 *PWR*—The emergency core cooling in a PWR operates as part of the safety injection system described in Chapter 2. As long as the charging pumps of the chemical and volume control system have adequate capacity, and are centrifugal pumps, they are used as the high-pressure injection pumps. When the required cooling flow rates are too large for the charging pumps, then independent pumps of adequate size are used. Figure 9-1 shows a plant rated at about 900 MWe, in which the CVCS charging pumps are used in the safety injection system. Figure 8-2 shows a plant rated at 1100 MWe in which large, high-pressure safety injection pumps have been furnished and there is no interconnection with the CVCS system. In both Figs. 8-2 and 9-1, the low-pressure end of the cooling operation is served by the RHR pumps and exchangers. Note, however, that in Fig. 8-2, the containment spray pumps are shown and interconnected with the RHR exchangers. CVCS charging pumps will be described in Chapter 10. A description of high-pressure safety injection pumps follows.

Safety Injection Pump. The safety injection pump

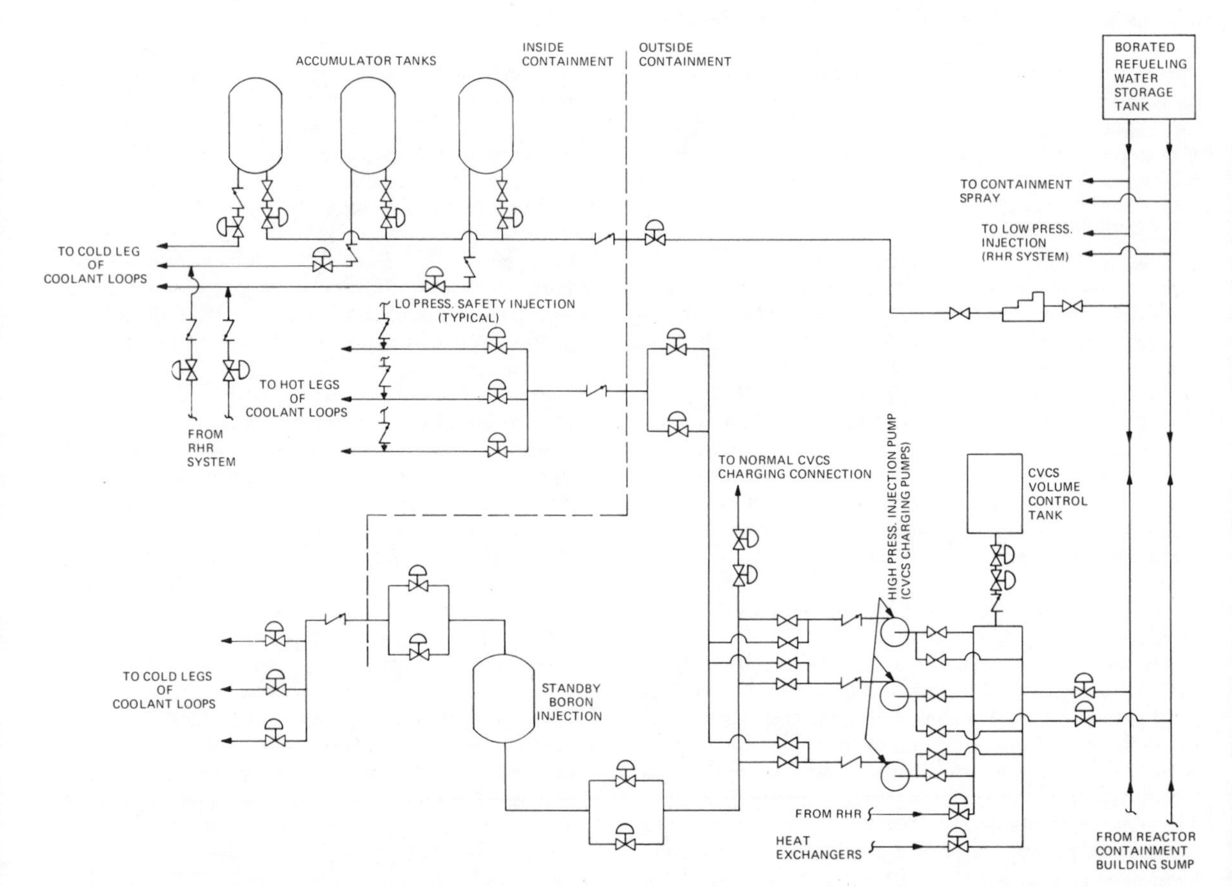

Courtesy, Westinghouse Electric Corp.

Fig. 9-1. Safety Injection System for a three-loop PWR. In this system the centrifugal CVCS charging pumps are used for high-pressure injection. Note the pump-through standby boron-injection tank for reserve reactor-shutdown capability.

is a stainless steel, multistage centrifugal pump designed in conformance with Class 2, of Section III of the Code. Construction, materials, nondestructive examination, and performance testing are the same as for the boron injection pump described earlier in the chapter. Close attention must be given to the system arrangement and NPSH at the pump inlet when recirculating water from the reactor building sump.

Safety Injection Tanks. These tanks have two other names that are used by the vendors—accumulator tanks or core flooding tanks. The tanks are pressurized with nitrogen. Tank volumes run from 1400 ft^3 to 2000 ft^3, according to reactor rating, and are approximately half full of liquid. The tanks operate from about 800 psi down to 350 psi, when the low pressure injection pumps come on. Since the tanks are isolated from the primary system by a check valve and an automatic isolation valve, design pressure is tank operating pressure plus a 10 or 15 percent margin. Design temperature is 175 degrees F. Design code is Class 3, of Section III of the Code. Material is stainless steel SA-240, usually Type 304. Construction is all welded. Inspection of fabrication and qualification of inspectors should be to the standards of Class 1 rather than Class 3. This is a deliberate upgrading of the vessel because of its emergency service duty and inaccessibility. The tanks are used in a vertical configuration rather than horizontal, because of greater ease in placing them in the reactor building.

9.2.3 *System Design*—The piping systems for emergency core cooling are designed to Class 2 standards with inspection of fabrication to Class 1 standards of Section III. All operating valves must be controlled remotely. Valves must fail into the position required to set up the flow path for the emergency cooling situation. Flow paths must be provided to test operation of each pump and each system path at any time. The system test should go as close to the reactor as practicable. Each system path should be physically separated from the other, and protected, so that a physical failure of one path will not affect operation of the other system.

9.3 Containment Cooling

As a back-up to the emergency core cooling systems, containment cooling systems are provided. In a BWR, the containment cooling mode of the RHR supplies two parallel 100 percent capacity systems to cool containment by direct water spray into the containment building. The water supply is taken from the pressure suppression pool. Operation of the RHR service water system rejects the heat energy to the final heat sink. The RHR system equipment has been described in Chapter 8.

In the PWR a recirculating air cooling system and a back-up direct spray system are provided. Each system has both cooling and I_2^{131} scavenging capability. The air recirculation system uses an activated charcoal filter to adsorb the halogen; the spray system uses a sodium solution, thiosulfate or hydroxide, as a scavenger to form the soluble sodium iodide. Containment spray, as shown in Fig. 8-2, is taken initially from the borated water storage tank; after exhaustion of that water supply, cooling is continued by recirculation from the reactor building sump. Each pump in Fig. 8-2 is 50 percent capacity.

9.3.1 *Containment Spray Pump*—This is a single stage centrifugal pump of all stainless steel. Pump design and materials of construction are the same as those of the residual heat removal pumps described in Chapter 8. Fabrication inspection and inspector qualification should be the same as for the safety injection pumps just described. The pump operates at both atmospheric pressure and containment design pressure. Additionally, it must be assumed that the spray reaches equilibrium temperature with the containment steam and the pump must operate at these high temperatures. Bearings and seals should be investigated to see if they require cooling. NPSH should also be checked at the high temperatures. Design pressure should be pump shut-off head, plus static head, plus a minimum of 50 ft. The pump should conform to Class 2 of Section III; certified performance curves should be required at both room temperature and the highest operating temperature. Since this is an emergency service pump, actual NPSH should be duplicated in the performance tests even though the unit may be a standard production pump.

9.3.2 *System Design*—The spray system is all stainless steel to preserve the purity of the borated refueling water. It is cheaper, over the life of the plant, to use corrosion resistant materials initially, than to use carbon-steel piping and equipment in this system and repurify the water each time a routine test is performed. A minimum of two 50 percent pumps should be supplied so that if one unit fails, a single unit plus the air-cooling system still has 150 percent of the required capacity. In a similar manner at least two recirculation fans should be used. Note that if each system has two 50 percent pumping units, it is possible to lose one unit from each system and still have a total capacity of 100 percent available. Thus, it takes a double failure to reduce the cooling to 100 percent of requirements.

System design conforms to Class 2 of the Code. Design temperature should be 50 degrees F, plus the saturated steam temperature corresponding to containment gage pressure. Design pressure on the pump discharge should be pump shut-off head, plus static head, plus 10 percent of the design dynamic head. Design pressure on the suction side of the pump should be static head of the refueling water storage tank, or the sprays, whichever is higher plus 15 percent.

Redundant parts of the system should be separated

from each other and physically protected so that no single accident can disable both portions of the system.

9.4 PWR Air Cooling

This system is used for cooling of the PWR containment structure during normal operation. It thus has the advantage of continuous operation so that any failure or abnormal performance will be noted immediately. Two important points must be noted and accommodated:

a. The system must withstand any pressure transients that may occur between the containment structure atmosphere and the inside of the air-handling system and equipment during the pressure rise immediately after an accident.

b. The system must handle air at atmospheric pressure and at containment design pressure.

9.4.1 *Air Handling Equipment*—Pressure transients arise due to the complex duct paths resulting from the limited space available inside containment. Overpressure protection is furnished by weighted vents which open when the pressure differential exceeds design values. A typical, external-duct design, pressure value is 7 inches of H_2O. The actual air handler should be multifan—units with four and five fans are normal. Fans ought to be two-speed so as to handle the higher-density air; they must be sized so that no single fan handles more than 50 percent of the emergency load. Cooling coils in the air handler depend on the water supply available at the site. Under normal conditions the cooling load is handled by the reactor building closed circuit cooling system. Because of the emergency service requirements, the closed circuit cooling system is backed-up by its heat sink. If the heat sink is fresh water, as from a river, lake, or cooling tower, the air handler cooling coil has special connections, as shown in Fig. 9-2, which permit direct emergency cooling by the nuclear service water system. If the heat sink is sea water, then the air handler should have a second set of special coils, sized for emergency service only, made of a copper base alloy. The alloy must be chosen for resistance to sea water under both dynamic and stagnant conditions. A good starting selection is 70-30 cupro-nickel alloy which has excellent "textbook" resistance to sea water. The alloy selection must be checked either by local experience with the actual site sea water or by corrosion tests in aerated stagnant and flowing water. Both conditions must be checked to insure that the material is suitable for all service conditions.

Ductwork should be heavily galvanized sheet metal.

9.5 Heat Sinks

Each of the previously described cooling systems requires either a closed loop cooling system or a

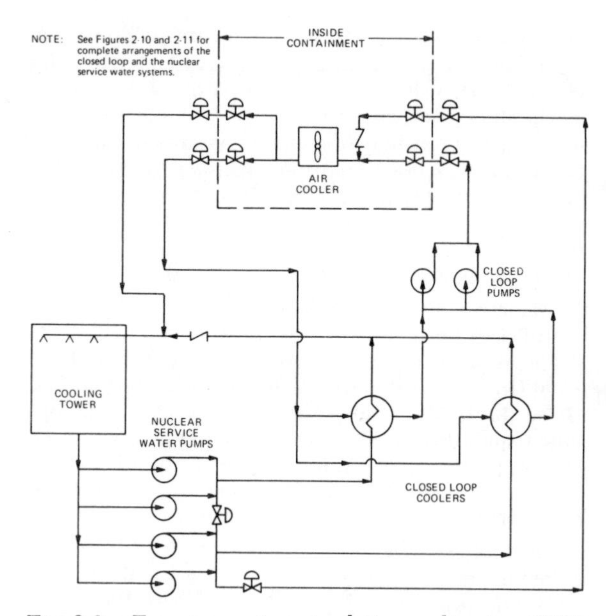

Fig. 9-2. Emergency air recirculation cooling in a PWR containment structure. This closed cooling loop has only one set of piping; redundancy is furnished by a spray system. Note the back-up water supply directly from the nuclear service water system. The cooling tower has been added for convenience to close the loop with a heat sink.

service water system, or both, to continue cooling. These systems and their components are described in Chapter 13, Closed Loops and Nuclear Service Water Systems.

9.6 Containment Penetration Systems

In Chapter 5, Figs. 5-12 and 5-13 show penetration sealing systems required on PWRs. These systems are required because a PWR containment structure has no secondary building enveloping it as the reactor building envelops the containment structure of a BWR.

9.6.1 *Penetration Pressurization*—The system is designed in accordance with Section VIII, Division 1 of the ASME Code and USAS B31.1.1.0 for the piping. All piping and air receivers are constructed of carbon steel. Piping should be seamless; welded piping is not recommended due to the additional cost of inspecting the longitudinal weld in order to guarantee pipe integrity. Pipe routing should be correlated with plant location to determine if the need for Charpy impact tests and nil ductility transition temperature definition exist. It is entirely possible that plants located in the northern half of the United States may require low-temperature steel piping from ASTM Spec. A-351. Design pressure for piping and receivers should be the same as the system that supplies the air. Each receiver should be sized to supply air for not less than four hours. Leakage shall be calculated on the containment leakage specification, i.e.: 0.1 percent or 0.2 percent per day of containment free volume, as specified. Further, since this pressurization air is injected between two seals, both of which can leak the maximum amount, the receiver air-supply must handle a double leak. If the containment leakage specification is 0.2 percent per day, a receiver sized for 4 hours must supply the following free volume:

$$\text{ft}^3\,(\text{Containment free volume}) \times .002 \times \frac{4}{24} \times 2 = \text{cu ft}$$

The back-up nitrogen gas supply should be sized in the same manner as each air receiver, but it should be sufficient for 24 hours.

The normal air supply to the system should be taken from the instrument air supply system, backed-up by the service air system. In no case should the dew point of the supply air be higher than −40 degrees F.

9.7 Common Requirements

Since all systems described in this chapter are required in an emergency, all equipment; supports; and protection must conform to Seismic Class I requirements. Additionally, the specifications must require certified raw material analyses and documented test and inspection acceptances. All pumping equipment should have certified performance curves; catalog curves are not acceptable for this equipment.

Specifications should require that not less than 10 percent of the welds be radiographed or have an analogous recorded volumetric examination on piping and pressure vessels. Large vertical storage tanks used for emergency water supplies should have at least 2 percent of normal welds plus 100 percent of all three- and four-plate intersections (shown in Fig. 9-3) radiographed. All inspector qualifications should be documented and preserved along with the radiographs.

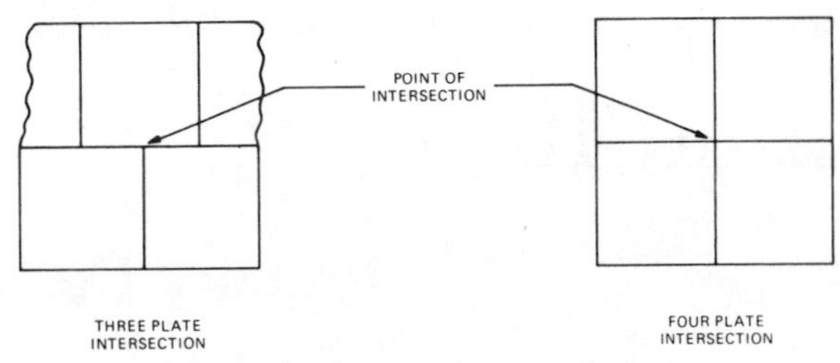

Fig. 9-3. Three and four plate intersections in large vertical tanks.

All systems must be designed so that each active component, such as a pump or a valve, can be tested during operation. Additionally, the system must be capable of being tested at any time. The test limits should go as close to the reactor as practicable. The system test should start with a simulated alarm signal, and exercise the complete instrumentation system and the process system.

CHAPTER 10

Primary Coolant Purification

10.1 System Function

All primary coolant systems must be maintained at a high level of purity in order to operate properly. Typical water conditions in a PWR primary loop are:

Conductivity	<1 micromho/cm^3
pH	4.5 to 10.0 according to boric acid concentration
O_2, ppm, max	0.1
Cl^-, ppm, max	0.15
F^-, ppm, max	0.1
H_2, cc(STP)/kgH_2O	25-35
Total suspended solids, ppm, max	1.0

BWR water conditions are slightly different, viz:

Conductivity	<1 micromho/cm^3
Cl^-, ppm, max	0.1

There are no oxygen and hydrogen requirements in the BWR because they are continuously removed by the turbine vacuum system. No limitation, as such, is given for fluorides by the vendor, but they should not be permitted to rise above 0.1 ppm. Suspended solids should not be higher than 1.0 ppm, as given for the PWR.

The purification systems remove products of corrosion and fission products, thereby controlling overall plant activity levels. In a PWR, the purification system has the following additional functions:

a. Acts as an expansion volume during plant heat-up

b. Supplies and controls the primary coolant boric acid for long-term reactivity control.

c. Controls coolant pH by addition of lithium hydroxide

d. Controls coolant oxygen by addition of hydrazene

e. Assists in control of radiolytic decomposition of the coolant by saturating the coolant with hydrogen as it passes through the volume-control tank

f. Supplies the additional water needed during plant cooldown.

The systems have several names according to the reactor vendor. In a BWR, it is known as the Reactor Water Cleanup System. In a PWR, it may be called the Chemical and Volume Control System or the Make-up and Purification System. The reactor water cleanup system in a BWR operates on the recirculation system, as shown in Fig. 3-12. In addition, there is a full-flow condensate-polishing unit in the feedwater train after the first or second feedwater heater, as shown in Fig. 3-1. The condensate-polishing unit in a BWR plant is usually the same type demineralizer as in the reactor water cleanup system, for reasons of economy, except when the condenser has seawater.

In a PWR, the system may have two additional duties in emergencies, not given in the above list. Its charging pumps can act as the high-pressure safety injection pumps or as the standby, emergency boron-injection pumps; both of these emergency duties are shown in Fig. 9-1. It is possible, therefore, to have a system of mixed quality levels according to the uses of the system.

10.1.1 *Boron Management*—Boron management is the general term applied to all the operations performed in varying the dissolved boron content of the primary coolant in a Pressurized Water Reactor. Boron addition is accomplished by injection of high-concentration boric acid through the charging pumps. Boron removal may be accomplished rapidly, by a bleed and feed system, feeding in pure water; or slowly, by a deborating demineralizer.

Figure 10-1 is a simplified diagram of a basic pressurized water reactor chemical and volume control system in which minor boron removal is by means of a demineralizer. Boron is added to the system by making up fresh boric acid in the make-up tank and injecting it into the system. In this system, the boron withdrawn from solution by the deborating demineralizer is permanently lost because operating economics do not favor resin regeneration. An alternate method, which has a higher capital cost but a lower operating cost, uses an evaporative system to recover both the boric acid and pure water, for reuse. In this system, shown conceptually in Fig. 10-2, the influent is first passed through a demineralizer then goes to the evaporator system. The evaporator condensate is upgraded to primary coolant quality water by another demineralization treatment, and is then returned to make-up. Concentrated boric acid is returned to the concentrated boric acid tank for reuse. The use of evaporative systems is favored because of the decrease in the amount of radioactive wastes to be handled.

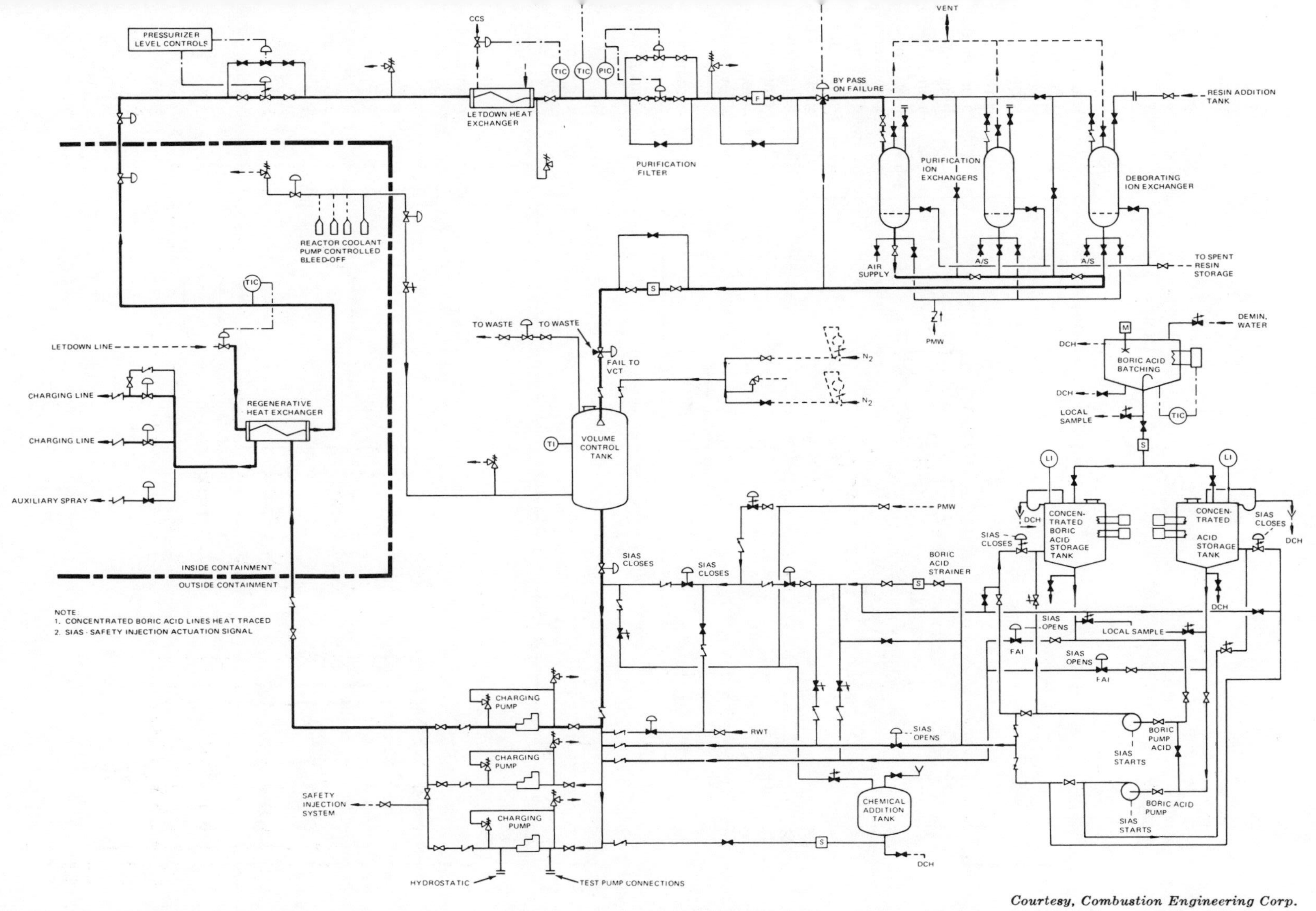

Courtesy, Combustion Engineering Corp.

Fig. 10-1. Typical chemical and volume control system. Minor boron removal is by the deborating demineralizer; major removal by a feed and bleed system to an evaporative boric acid recovery system. Back-up shut-down boron supply is stored in the concentrated boric acid tanks.

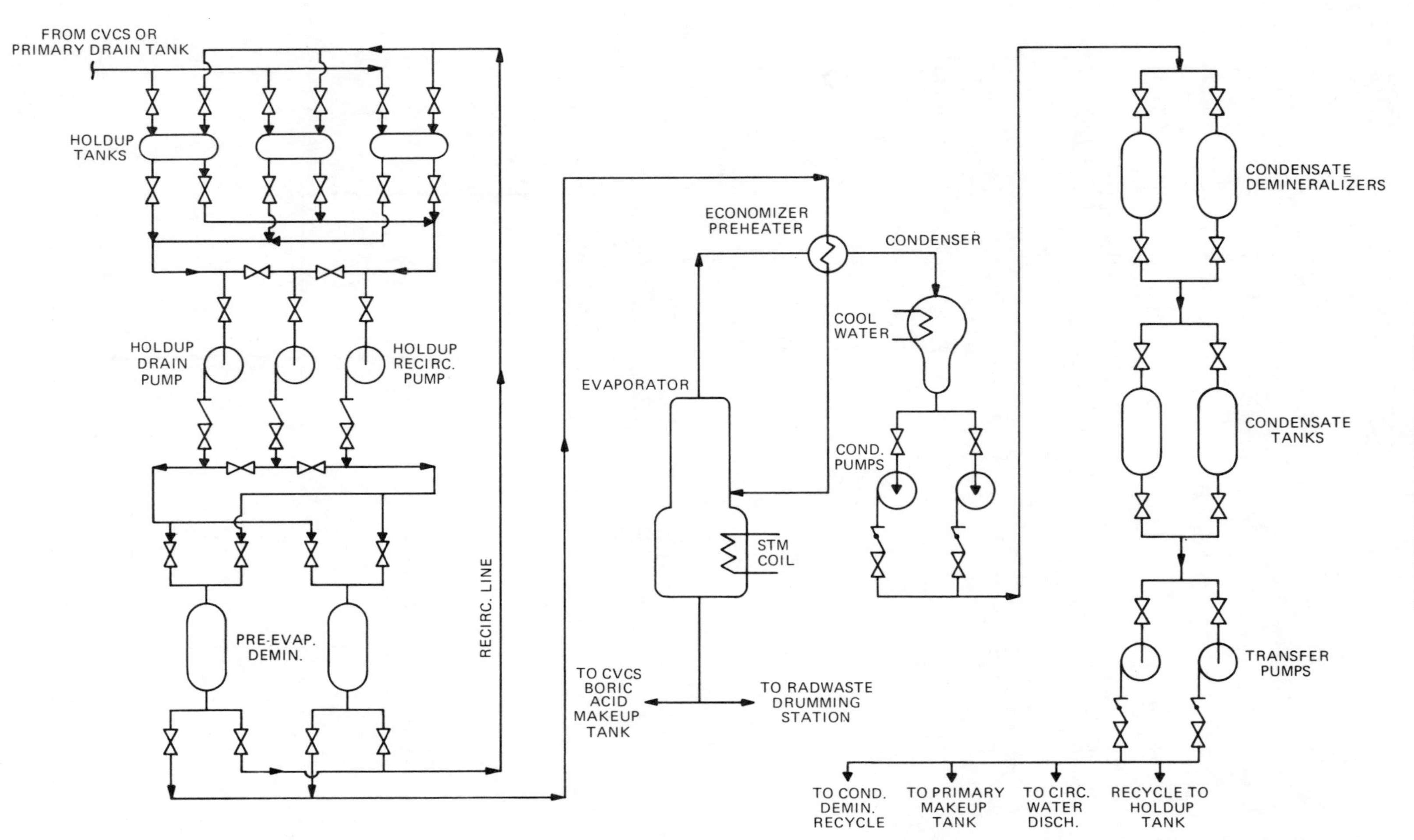

Fig. 10-2. Diagrammatic arrangement of an evaporative boric acid recovery plant. Distillate from this plant is recycled as primary system make-up; concentrate is recycled to the concentrated boric acid storage tank. This system will be required for any "near zero release" PWR plant. (See Chapter 12.)

The boric acid evaporation system will probably be used exclusively in the future, because it becomes a natural part of the "zero release" scheme described later, in Chapter 12.

The only emergency service required of the boron management system is when, as in Fig. 10-1, the concentrated boric acid storage tanks are used as the poison source for the back-up shut-down system. Note in Fig. 9-1, that even though the CVCS charging pumps are used for emergency boron injection, a special boric acid storage tank is used and the boron management system is not involved in the emergency. Figures 9-1 and 10-1 reflect different philosophies by two system designers; both approaches are acceptable.

The accepted boric acid make-up and storage-system concentration has been 12 percent. This required that all storage tanks and piping be kept above 120 degrees F to prevent precipitation of boric acid crystals in the process system. One designer is now changing his philosophy and reducing the concentration of his charging system to 4 percent so that his system need not be heat-traced. He has checked the allowable cooling rate of his primary system and found that with the same size charging pumps, he can add dissolved boron to the primary system fast enough to maintain the reactor subcritical as it cools down to room temperature.[1] This designer, however, keeps his emergency injection boric acid in a separate tank at 12 percent concentration and pumps through the tank when he has to inject the back-up shut-down boric acid. Pumping through the tank with low-concentration water insures that any boric acid crystals that may be present will dissolve and not interfere with system operation.

10.2 Demineralizers

The demineralizers are always the polishing type with a mixed bed of anion and cation resin. Two types of units are used—a bed of granular resin (a minimum of 36-inches deep), and a unit that uses ground resin in a manner similar to a filter-aid type filter. In both types of units the resin is used for only one cycle and is then sluiced to radioactive waste. It has been found more economical to replace the resin than to regenerate it and then process the regeneration effluents in a radioactive waste system. The radioactive impurities are fixed on a solid bed and, if not regenerated, are in a suitable physical form for shipping as a spent resin, or for inclusion as fine aggregate in a concrete shipping package. In any case, the resin eventually suffers radiation damage and must be replaced.

Since the radioactive impurities are almost all cations such as Fe^{+++}, Cr^{++++}, Ni^{++}, etc., some vendors are working on a deep bed system with a cation precolumn followed by a mixed bed column. In this system the cation precolumn removes the bulk of the radioactive impurities and takes almost all the damage from radiation so a cheaper, lower grade cation resin is used. The mixed bed uses expensive, nuclear grade resin.

10.2.1 *Deep Bed Demineralizer*—The deep bed demineralizer is designed for a flow rate of 20 to 50 gpm per square foot of bed area. Since the water purity maintained in primary systems makes it almost impossible to calculate a bed size properly, the approach used is to design a bed with adequate cross-section for the flow required, and sufficient depth to insure proper operation. This type of design produces a total bed-volume that may have a theoretical life of over two years. Actual bed-depth varies from 36 to 66 inches.

The demineralizer unit is made of Type 304 stainless steel, and conforms to Class 3, of Section III of the ASME Code. Design temperature need never be higher than 175 degrees F and may be as low as 150 degrees F because the resin cannot accept temperatures higher than 140 degrees F. Addition and discharge of resin is by sluicing of a resin slurry. An air connection should be provided at the bottom of the vessel so that the bed may be reset at periodic intervals to prevent channeling in the unit. The vessel should have 50 percent freeboard above the bed so that it may be reset properly.

Design pressure is the option of the system designer. In Fig. 10-1, the high-pressure primary coolant passes through a pressure reducing station, on through the demineralizers, and then into the volume control tank. The system designer then will select his system pressure levels and design the demineralizer accordingly.

The slurry is best handled with 5 to 10 percent solids, and velocities of 5 to 10 ft per sec. Branches should be taken off with laterals in the direction of flow, so that the slurry does not have to take more than a 45-degree turn. Branches in horizontal runs must always exit out of the top quadrant of the pipe. Direction changes requiring elbows should be made with five-diameter bends, or greater, whenever possible. In close quarters, direction changes may be reduced to as small as a long-radius elbow, but should never be tighter. Fresh resin must always be handled in stainless steel to preserve the resin capacity for the demineralizer; spent resin may be handled in either stainless steel or carbon steel. Valves should be ball type.

10.2.2 *Ground Resin Demineralizers*—A ground resin bed demineralizer is supplied with a Boiling Water Reactor; it operates as a combined filter and demineralizer and can require bed replacement either because the bed is chemically exhausted from demin-

[1] The increase in water density due to cooling has the effect of adding positive reactivity to the core in the same fashion as withdrawing a control rod. Hence, boric acid must be added to balance this effect as coolant temperature drops.

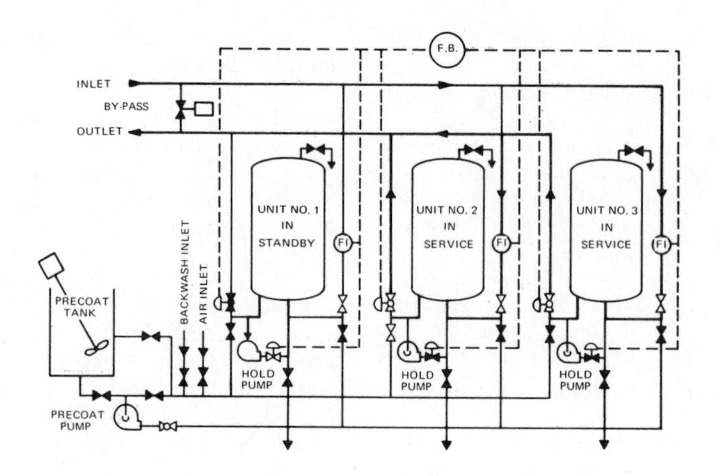

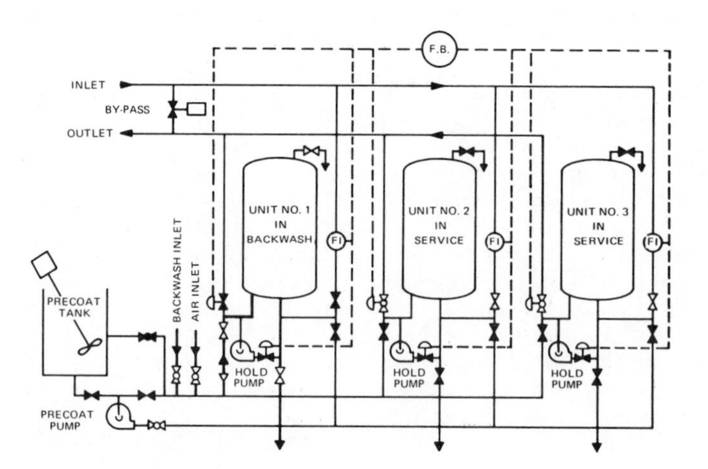

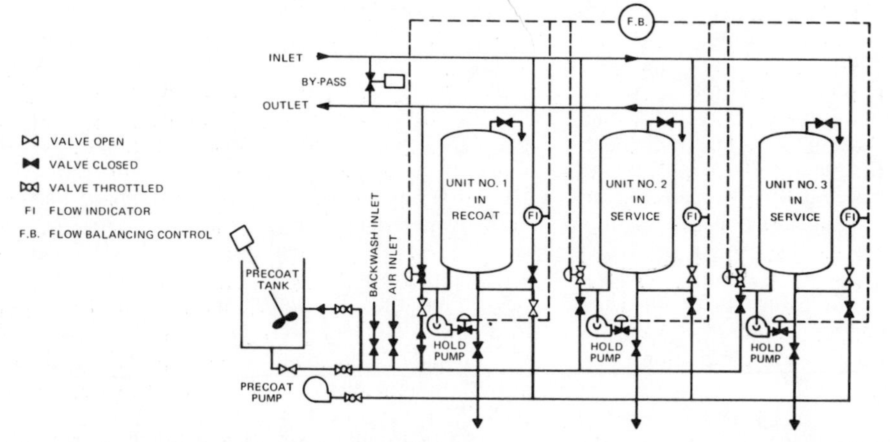

Courtesy, Graver Water Conditioning Co.

Fig. 10-3. Unit Nos. 2 and 3 of this ground bed demineralizer plant are in service, while the holding pump holds Unit No. 1 on standby (Top) until needed; Unit No. 1 being backwashed (Center); and Unit No. 1 being precoated with resin (Bottom).

eralization or it is physically dirty from filtration. It is, therefore, monitored by both a conductivity meter and a differential pressure gage. The plant requires a resin mixing tank, precoat pumps, a filter/demineralizer unit, a backwash water supply, compressed air, and a holding pump, all as shown in Fig. 10-3. A resin slurry is prepared in the mixing tank, using nuclear grade, ground, mixed bed resin, and demineralized water or turbine condensate. The slurry is applied to the filter/demineralizer unit (coating filter cartridges) with the precoat pump. Coating thickness may vary between 3/16 inch and 1/2 inch. If the unit is to be held on standby before use, the precoat layer is held in place on the cartridges with the unit's holding pump which simply recirculates demineralized water through the unit. Holdup flow is about 1/16 to 1/8 gpm per square foot of filter surface. The plant is arranged so that on low flows the hold-up pump comes on automatically, to keep the bed in place.

Plants are designed as multibed plants with one or two more filter/demineralizer units than are needed to handle the flow. In this manner, the extra unit (s) is being recoated while the remainder of the units are on stream; thus, continual operation is achieved. Average recoating time is of the order of 45 minutes to 1 hour. An individual filter/demineralizer unit may handle up to 5000 gpm; average flow rates in a condensate polishing unit are of the order of 4 to 5 gpm per sq ft of bed surface; in the reactor water cleanup system unit of a BWR, flow rates are lower—down to 1 or 2 gpm per sq ft of bed surface.

A plant may be manual, automatic, or semi-automatic. In a manual plant, all operations except holding are manual. In an automatic plant, all operations are performed automatically and a signal is sounded at the initiation of major steps such as "recoat" or "in service." In a semi-automatic plant, recoating is initiated manually by pressing a button, after which recoating is under the control of a sequence timer and a stepping switch. In a semi-automatic plant the unit is placed into service automatically, and withdrawn automatically, sounding the alarm referred to above. At the present time the semiautomatic plants are the most popular because it is easier to monitor the recoating process when it must be operator-initiated.

The filter/demineralizer unit is a vertical cylindrical tank with an entrance manhole in the top. The tank has a false bottom which is used as a tube sheet for the precoat cartridges. The unit shown in Fig. 10-4 is an early unit used in fossil powerplant work. Each cartridge has a nylon or stainless steel base wound on a stainless steel core. The vessel is carbon steel, usually SA-515 or 516, lined internally with a multilayer coating of baked phenolic, 8- to 10-mils-thick. The reactor water cleanup unit is built in accordance with Class 3, of Section III of the ASME Code; the condensate polishing unit is Section VIII

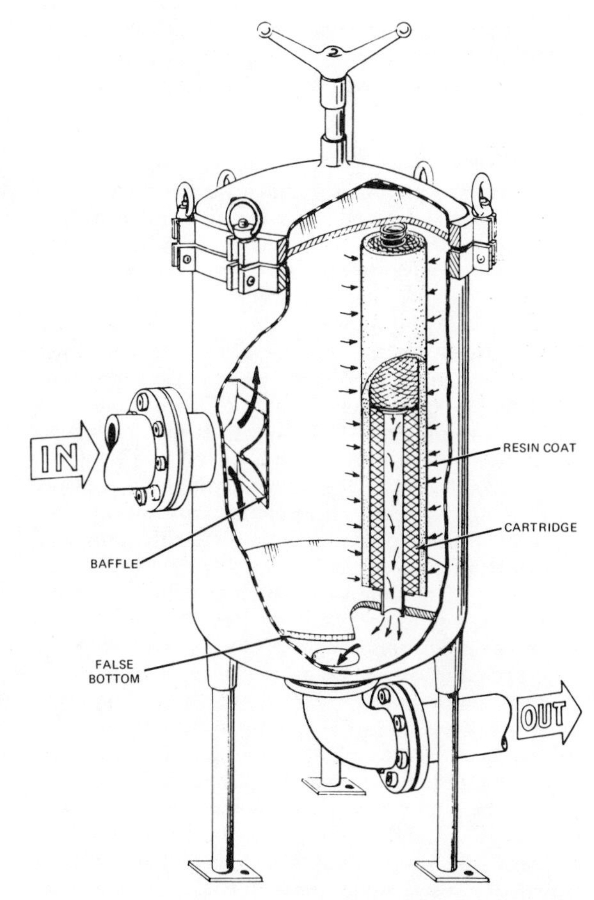

Courtesy, Graver Water Conditioning Co.

Fig. 10-4. Cutaway drawing of a typical ground resin filter/demineralizer unit.

Division 1. Design conditions are usually 1400 psig and 150 degrees F for the reactor water cleanup system; 150 psig and 150 degrees F for condensate polishing. The manhole is designed so that when the unit is installed in a concrete shielding cell, directly beneath a removable ceiling plug, the manhole cover can be removed from the floor above, after which all the cartridges are removable. The necessary maintenance tools are provided by the manufacturer and should be specifically required by the purchase specification.

Strong base and acid resins are used, ground to about 325 mesh. The resin mix is 3:1 cation to anion ratio. A single coating runs 4 to 5 weeks before recoating is required. The resin is used for one cycle only, and then transferred to waste. Clean pressure drop of a fresh coating is 3 to 5 psig with the system designed to accept a dirty pressure drop of between 25 and 30 psi. In cases where the ion exchange capacity is exhausted at low-pressure differentials, the reactor operator has the option of applying an additional resin coating to the unit, or backwashing and recoating. Resin handling is by slurrying, in accordance with standard chemical engineering practice for filter aid.

10.3 Filters

Filters, except for the filter/demineralizer previously described, are of the disposable cartridge-type arranged for remote opening and cartridge replacement. Filter casings are austenitic stainless steel, usually Type 304, built in accordance with Section III Class 3 of the ASME Code. Construction is all-welded except for the head, with connections which are for welding rather than flanged. Cartridges are sized for 98-percent retention of 25 micron particles, and a clean pressure drop of 3 to 6 psi. The filters are run to a dirty pressure drop of 20 to 25 psi, depending both on the engineer and the total pressure drop of the system. The procurement specification must give clean pressure drop for sizing purposes, and dirty pressure drop for cartridge strength. Cartridges are of organic yarns, such as cellulose, nylon, or dacron, wound on a stainless steel wire-mesh core. In every application using a disposable cartridge, the total accumulated radiation dosage must be checked against the estimated life of the cartridge. By-passes must be provided in the piping to permit cartridge replacement during plant operation. Remote tools should be supplied, or definitive design criteria specified by the filter vendor. This includes tools for opening the filter and removing or replacing the cartridges. The plant design engineer must provide shielded access, usually from above, and crane service. Cartridge transport facilities, in the form of a general service or special-purpose shielding cask, must be provided by either the design engineer or the owner.

10.4 Heat Exchangers

Heat exchangers in reactor purification systems have a variety of names depending on the purpose and the vendor. The three most common names are the regenerative and the non-regenerative, or letdown, heat exchangers. These are always found in tandem on the high-temperature, high-pressure influent line to the purification system. The regenerative heat exchanger acts as an economizer, transferring heat energy from the hot, primary coolant bleed to the cooled, purified primary coolant feed. Excess heat energy remaining in the bleed stream is removed by the non-regenerative, or letdown heat exchanger which transfers it to the Closed Loop Component Cooling System. Excess heat is that which must be removed to reduce the fluid temperature to levels acceptable to the demineralizer resins—120 degrees F to 140 degrees F, depending on the resins used. Other names for exchangers in this system are excess letdown heat exchanger and seal water cooler. The units vary in size with the regenerative, non-regenerative, and letdown heat exchangers handling full system flow—as high as 150 to 200 gpm—and the seal water coolers handling as high as 40 gpm, depending on the amount of water bled from the primary pump seals.

Design pressures and temperatures depend on the exchanger application so no numbers can be given. Exchanger types are either shell and tube, or shell and spiral. Because there are so many different designs, some general guides will be given. The heat exchanger must be Class 3, Section III of the ASME Code. Heat transfer tubes are always austenitic stainless steel. In a PWR, primary coolant is handled in austenitic stainless steel. In a BWR, primary coolant is handled in carbon steel, except for the heat transfer tubes. The tube-to-tubesheet joint should have a light contact roll followed by welding. The tubesheet should be a carbon steel forging clad with stainless steel to accept the tube-to-tubesheet weld. TEMA Class R, is preferred, although some of the vendors offer Class C. Fouling factors of .0003 for primary coolant and .0005 for closed-circuit cooling water are adequate, although a great many engineers prefer a fouling factor of .0005 for primary coolant.

Heat loads generally require multishell units so it is not unusual to see two- and three-shell exchangers. As may be expected, both horizontal and vertical equipment arrangements are used. Access in the form of inspection ports and removable bonnets or channels must be provided to both ends of the tubes for inspection and repair. Inspection ports should be provided along the length of the shell to permit inspection of the tube bundle. The favored exchanger design is a U-tube type with a single pass shell because of initial low cost, and automatic provision for differential thermal expansion.

10.5 Pumps

The purification systems use both centrifugal and reciprocating pumps. As with the heat exchangers, the same pump has different names according to the individual vendor. In a PWR, the pump injecting purified water into the primary system may be called the "charging pump" or the "make-up pump"; it may be either centrifugal or reciprocating. For instance, in a four-loop plant, Westinghouse furnishes them as two centrifugal units and one reciprocating unit. The charging pump always has the emergency duty of injecting high-concentration boric acid in the PWR as the redundant shut-down system, to comply with Design Criterion 26. The pump may also act as the high-pressure safety injection pump as described in Chapter 9. The charging pump is all stainless steel, designed in accordance with Seismic Class 1; and Class 2, of Section III of the ASME Code. The pump is Seismic Class 1 because of its emergency duty requirements; if the system is arranged so that the pump has no emergency service requirements it need not be designed to operate in seismic shock but only to retain the integrity of the pressure envelope. Typical dynamic head for charging pump design is 5700 ft of water because it must pump water from the low pressure (one or two psig) volume control tank, into the primary system at about 2285 psig.

In the BWR, the pump is used strictly for recirculation and supplies only friction losses, as shown in Fig. 3-12. Note that the pump is located upstream of the filter/demineralizer units so it need not preserve demineralizer purity; as a result, this pump has a cast carbon steel casing and austenitic stainless steel impeller and trim. A typical set of design conditions is 1400 psig and 570 degrees F; and rating is 180 gpm at 500 ft.

In the PWR, the charging pump is all stainless steel. Both the BWR reactor water clean-up pump and the PWR centrifugal charging pump use mechanical seals—two seals in tandem—with the conventional graphite running against stainless steel. A major difference in the two units is the design temperature; the BWR pump is designed for 550 to 575 degrees F; and the PWR pump for 150 to 200 degrees F. Thus, the BWR pump requires bearing and seal cooling, while the PWR pump does not.

The reciprocating charging pump has a forged, stainless steel body—Type 304 or 316, with a 17-4 PH piston. The stuffing box uses Teflon or graphited asbestos packing rings; if Teflon is used, the total radiation dosage on the Teflon must be checked over the longest anticipated refueling cycle. The packing should be changed at least every refueling and the theoretical life of the Teflon should be at least twice the total anticipated radiation dosage between refuelings. The pump drive should be through a remotely controlled variable-speed gear box to permit injection variation. The pump should have double inlet and discharge check valves of stainless steel, with a cobalt-tungsten hard-faced trim. It should have a relief valve with a capacity equal to 100 percent of pump rating.

Manufacturers are hesitant about performing an adequate seismic dynamic analysis on pumps as small as these; therefore, it is suggested that the procurement specification allow the option of either a dynamic analysis or a vibration test, to demonstrate seismic adequacy.

The remainder of the pumps in a Chemical and Volume Control System handle boric acid. The pumps are centrifugal type when used for transfer operations, and reciprocating type when used for metering operations. Metering pumps use the same construction as the reciprocating charging pumps. The centrifugal pumps use the same materials of construction as the larger charging pumps, but canned pumps are often used. The main reason for the use of a canned pump in lieu of a conventional centrifugal pump with a mechanical seal, is that the cost of maintenance on a small conventional pump makes the more expensive canned pump a better investment over the life of the plant.

Pumps handling boric acid must either be heated by hot water jacketing or have provisions for flushing the pump with pure water at shutdown, to prevent boric acid crystallization. The preferred method is pure water flushing.

10.6 Tanks

Tanks are always stainless steel or stainless clad carbon steel both for corrosion resistance and fluid purity. They are either horizontal or vertical, depending on equipment arrangement. All tanks except the volume control tank and the chemical mixing tank are atmospheric. The volume control or make-up tank, as it is sometimes called, is designed for 75 to 100 psig internal pressure and 15 psig external pressure. Design temperatures are usually about 250 degrees F, even though most tanks are atmospheric.

The volume-control tank operates at 2 to 3 psig with a hydrogen atmosphere above the liquid surface. The hydrogen dissolved into the water in the volume control tank retards the radiolytic decomposition of the primary cooling water as it passes through the core.

Tanks storing or preparing boric acid solutions must be heated to prevent crystallization of boric acid. Operating temperatures of boric tanks vary from 140 to 175 degrees F, depending on the concentration of the acid being stored. Tanks are designed in accordance with Section III, Class 3 of the ASME Code. Tanks storing boric acid for emergency shutdown service, as in Fig. 10-1, must also be Seismic Class I. Construction is all welded. The chemical mixing tank is constructed to Section VIII, Division I of the ASME Code.

10.7 Piping

Piping in the purification systems starts at the second valve from the primary system, none of it is Class 1. That portion of the piping required for emergency service must conform to Seismic Class 1 and ASME Code, Section III Class 2. The remainder of the piping is ASME Code Section III, Class 3 and Seismic Class II. Emergency piping in this system belongs to the back-up boron injection shutdown subsystem and interconnecting piping when the charging pumps are used for high-pressure safety injection. Design pressures and temperatures vary widely from 2750 psig and 650 degrees F, down to 150 psig and 250 degrees F. Material is always stainless steel, Type 304 or 316. The Type 316 stainless steel is used in the more severe services to take advantage of its higher strength at elevated temperatures; for instance, at 650 degrees F in Class 1 and 2 piping specifications. ASME SA-376 Type 304 stainless steel, has an allowable stress of 14,300 psi, and Type 316 has 16,000 psi. The actual selection of the alloy must be based on economics because either alloy is suitable. An all-welded system should be used. Boric acid lines should be warmed with electrical heat tracing; steam tracing requires too much maintenance in areas of zero or limited personnel access, to be practical in this application.

Valves should be stainless steel with double packing, and lantern ring leak-off as previously described in Chapter 7. Some of the valves in this system will be as small as 1½ or 2 inches in size. Valve construction in the small sizes must be carefully checked. A popular design uses a screwed bonnet; care must be taken by the use of hardness differentials or appropriate coatings to insure that the bonnet will not gall in the valve body. Packing should be an Inconel-wire-reinforced graphited asbestos.

10.8 Boric Acid Recovery

Boric acid is recovered by concentration of dilute boric acid to about 12 percent in an evaporator plant. The plant is purchased as a package consisting of the evaporator, preheater, condenser, piping, and controls. The boron-free distillate, after a demineralization treatment, is primary water quality. The concentrate or evaporator bottoms is concentrated boric acid and is used until its radioactivity level no longer permits. The condensate tanks, shown in Fig. 10-2, permit sampling of a condensate batch to determine its disposition. This permits the operator to make an informed decision and to do any recycling that may be necessary before reuse.

The plant is all stainless steel and may be skid-mounted if desired. The evaporator is a tray-type tower. If a feed tank and pump are furnished, then the pump is the canned type. All condensate pumps have mechanical seals. Plant capacities range from 5 to 20 gpm feed rate. Vessels, piping, pumps, and valves conform to Class 3, Section III of the ASME Code and Seismic Class II. All concentrated acid tanks and piping require heat and insulation to prevent unwanted crystallization and the temperature of the system is usually kept at 150 degrees F which provides a 40-degree F—margin above saturation temperature.

CHAPTER 11

Spent Fuel Storage Facilities

11.1 Facility Description

Spent fuel removed from the reactor is stored in pools filled with demineralized water until such time as it has decayed sufficiently so that the activity level and heat output of the spent fuel are low enough to permit shipment to a reprocessing facility. This storage period varies from 90 to 360 days, depending on the policies and economics of the utility.

Water is used in the pools because it provides cooling, shielding, and visibility, and in a PWR it is borated as an added protection against accidental criticality. The fuel elements are stored in metallic frame structures that keep the elements in a non-critical geometric array and provide adequate flow paths for convective cooling. The storage racks are supplied by the reactor vendor and are always of a corrosion resistant material, usually austenitic stainless steel.

In a PWR, the fuel storage pool is always in a separate building which is connected to the reactor building by a transfer tube, as shown in Fig. 2-15. In a BWR, the storage pool is in the reactor building and is accessible from the reactor as soon as the drywell cover has been removed, and the seal between the reactor and drywell has been installed. The storage pools, themselves, are reinforced concrete lined with stainless steel. The metal lining is not less than 3/16-inch for reasons of stability during construction —solid stainless steel or stainless clad steel is an economic decision, taking into account all costs of material and labor. The surge tanks shown in Figs. 2- 9 and 3-21 are arranged so that the pool water overflows over a skimmer edge into the surge. Thus, the water level in the fuel storage pool is always constant and the inevitable evaporative water loss shows up as a level change, greatly magnified, in the surge tank where it can be measured.

The pool is usually designed to accept two core reload batches plus a complete core. A core reload batch is defined as the number of fuel-element assemblies changed in one refueling. Thus, if for some unexpected reason the plant has an old batch that has not yet been shipped, has just been refueled and then must be completely unloaded, there is sufficient room to store all the fuel. In a BWR, all the fuel is stored in a single pool because it is too expensive to sectionalize the pool. In a PWR it is common practice to compartmentalize the pool so as to make the fuel transfer device (shown in Fig. 2-15) furnished by the reactor vendor, accessible for maintenance.

The fuel storage cooling system is usually sized on the heat load produced by a core reload, five or six days after shutdown. Decay heat production falls off exponentially according to a variation of the Wigner Way equation which can be expressed as follows:

where

P_t = Power level at time, t, after shutdown, Megawatts thermal

P_o = Core operating level, Megawatts thermal

$T - T_o$ = time after shutdown, days

T = time after startup, days

T_o = operating time, days.

$$P_t = 5.9 \times 10^{-3} P_o \left[(T - T_o)^{-0.2} - T^{-0.2} \right]$$

The cooling system is sized so that there will be a net pool temperature rise for the first three or four days after reloading, followed by a return to normal pool temperature. The temperature rise occurs because actual fuel removal may take place in 3 or 3½ days instead of the design, 5 or 6 days. The pool temperature rise can be as high as 8 to 10 degrees F; it takes advantage of the heat absorption capability of the pool water to reduce the size of the cooling equipment. If the load is higher than normal due to unexpected events, connections are provided to permit the use of the RHR cooling capacity in the fuel pool. A typical set of design conditions will maintain the storage pool at 115 degrees F with ⅓ of a core transferred 150 hours after shutdown; the fuel elements will have operated 930 full-power days.

An important requirement of the spent fuel storage pool is sufficient purity and clarity of water to minimize corrosion of the stored fuel elements and permit clear vision to all parts of the pool. To achieve this, all installations have filtering and purification equipment, as shown in Figs. 2-9 and 3-21. Purification equipment should be sized to pass the complete volume of the storage pool in 36 to 48 hours. This will require flow rates of the order of 150 to 200 gpm. The purification plant should be arranged so that water can be filtered, even though it may not require demineralization.

11.2 Water Requirements

The boron requirements of the fuel storage pool for PWRs vary from manufacturer to manufacturer but are in the range of 2100 to 2500 ppm of boron. This corresponds to a range of 12,000 to 13,000 ppm of boric acid.

A definition of suitable water clarity that has sometimes been used is the following: the water shall be clear enough to read letters ¼-inch high under 25 ft of water, using binoculars and a viewing aid to avoid surface-breaking aberrations. Total nondissolved solids should be less than 1.0 ppm. Make-up water, before the addition of boron compounds, should have the following characteristics:

Conductivity	1.0 μ mho/cm at 25°C, max
pH	6.0 to 8.0
Oxygen	0.1 ppm, max
Chlorides	0.15 ppm, max
Fluorides	0.10 ppm, max
CO_2	2.0 ppm, max
Total undissolved solids	0.5 ppm, max

During operation, pool water conductivity may be permitted to rise to 1.5 μ mho/cm, but demineralizer outlet should never be permitted to go higher than 1.0 μ mho. The halides—chloride and fluoride—should be kept at the inlet conditions.

11.3 Seismic Classification

Cooling and purification equipment and piping are Class II seismic design. The actual storage pool, itself, is Class I seismic to insure adequate water for cooling and shielding. No penetrations are permitted below the waterline so that accidental piping ruptures cannot drain the pool.

Equipment and structures above the open pool must be carefully examined. The design engineer has two alternatives available: In the first alternate the structure, crane, and equipment may all be Class I seismic and carry the corresponding cost; in the second alternate the structure, crane, etc., may be Class II if the design is such that under *no* condition can a Class II item fail, and damage a Class I item or a fuel element. The choice is not a simple one to make and each design requires a careful evaluation for a correct decision.

11.4 Heat Exchangers

The heat exchangers are shell and tube type with two pass U-tubes and a single pass shell. Fuel pool water is in the tubes and closed circuit cooling water is in the shell. Tubes and channel are stainless steel while the shell, baffles and tie rods are carbon steel. The unit may be of bolted or welded construction. The author prefers bolted construction so all parts of the unit are accessible for inspection. The tubesheet is carbon steel clad with stainless steel. The tube-to-tubesheet joint is welded, since the pool water is potentially radioactive. Gaskets are stainless steel, spiral metallic gaskets with a low chloride (less than 200 ppm) asbestos filler. Design codes are TEMA Class R, and Section III Class 3 of the ASME Code. Typical design conditions are 100 psig and 175 degrees F tube side, and 75 psig and 150 degrees F shell side. Each unit is sized for 50 percent of the design load. Water temperature ranges are about 105 degrees F into the pool, and 115 degrees F out of the pool.

In the current 1100 MWe plants, thermal rating is 3400 MWt. One third of the core, a single reload batch, is 1100 MWt. Normal refueling time is between 10 and 12 months; for this illustration 310 days will be used, so one batch or ⅓ of a full core runs three complete fuel cycles or 930 days of operating time. If 150 hours after shutdown (6.25 days) is used as the design rating for the heat exchangers, then the design heat load, P_t, will be:

$$P_t = 5.9 \times 10^{-3} (1100) [(6.25)^{-0.2} - (936.25)^{-0.2}]$$
$$P_t = 6.49 [(6.25)^{-0.2} - (936.25)^{-0.2}]$$
$$P_t = 6.49 [.693 - .254]$$
$$P_t = 2.85 \text{ Megawatts thermal}$$

Water flow through the tubes will be:

where

Q = heat load, Btu/hr
C_p = specific heat, Btu/lb/°F
w = lbs water per hour
Δt = temperature change, °F

$$Q = C_p w \, \Delta t$$

$$2.85 \times 3.413 \times 10^6 = 1 \times (115\text{-}105) \times w$$

$$w = 2.85 \times 3.413 \times 10^5 \text{ lbs per hour}$$

$$\text{Vol} = \frac{2.85 \times 3.413 \times 10^5}{8.33 \times 60} = 1980 \text{ gpm}$$

Thus heat loads in an individual unit will be a maximum of 1.45 MW (4.95 × 10 Btu/hr). Water flow through the unit will be selected to give the most economical design when considering the temperature of the final heat sink.

11.5 Pumps

Two 50-percent capacity pumps are used for the spent fuel coolant circulation; some engineering companies, as a standard practice, specify each pump at the calculated head with 60-percent volume. There is no objection to this practice as long as the motor is adequate. The pump for a PWR is of stainless steel construction with mechanical seals. For a BWR pump, the casing is carbon steel with stainless steel wear rings and shaft. Seal material should be Stellite against graphite with all other metallic parts of stainless steel. Static seals in the assembly should be neoprene, Hycar or Hypalon; ethylene-propylene terpolymer, if available for a reasonable price, has a life in radiation fields at least ten times longer than the preceding elastomers. Casing gaskets should be the same type of spiral metallic gaskets described for the heat exchanger. The pump design should permit removal of the impeller without disturbing the motor, the pump casing, or the piping connections. The pump conforms to Class 3, of Section III of the ASME Code.

The prime purpose of the borated water pump shown in Fig. 2-9 is to permit purification of the water in the refueling canal during refueling operations. The pump has the same specifications as the coolant circulation pump. Its capacity is such that it can supply the system demineralizer only; there is no cooling requirement for this pump.

11.6 Demineralizers

Demineralizers are the mixed-bed type with nuclear grade resins. The best choice is a deep bed demineralizer for scavenging service; this type is actually used in PWR installations. In BWR facilities the presence of large powdered resin filter/demineralizers for condensate and primary purification make the use of a similar design economically imperative in the spent fuel system. The use of a cheaper deep bed unit in a BWR would add great cost to the radioactive waste system and require that two different forms of resin be stocked. The equipment should be arranged so that one resin preparation installation serves the spent fuel and primary water filter/demineralizers.

Specifications for fuel pool demineralizers are the same as for the primary system equipment described in Chapter 10, but modified for pressure rating and quality level. Quality is discussed later, in Chapter 14.

11.7 Piping and Valves

Piping is all welded and conforms to Class 3, Section III of the ASME Code. PWRs use Type 304 stainless steel SA-312 or SA-376 or SA-358, while BWRs use carbon steel SA-106 or SA-155. Valve materials agree with the piping and have hard-faced

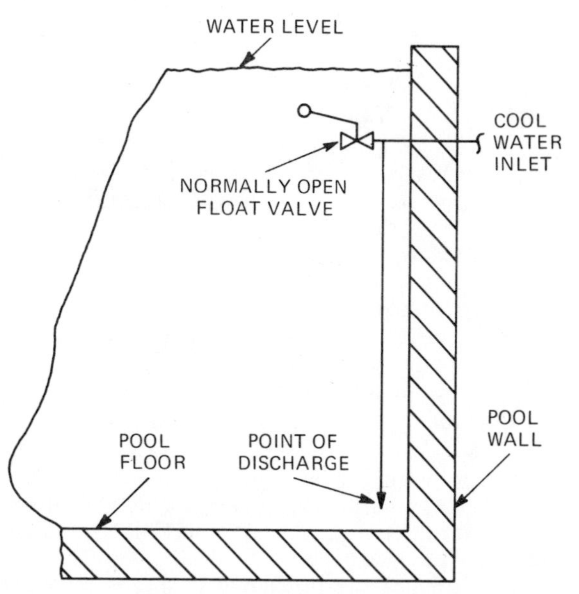

Fig. 11-1. A fuel pool siphon break using a normally open float valve. A low waterlevel permits the float to drop, opening the valve venting the pipeline.

alloy trim. Valves require either a back seat or gland leak-off to prevent leakage to the atmosphere; the author specifies both. Piping entering the pool always penetrates the pool wall high enough to prevent serious loss of water due to an external rupture; the penetration is usually between 6 inches and 36 inches below the water surface. Piping internal to the pool, carries the cool water down into the pool and discharges it just above the bottom. A siphon break, one arrangement of which is shown in Fig. 11-1, is required to prevent loss of water after an external break. Pool water discharge is taken off near the top to assist natural convection cooling. A variation of piping that has been used with good results is to return a portion of the purified and filtered flow through an inlet scupper in the opposite side of the pool from the outlet skimmer. This causes a definite flow pattern across the pool surface and assists in keeping the surface clean of floating dusts.

CHAPTER 12

Radioactive Waste Handling

12.1 System Description

Radioactive waste must be handled to the satisfaction of all regulatory bodies. In general, this means that all solid, liquid, and gaseous wastes must not exceed the limits prescribed in Part 10, Chapter 20, of the Code of Federal Regulations (10CFR20). This section of the regulations imposes instantaneous limits and integrated limits per year, for individual isotopes and collective discharges. The Federal Government has issued guidelines for design objectives for control of radioactive discharges to the environment that require the design basis to be effectively 10^{-5} times the limits given in 10CFR20. Details of these design objectives are given in Appendix I of Part 10, Chapter 50, of the Code of Federal Regulations.

The plant design for liquid and gaseous waste processing must have a theoretical basis and the base usually chosen is the failure of 1 percent of the fuel. It is appropriate to note here that no nuclear installation has ever approached this limit. Liquid wastes originate from demineralizer wastes, valve and pump seal leaks, laboratory processing fluids, "hot" shower and laundry drains, equipment drains, equipment washdown operations, and in the case of a PWR, primary coolant excessive volume changes and the boron management system.

Gaseous wastes originate mainly from dissolved gases in the coolant, radiolytic decomposition of water, gases added to the plant such as N_2 and H_2, vent air circulated through storage tanks, and a minute amount of fission product gases that might leak out of one or two defective fuel-rods in a core. In a PWR, gas volumes are relatively small, because the primary system is sealed, and the only gas is in the CVCS volume control tank where the coolant is sprayed into the hydrogen atmosphere at a relatively low pressure.

In a BWR, gas volumes are larger because the feedwater (turbine condensate) is being continuously deareated and all deareator off-gas must be treated as if it were radioactive.

Solid wastes arise from a variety of miscellaneous sources. There are three "regular" sources: spent resin from the various demineralizers, used filter cartridges, and prepared radioactive concrete. In addition, there are such things as blotters, worn clothing, shoe covers, gloves, broken tools, decontamination materials, worn-out equipment and other, miscellaneous solid items.

Waste treatment philosophy is basically simple: Nothing must leave the site in any form—solid, liquid, or gas—unless it has been properly monitored and the method of disposal specified. Thus, regulations are always observed.

Gaseous wastes are handled in two ways: 1. Very low activity gases, such as may be expected from storage tank vent systems, may be monitored for activity; diluted with copious volumes of outside air; monitored again, and then released at high velocity from tall stacks, high in the air (300 ft, or more). The height and velocity permit the gases to be further diluted before they sink to the ground: 2. Higher level wastes, such as may come from the gases evolved from primary coolant in a PWR, are compressed and stored in decay tanks for 30 to 45 days before their controlled release to the atmosphere. The release for any gases is through three filters in tandem—fiber glass "air conditioning" filter, high-efficiency particulate filter, and finally a halogen adsorbing filter.

Gaseous wastes from a BWR must be processed on a continuous, rather than a batch, basis because of the quantity involved. Since present-day BWRs are direct cycle machines it follows that the gases removed from the main condenser by the air ejector will contain any gases that may have been released or formed in the reactor. Thus, there will be oxygen and hydrogen formed by radiolytic decomposition of water, N_2^{13}, N_2^{16}, O_2^{19} as well as any trace amounts of xenon and krypton released from fuel rod surface contamination. A typical BWR gaseous waste system before publication of the Appendix I design objectives is shown in Fig. 12-1; its description follows. A suitably sized path was provided to hold up the gas for at least 30 minutes to permit the active isotopes to decay before release. The noble gases, xenon and krypton, decay to solid daughters and are removed by particulate filters. The other main isotopes decay by at least 3 half-lives and are reduced to ⅛ of the original content. The gas is filtered by high-efficiency particulate filters before release. Radiation monitors measure gas activity continuously and are able to seal the plant exit before excess activity reaches the stack. This is achieved by monitoring at the inlet to the gas holdup section. The gas is diluted in the stack

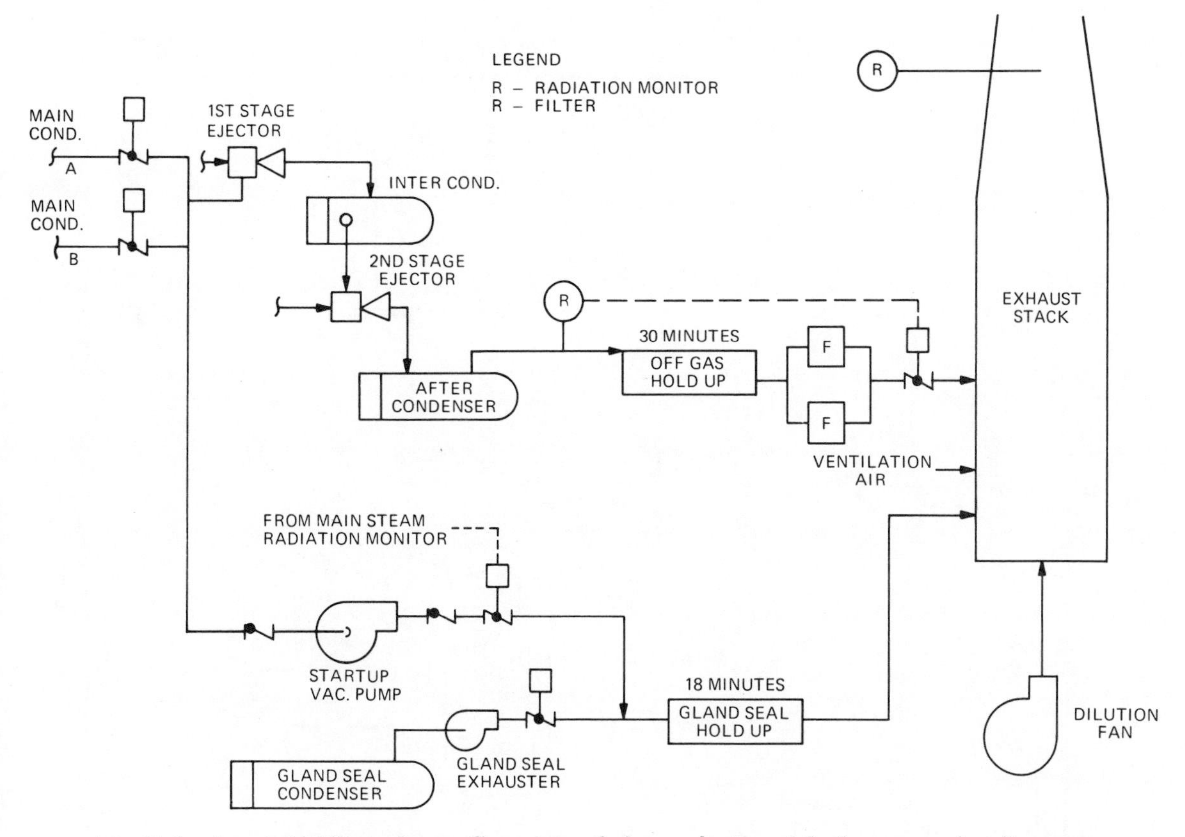

Fig. 12-1. A typical BWR condenser off-gas system before application of the "near zero release" concept. See Fig. 12-4 for the same system arranged for "near zero release."

for control of activity per unit volume, as well as to assure that the hydrogen content of the exhaust gas is below the explosive limit. The revised system is described in Paragraph 12.2, Near Zero Release.

Liquid waste handling has two goals, first to release innocuous wastes to the environs after suitable monitoring, and second to reduce the volume of higher-level waste by recovering pure water for reuse and converting the concentrated active waste to a stable form for long-term storage before release from the facility. Liquid waste systems differ with each engineering company, but all have the same basic principles. Wastes are classified as "clean" or "dirty" according to the chemical content of the liquid. Those wastes originating in the primary system or other pure water systems are "clean," but wastes originating in the chemistry laboratories, the laundry, or the showers are "dirty." A further classification is sometimes made that separates liquids containing detergents from other chemically contaminated liquids, because they require different treatment. Wastes are collected in holdup tanks for classification in individual batches. One batch may be chemically contaminated but radioactively innocuous so that after pH adjustment, it may be discharged from the facility; another batch may be held for 30 to 60 days and then discharged; a third batch may be treated by evaporation and demineralization to recycle water and reduce the volume of the active fluid. The active concentrate is then fixed as concrete in a drum, and is buried.

Equipment is usually provided in plants to process all liquid wastes originating in clean systems and to recover as much water for recycle as possible. Treatment today usually consists of batch evaporation, with the condensate returning to the process after sampling, and further purification by demineralization. Several years ago "clean" wastes were treated by demineralization directly instead of by evaporation, but operating costs were too high and evaporation has replaced demineralization.

In all cases, waste is filtered for particulate removal at least once before discharge from the facility. Normal procedure is to filter the waste as early as possible, to avoid particulate contamination in storage tanks.

A large source of liquid waste is in the slurrying of spent resins from demineralizers to spent resin

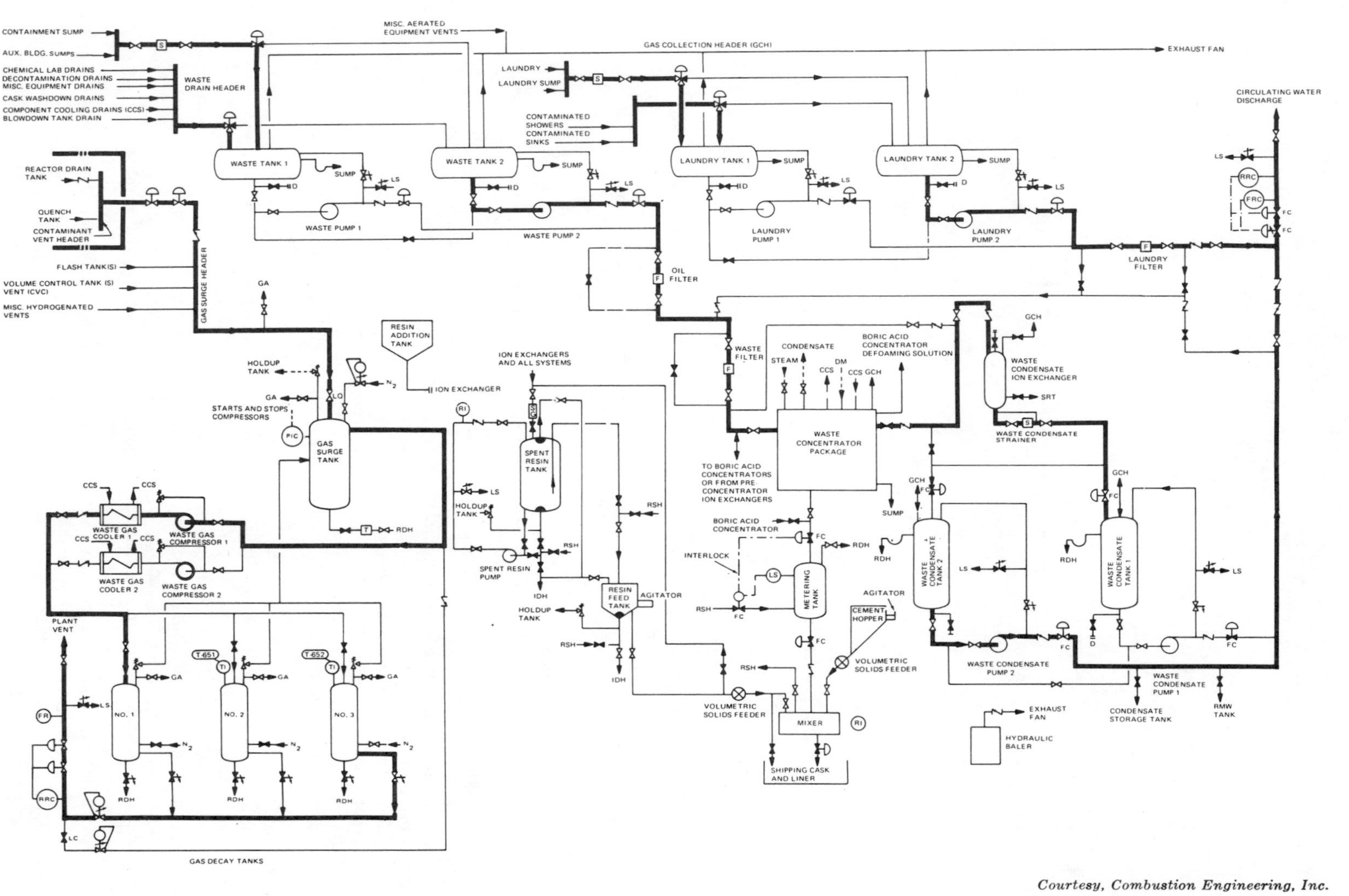

Courtesy, Combustion Engineering, Inc.

Fig. 12-2. Complete gaseous and liquid radioactive waste system for a PWR. This system is capable of recovering water for recycling; nothing is released without radiation monitoring.

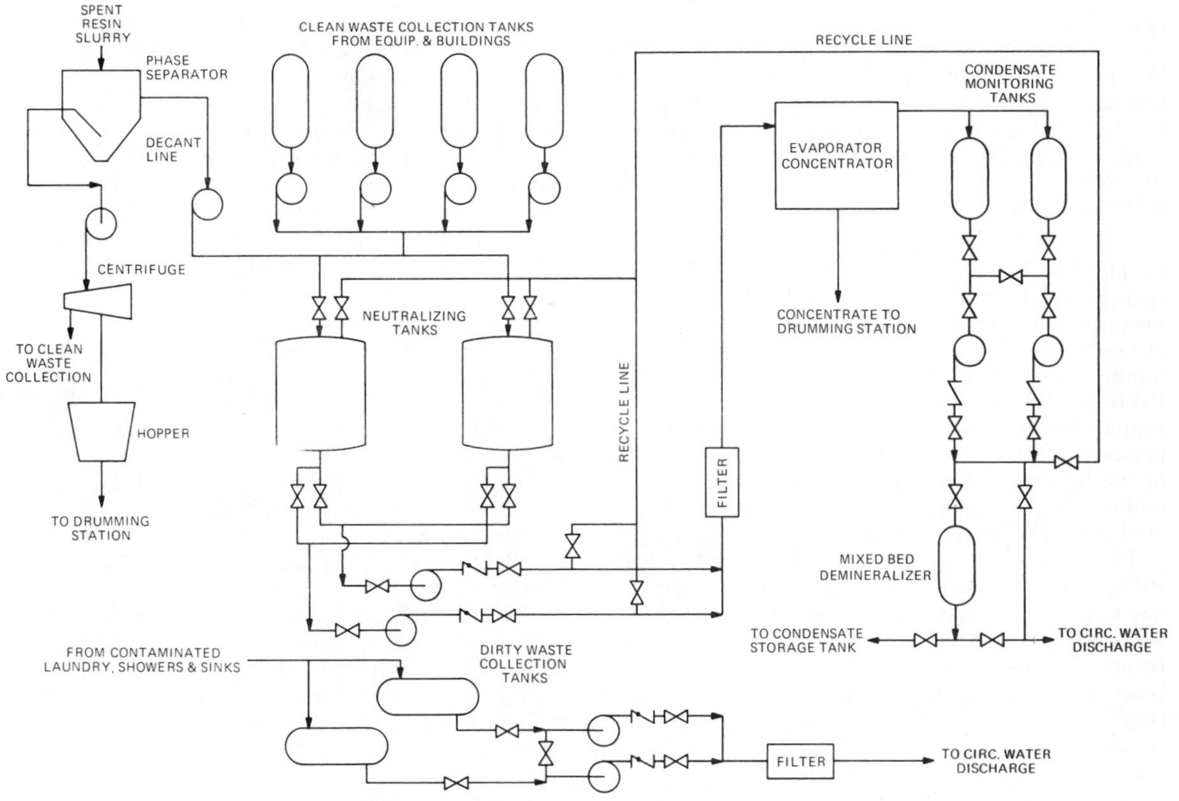

Fig. 12-3. Radwaste system for a BWR. The sludge from the phase separator and the concentrator bottoms can be mixed together in the same cement mixer for final disposal.

storage. This water, after separation from the slurry, is sampled and its disposition then determined. Water quantities and slurry-handling facilities differ for bead resins and powdered resins. Water can readily be separated from bead resins by overflowing out of a large, cross section area tank; a settling tank called a "phase separator" is needed for powdered resin separation and clear liquid is removed by decanting.

The basic philosophy in fluid waste handling is that it must be handled in individual batches which keep their identity, and whose level of radioactivity can be measured at any time. Any release to the environment must be under carefully controlled conditions such that 10CFR20, as modified by Appendix I of 10CFR50, is never violated. Continuous recorders must monitor the activity level of all potentially radioactive exits from the facility and the records must be preserved. Individual system details may vary but economics dictate that as much "clean" water as is practicable be recycled as primary coolant makeup. The alternate to recycling is extensive tankage, together with the building required to house it. Figure 12-2 shows a typical liquid waste system for a PWR, and in Fig. 12-3 an analogous system for a BWR is shown.

Solid and high-level liquid wastes are removed from the facility in discrete controlled shipments. Solids such as shoes, gloves, paper, clothes, etc., are compressed by a hydraulic baler, into as small a volume as possible and then shipped. Spent resins and high-level liquid wastes such as waste evaporator bottoms are mixed with sand and cement and cast into suitable containers, such as steel drums, which are then shipped after the concrete has cured and decayed sufficiently to permit shipment.

Shipment is made to a Federal or government-authorized burial ground where, for a fee, the waste is buried and kept under continuous surveillance thereafter. A Federal burial ground is located at Oak Ridge, Tenn.; a private facility is located near Buffalo, N.Y. and is operated by Nuclear Fuel Services, Inc.

12.2 Near Zero Release

In 1970, the "near zero release" concept was formed in answer to the demands of licensing action intervenors. The concept answers the AEC requirement that activity releases be "as low as practicable." The term has been defined as 10^{-5} times 10CFR20 limits.

The equipment used for liquid waste handling is essentially the same in all systems as before the

"Zero Release" concept. There is greater discrimination in collection of waste with more consideration being given to future processing that will be required. Now, not only is there consideration given to clean and dirty waste liquid, and active and inactive waste, but also to the degree of activity anticipated. For instance, water used to sluice spent resin to a holdup tank can be reasonably cleaned for reuse by filtration. Greater use is made of evaporators so that a maximum of water is recovered and recycled. Demineralization is reserved for evaporator distillate cleanup and water only slightly contaminated. All liquid waste originating in the primary system of a PWR is processed through the evaporator. In the evaporation process, all tritium dissolved in the water passes overhead with the distillate, and is returned to the primary system. The remainder in the evaporator bottoms is fixed into concrete; thus, no tritiated water is released to the environment.

12.2.1 *PWR*—New techniques are being used in PWR gaseous waste handling. The first step in waste-gas handling is catalytic recombination of hydrogen and oxygen to recover radiolytic hydrogen or tritium. By installing surge tanks downstream of the recombiner, effluent gas can be monitored for hydrogen and recycled as required, until all the hydrogen has been combined into water. The remaining gas stream has traces of xenon and krypton that are the sources of radioactivity. One scheme simply compresses the gas and stores it for the life of the plant. The other scheme removes the noble gases, stores them, and releases the remaining innocuous gases. Two techniques have been proposed thus far, for separation. In one method, the gases are condensed, the xenon and krypton separated by cryogenic distillation, and then packaged into gas bottles. In the second method, the xenon and krypton are separated by absorption with Refrigerant-12, separated by distillation, and packaged into gas bottles. This process is also carried out at subzero temperatures. As of mid-1971, these plants are in the design stage; any results are prototypical. The designs are based on well-known chemical engineering principles. Cryogenic distillation and absorption have been used successfully for many years.

12.2.2 *BWR*—The original BWR off-gas system shown in Fig. 12-1, has evolved for the more rigid design objectives, into the system shown in Fig. 12-4. A catalytic recombiner has been added to combine hydrogen into water, after which the water is removed by condensing. Following a 30-minute delay, additional moisture is removed by condensation at a few degrees above the freezing point of water. Following the cold condenser is a charcoal adsorption system which adds further delay before release from the plant.

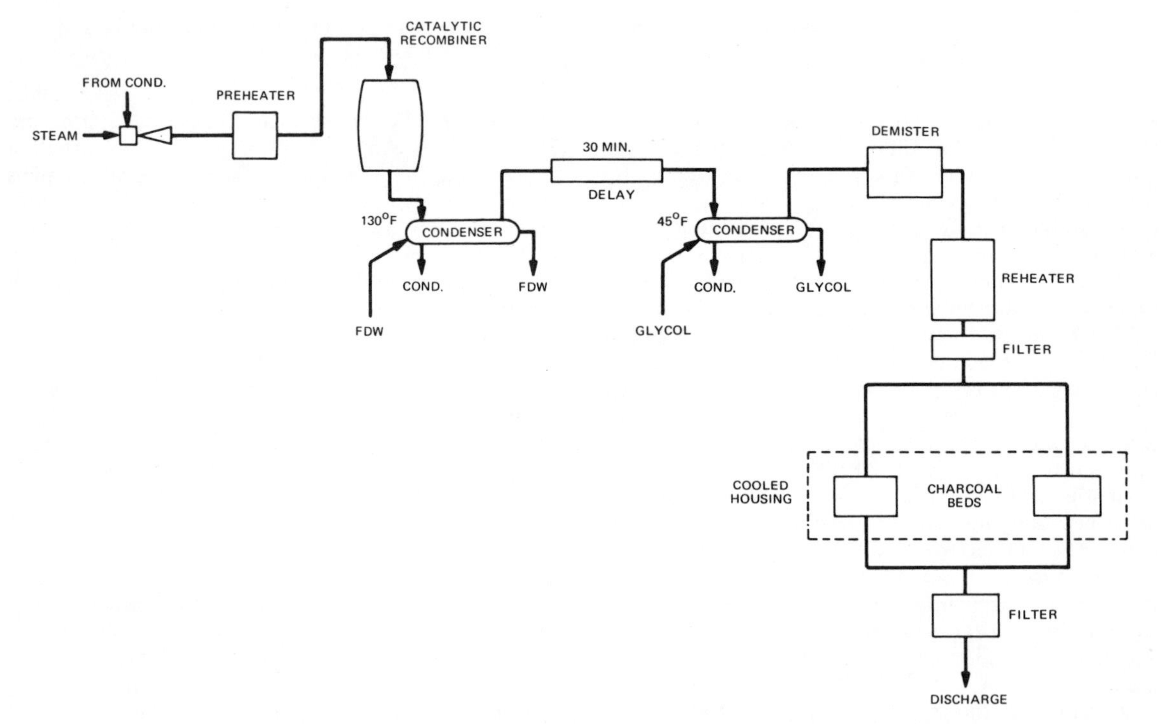

Courtesy, General Electric Co.

Fig. 12-4. A BWR condenser off-gas system designed for "near zero release." This system has hydrogen removal and low temperature charcoal adsorption beds.

As a result of this additional processing, continuous discharges are 1 percent, or less, than 10CFR20 limits, and high-level discharges are in the form of solids that are held on site or shipped to authorized storage depots.

12.3 Seismic Classification

There is a fair degree of discussion underway concerning the seismic classification of radioactive waste systems. It is the author's opinion that the storage portions of these systems can come under the description of "uncontrolled release of large amounts of radioactivity" if they fail. Thus, all tanks, together with their first shut-off valves would be Seismic Class I at all times. Such additional items as the waste evaporator, gas compressors, and feed tanks must be examined from the point of view of the amount of activity they contain and a health physicist must be consulted about the applicability of the uncontrolled activity release criteria. The furthest the items should go in seismic design is operational failure while retaining the integrity of their pressure envelopes.

12.4 Radiation Effects

After the first few months of operation, all parts of the radioactive waste handling systems are exposed to gamma radiation which, although it may vary in intensity, never ceases. Materials of construction must be chosen with as much attention given to radiation damage considerations as to chemical damage considerations. This portion of the plant sees complete process systems devoted to handling radioactive gases which require soft seated valves to insure a tight shut-off. Metals offer no problem to gamma irradiation. Organics pose a serious problem, and the fluorinated hydrocarbons are either to be avoided completely, or used under carefully controlled conditions so they do not exceed their tolerance levels.

The energy radiated during a chain reaction or by a material that has been activated by neutron bombardment, is usually expressed in "roentgens per hour" (r/hr). The roentgen unit is equal to the amount of energy passing through one cm^3 of air at 0°C and 760 mms Hg absolute pressure, sufficient to produce one statcoulomb of electricity or its equivalent, 83.8 ergs per gram. Because 83.8 ergs per gram is an awkward number to manipulate, the industry defined another unit, the "rad," and made it equal 100 ergs per gram so that it would be easier to use in calculations; 1 rad equals 1.193 roentgens. The table of threshold damage is expressed in rads.

Table 12-1, "Functional Damage Thresholds of Organic Materials Due to Radiation" lists the most common of the organic elastomers and thermosetting materials in ascending order of resistance to radiation. Above these total integrated doses, permanent damage can be expected to occur; the damage may take the form of hardening, swelling, softening, or degradation. The organics should only be applied with a complete knowledge of the anticipated radiation levels and a specific exposure period, so that the total integrated dose can be calculated. The author prefers a minimum safety factor of 2.5 on integrated dosage so that if a material has a threshold value of 5×10^7 rads, it would never be designed to go above 2×10^7. Thus, in an ambient of 1000 rads per hour it takes 20,000 hours; over two years, to reach a total dosage of 2×10^7 with an additional 3 years available if, for any reason, the replacement could not be made on schedule.

TABLE 12-1. Functional Damage Thresholds of Organic Materials Due to Radiation

Material	*Integrated Dosage, Rads*
TFE Resins	5.0×10^5
Kel-F	8.0×10^5
PVC	9.0×10^5
Butyl Rubber	1.0×10^6
Nylon	1.0×10^6
Polypropylene	1.5×10^6
FEP Resins	5.0×10^6
Silicone Rubber	9.0×10^6
Hypalon	9.5×10^6
Viton	9.5×10^6
Nitrile	1.0×10^7
Neoprene	1.0×10^7
Polyethylene	5.0×10^7
SBR Rubber	8.0×10^7
Natural Rubber	1.5×10^8
Ethylene-Propylene Rubber	1.5×10^8
Epoxy	2.0×10^8
Polyester	2.0×10^8
Phenolic	2.0×10^8
Penton	3.0×10^8
Saran	4.0×10^8
Kynar	5.0×10^8
Filled Silicone Resins	8.0×10^8
Polystyrene	1.5×10^9

12.5 Waste Solidification

As stated earlier, spent resins and liquid waste concentrates are shipped off-site in the solid form. The goal of solidification is to prepare a package that will resist physical damage and chemical action for as long as it is necessary to permit the waste activity to decay to acceptable levels. Activity levels are dependent only on the amount of activity put in the package at the facility. It follows, therefore, that by using a batching process and measuring the activity level of each batch, the total activity in each shipping package can be controlled. In the following discussion the terms "hot" and "cold" refer to the presence or absence of radioactivity, not temperature level. In preparing hot concrete, three constituents are used—fine aggregate, water, and cement. Of these three constituents, the facility waste products contribute to the fine aggregate and the water. So in furnishing a fine aggregate for the hot concrete, provision must be made for dilution of the hot spent resin with cold sand, and for diluting hot waste concentrates with cold water.

The solidification facility must have batching tanks for spent resins and these tanks must also have redundant capabilities for accurate weight measurement. The batch preparation procedure is as follows:

Shipping regulations, shipping casks, shielding capability and handling facilities (cranes, trucks, etc.) determine the maximum weight, and curies that can be handled in a single package. Laboratory testing has determined the proportions of each constituent required in the concrete. Working within these constraints, the facility engineers determine the amounts of spent resin, sand, waste concentrate, water, and cement for each concrete package. It must be noted that usual measurements of activity level require that either spent resin or waste concentrate be added to the package; combinations of the two become uneconomical. The ingredients are added to a cement mixer, mixed and discharged into the shipping package. The shipping package is given permanent identification and records are prepared, keyed to the identification and describing the amount of each constituent in the package, the number of curies of activity in each constituent, and the date of filling. The radiation level at the outside of the freshly poured package is also recorded. The package is then cured and moved to a storage area in the solidification facility until the activity level decays to the level set for shipment.

Spent resin is mixed in the storage tank to insure uniformity of distribution, its activity level measured, and the batch size determined. The resin is then transferred to a batching tank and dewatered. Batch measurement is best determined by weighing after dewatering is completed. Transfer from the batching tank to the mixer is by a solids conveyor.

Liquid waste concentrates are treated similarly to the solid waste concentrates. The storage tank activity level is measured and the batch size determined. Batching may be accomplished either by a batch tank or by a meter with an automatic cut-off. Transfer is by pumping.

Since these operations manipulate radioactive material, all activities are controlled remotely in back of biological shield walls. Control is best described as manual remote. The only automatic devices incorporated in the facilities are batching cut-offs and safety devices. All operations are initiated manually. Each piece of radioactive equipment is installed in its own concrete cell; the concrete is used for shielding. The building must be arranged with an operating gallery that has all operating controls and facilities that permit the operator to observe the operations. Viewing facilities may be shielding windows, mirrors, periscopes, closed-circuit television, or any convenient economic combination of the four. Individual cells are recommended so that when maintenance is required on a machine, only that one cell must be decontaminated rather than a large operating line. A building design philosophy must be evolved that provides for initial equipment installation and later replacement. Two equipment removal methods are prevalent. In one method, hatches are provided in the cell ceiling and the equipment is removed through the hatch. The other method has one cell wall partially constructed of removable concrete blocks. Large, heavy equipment is usually removed by cranes through hatches. Small equipment, like pumps, is removed through openings created by the removal of some of these concrete blocks.

Waste solidification equipment is still evolving. Installations use screw-type conveyors and slurries to transfer spent resins. Concrete is mixed in conventional concrete mixers and NAUTA mixers; both types are made of stainless steel to permit decontamination. The NAUTA mixer is easier to discharge and clean; its design can be closer to the normal industrial product.

The solid shipping package is either a steel drum or a concrete box; some plants prepare to handle both. Handling depends on the package; a drum is usually moved and stored on a motorized roller conveyor with the drum on end. The filled drum is then stored on the conveyor until it is shipped from the facility. Concrete boxes are usually handled by a combination of a wheeled dolly and a bridge crane. The concrete box is moved from the filling room to the storage room on the dolly; it is then put into storage and handled thereafter with the bridge crane. The building is so arranged that the shipping vehicle (usually a low-boy trailer and tractor) can come indoors for loading by the crane.

12.6 Equipment

12.6.1 *Pumps*—Pumps are direct motor-driven centrifugal pumps of all stainless steel construction. Shafts are sealed with mechanical seals using graphite running against Stellite or stainless steel. Stationary seals should preferably be ethylene-propylene or neoprene. Casing gaskets should be stainless steel spiral metallic with a low-chloride asbestos filler. Impellers, in pumps handling clear liquids, may be either open or closed; impellers in pumps handling slurries (resins), should be open. Pumps handling slurries should not require wear rings nor be sensitive to increased clearance, due to wear, between the impeller and casing. Impellers and seals should be removable, by means of spacer couplings and back pull-out design, without disturbing either the motor or the casing pipe connections. Consideration should be given to the use of vertical in-line pumps as well as horizontal pumps. If the design is such as to call for the use of a vertical sump pump in either a tank or a building sump, then the pump should have an open impeller and water-lubricated graphite sleeve bearings.

The pumps should conform to Class 3, of Section III of the ASME Code. Operational failure of waste system pumps in an earthquake can be accepted because it does not create any hazard such as might follow a loss of cooling.

12.6.2 *Heat Exchangers*—Heat exchangers, if required, are of the U-tube type with welded tube-to-tubesheet joints. They should conform to Section III, Class 3 of the ASME Code, and Class R, of the TEMA Standards. Process fluid should always be in the tubes to minimize the volume of the active fluid and take advantage of any shielding that might be available from cooling water in the shell. Specific materials of construction cannot be given for these exchangers because the fluid chemistry of each application varies. Some guidelines can be given. Cooling water can be handled in carbon steel if it is fresh water, or a copper base alloy if it is sea water. The process side may be carbon steel, stainless steel, nickel-chrome-iron alloy, nickel-iron-chrome alloy, or even a copper base alloy. Particular care must be exercised in the choice of the tube side material to insure that it is compatible with all the anticipated chemical environments it will see. The construction most used so far, has seamless stainless-steel tubes, carbon-steel shell, baffles, tie rods, and spacers. The bonnet and tubesheet, depending only on economics, are stainless steel or stainless clad carbon steel.

12.6.3 *Demineralizers*—Demineralizers should have mixed resin beds and be of the same type (bead resin or powdered resin) as the rest of the demineralizers in the plant so that the radioactive waste system only has to handle one type of resin. The bead resin vessel and internals should be of stainless steel. Construction should be all welded and be in conformance with Section III, Class 3, of the ASME Code. Connections should be welded.

The reader is referred to the discussion in Chapter 10 on ground resin demineralizers, as used in BWRs. A unit in a radioactive waste system will have the same specifications as previously described. Design temperature, pressure and flow rate will be different —in all probability, flow rates will be substantially smaller. The engineer should examine the possibility of using the same size demineralizer/filter as in the Reactor Water Cleanup System. If that is not practical, he should then look at the practicality of using the same size internal elements as in the RWCU System. The next possibility for cost saving is in the operating cycle. There is no need to provide continuous operation of the filter/demineralizer units. Judicious sizing of storage tanks will permit flow to be either stopped or reduced while a unit is taken off stream and its resin replaced.

12.6.4 *Filters*—Filters should have the same specifications as those described in Chapter 10. Equipment size will be smaller; whenever possible this filter should use the same size cartridge assembly as in the units in the primary purification system and the spent fuel pool cooling system. These units will then be designed for 25-micron filtration.

12.6.5 *Evaporators*—The evaporators must be designed to handle low chemical purity, low activity level wastes originating as floor drains, laboratory drains, laundry drains, etc. The design of the evaporator must be integrated into the design of the overall facility and a procurement package scope defined. It is usual practice to purchase the evaporator and its condenser as a single procurement together with operating controls, as the minimum scope package. Individual choice then may include such things as a feed tank and pump, concentrate discharge pump, distillate receiving tank, gas stripper, and prepackaging such as skid mounting, piping, and electrical work.

A study is required to determine the quantity and characteristics of the influent. Normal evaporator ratings are in the range of 5 to 10 gallons per minute at steady-state operation. Foaming tendencies and solids content have to be determined. Several types of evaporators are available and suitable for the purpose; among these are tray type and film type equipment. Waste volume may be concentrated to any degree so long as the concentrate can be readily pumped at the lowest handling temperature in the facility. This temperature may be ambient or it may be higher requiring the use of heated insulated equipment. Usual practice is to assume ambient temperature handling of concentrate.

The complete evaporator must be designed to decontaminate the distillate. The degree of decontamination achieved is known as the decontamination factor (DF), which is defined as the quotient of the specific activity of the feed solution divided by the specific activity of the distillate where specific activity is in curies per unit volume. Activity levels are usually expressed as millicuries (mc) or microcuries (μc) per milliliter. The minimum decontamination factor that should be considered acceptable is 10^6 and every effort should be made to achieve 10^7. The maximum permissible concentration of unidentified isotopes permitted to be discharged to the surroundings (according to 10CFR20) is 3×10^{-8} μc per milliliter. Thus, if the initial solution has an activity level of 120μc/ml and it is processed with a DF of 10^7, the resulting distillate activity is $1.2 \times 10^{-5} \mu$c which is still too high for discharge. If the distillate is then passed through a mixed bed demineralizer once, which normally has a DF of 10^3, the resulting activity level wil be 1.2×10^{-8}, which indicates an acceptable activity level for discharge. The previous example may be acceptable in operation, but the margin of acceptability is too low for design objectives. In present plant designs, this demineralized distillate is used as primary system make-up so it need never be released from the plant.

Evaporators should be constructed of austenitic stainless steels in accordance with Class 3, of Section III of the ASME Code, and Seismic Class I. Steam is the most popular heating medium, and the tube bundle should conform to Class R, of the TEMA standards and have welded tube-to-tubesheet joints. The design must be such that both the tube bundle and the evaporator shell are completely drainable by gravity. The tube bundle should be removable and all tube ends available for plugging. Connections for process liquids should be weld ends, while steam and condensate connections may be flanged. The tube bundle mounting joint, which must be flanged to permit removal, is best designed with a male-female joint having a concentric ring finish. The tube bundle gasket and shell manhole gasket should be stainless steel spiral metallic with a low-chloride asbestos filter.

It must be remembered that all valving, except those that are used only after flushing and decontamination, must have operators to permit remote control.

12.6.6 *Phase Separators*—Phase separator is the usual name applied to the vessel used in a Boiling Water Reactor facility to separate the powdered spent resin from the slurrying liquid used during transportation of the resin from the filter/demineralizer to the spent resin storage tank. The separator processes the slurry by decantation of the supernatant fluid, thus concentrating the slurry from a concentration of approximately ½ percent to about 12 to 14 percent.

The phase separator is a vertical cylindrical vessel with a Vee bottom and a flat or dome top of austenitic stainless steel. Pressure is very low, so that design pressure should be only high enough to permit design and stamping under Section III, Class 3 of the ASME Code. Design temperature should be 150 degrees F. Seismic requirements are Class I. The vessel must be copiously vented at all times. All connections should be flanged with connections included for flushing and decontamination. A spacious entry manhole should be provided in the top. Each vessel should be sized to accept at least one filter/demineralizer resin charge complete with slurry water.

12.6.7 *Piping*—Piping is all welded, low pressure. System pressures should be carefully investigated because judicious design may possibly permit the use of Schedule 10 piping instead of Schedule 40. Design temperatures will vary from 150 degrees F in the ambient temperature portions of the system, up to as high as 250 degrees F or 275 degrees F for some of the piping around the evaporator. Piping code classification is nominally Class 3, of Section III of the ASME Code, but all design temperatures selected must be checked to insure that an allowable stress exists in the code for the material and temperature selected. Connections to equipment and valves should be welded. Piping between shut-off valves and storage vessels should be Seismic Class I; while other portions of the piping may be Seismic Class II.

Valving has to be for three services; one service is for clear liquids, a second service is for slurries of resin, and the third service is for gases, In clear liquid service, an all metal globe valve is satisfactory; construction should have hard surfaced seats and double stem packing with a lantern ring and gland leak-off. For resin slurry service, the preferred valve pattern is the ball type, and the body and ball should be of austenitic stainless steel. The valve seals must be ethylene-propylene elastomer. Three types of ball valves are available: one type in which the ball is withdrawn through the top of the valve; a second type in which the valve body and two end-pieces are held together by through bolts—access to the ball is achieved by removing the body from the line leaving the end pieces in place—and the third type requires that the valve be removed completely from the line and have one end dismantled for access to the ball. The preferred valve for nuclear plant use is the first type, with access to the ball through the top. The second type has a special application. If, for any reason, a change in material must be effected, a dissimilar metal weld can be avoided by simply making the end pieces of the appropriate material. The gas lines require a soft seated valve for tight shut-off. The most popular pattern is the Saunders patent diaphragm valve, because it is a packless valve. The diaphragm is ethylene-propylene reinforced with nylon, and an O-ring added to the stem acts as a backup packing if the diaphragm should fail.

All the valves should be austenitic stainless steel with weld ends. The radiation dosage on the ball valves and diaphragm valves must be checked, and a service life commensurate with dosage and scheduled shutdowns established during the design period. All valves that are used during processing must have power operators. Air operators are preferred to electric operators because they can be more easily made to fail to a safe position. The only valves that are hand-operated are those that are used during contact maintenance after decontamination.

CHAPTER 13

Closed Loops and Nuclear Service Water Systems

13.1 System Description

"Nuclear service water" is the term used to denote the final heat sink for waste heat loads such as bearing cooling, radwaste evaporator condensing, residual heat cooling, etc. It accepts auxiliary heat loads that cannot be economically recovered. The heat load, in all cases except one, is transferred from the point of heat generation to the nuclear service water system by a "closed loop." The exception is the RHR system of the General Electric BWR which normally uses nuclear service water directly in the RHR heat exchangers and pumps. When this is the situation, the closed loop cooling system has no emergency duties and none of the emergency design requirements subsequently described in this chapter need be included. The typical nuclear service water system uses the same water and heat sink as the turbine condenser circulating water system. The closed loop uses inhibited demineralized water.

It must be noted that with the increasing emphasis being placed on control of all plant discharges, the nuclear service water heat sink must be carefully selected and integrated into the plant design. A portion of the nuclear service water system, as well as the closed loop cooling system, is required for emergency cooling service, and so must conform to the design requirements for Seismic Class I. In all cases the nuclear service water system is kept physically separate from the turbine circulating water system; thus it has separate pumps, piping, power distribution, etc. Additionally, it must be built so that no single failure can prevent the system from accomplishing its emergency duty. Hence, redundancy is required in all pumps and piping. As with all emergency services, the redundant components must be kept physically separated from each other so that a single failure will not damage more than one component. This separation must be considered from the viewpoint of the results of an extraneous structure failing onto the emergency system. If the design should permit this to occur, then only one component can be allowed to fail.

Nuclear service water can be ocean, bay, river, or lake water—or even cooling tower water. As available sites with natural cooling are occupied or thermal discharge restrictions are implemented, other solutions pertinent to individual site locations will be developed. Whenever these solutions involve the use of water storage, as in a cooling tower basin, the design must insure that an adequate water supply is available for emergency cooling after the design basis earthquake. This may take the form of ensuring that the heat rejection device will survive the earthquake shock or of providing a water supply large enough to permit "once- through" cooling after the heat rejection device has failed. An economic study is required in each case to determine the best solution. The study must also take into account the availability of earthquake-resistant heat rejectors.

Closed cooling loops must guarantee service to emergency loads after the operational earthquake, the same as the service water system. This system, as does the service water system, must conform to the single failure criterion. It is mandatory to furnish at least two 100 percent heat exchangers. Various combinations of pumps are used with the exchangers —the most popular of which are three 100-percent units, thereby furnishing a 100-percent standby for either pump. Since this system is used during normal plant operation, one of the pumps must be periodically withdrawn from service for routine maintenance. The third pump provides protection from the single failure criterion when one pump is out for maintenance during plant operation. Other arrangements are possible with 25-, 33- or 50-percent pumps after studying the comparison of full load flow rates and emergency flow rates so long as the single failure criterion is observed. The closed loop serves all the auxiliary heat loads in the reactor building and the reactor auxiliary building. The system must be so arranged that, in an emergency, all nonessential loads can be isolated from the loop. There are many ways of accomplishing this but the most common method used is two separate loops—one for essential loads, the other for nonessential loads—originating from supply and return manifolds. It must be remembered that the essential loop and the manifolds must be duplicated so that one manifold, or one loop, can be permitted to fail and not interfere with operation. Each emergency load is supplied from each loop so that there is complete flexibility in the installation. The only emergency load actually located inside containment is the containment air cooling system in a PWR. For this load, the connection from each loop has its own containment penetration for both supply and return. As previously noted for service water, the pumps; exchangers; manifolds; and loops must be arranged and protected so that a single fault such as a power feeder failure or a structural failure cannot disable both loops.

The last point to be considered is an alternate supply of coolant to the containment air cooling system. Many times an emergency connection is supplied, with its own containment penetrations, from the nuclear service water system to the containment air cooling system. This has been done often enough to warrant serious consideration, but not often enough to make it an absolute requirement. For the present, a sensible guideline can be that if the plant is within 15 or 20 miles of a sizable population center the connection should be included in the original PSAR. If it has not been included and the AEC should, during review of the PSAR, require the connection to be added, it is a simple change to make; penetrations can be provided in the containment structure design for this possibility.

Figure 2-12 shows a typical nuclear service water system for a PWR serving all the reactor auxiliary loads, the emergency diesel generators, and acting as a second water source for containment cooling if the supply in Fig. 2-11 fails. In Fig. 2-11, the component cooling system is split into two parts—an essential half for emergency use and a nonessential portion which is cut off during emergency operation. The essential service portion is Seismic Class I; the nonessential loop is Seismic Class II. Figure 2-11 does not show the two header arrangement required to meet the single failure criterion.

Figures 3-15, 3-22, and 3-23A and B show the nuclear service water systems and the closed loop system of a BWR. Note that in the BWR of Fig. 3-15, even though the RHR System is serviced by its own nuclear service water system, the pump bearings and seals must still be cooled by the closed loop cooling system of Fig. 3-22 which, therefore, is upgraded to a Seismic Class I system. Figure 3-22, recognizing this upgrading, uses only one nuclear service water system for the plant. It is appropriate to reiterate here that if the water quality permits, the RHR pump cooling should be directly assumed by the nuclear service water system so that the BWR closed loop cooling system need not furnish any emergency services.

Code requirements for these systems are not yet clearly established. Because of the safety duties of these systems the trend is to specify the systems in accordance with Class 3, of Section III of the ASME Code. This is an upgrading from Section VIII and B31.1.0, in the areas of quality assurance and quality control. It is recommended that these systems be designed and built in conformance with Section III, Class 3 of the ASME Code.

13.2 Equipment

13.2.1 *Nuclear Service Water Pump*—The specifications for this pump depend almost entirely on the plant location. It may be a vertical pump taking its suction from a body of water or a horizontal pump taking suction from a cooling tower basin. Typically, the pump head will be of the order of 60 ft of water, plus any necessary static lift. In fresh water applications the pump is usually steel with bronze trim. Steel is needed instead of cast iron because of its resistance to mechanical shock from earthquakes. Vertical pumps should be self-lubricated and not require any separately mounted auxiliary lubricating pumps; whenever possible they should be water lubricated rather than oil lubricated. Mechanical seals are preferred to conventional stuffing boxes whenever the pump design permits the option.

Horizontal pumps are single stage, double suction machines horizontally split along the centerline. If the design has an end suction then it should have a spacer-type coupling and a back pull-out design so that all the pump internals can be removed without disturbing the casing pipe connections or the motor; this permits single craft pump maintenance.

If the pump will handle salt water, the starting design should be carbon steel with Niresist trim. The ASME Code has not previously had to recognize salt water service and is lacking adequate materials for the pump. In 1971, the only suitable material recognized by the Code was an aluminum bronze under Code Case 1233. This situation will change, but the designer must check the materials of construction for a salt water pump with the code during preparation of the specification. It is essential that the surrounding area be canvased for previous experience with materials of construction and full advantage taken of this experience. It cannot be too strongly emphasized that the best guide to materials of construction is the local experience.

13.2.2 *Closed Loop Cooling Water Pump*—The closed loop cooling water pump is a horizontal pump of cast carbon steel with stainless steel or bronze impeller and trim. It should have the same construction features as the horizontal nuclear service water pump. In typical designs total dynamic head is of the order of 75 ft of water. Since this is a closed loop there is essentially no static lift, only friction head. Also, the pump must conform to the standards of the Hydraulic Institute.

13.2.3 *Closed Loop Water Heat Exchanger*—This unit transfers the heat load from the Closed Loop Cooling System to the Nuclear Service Water System. Clean closed loop cooling water should be on the shell side of the unit while raw service water should be in the tubes. The exchanger should be in accordance with Section III, Class 3 of the ASME Code, and Class R of TEMA Standards. The shell, baffles, spacers, and tie rods should be carbon steel such as SA-515 or 516. Tubesheets should be carbon steel. Tubes, channels, and tubesheet cladding depend on the service water. Salt water should have cupro-nickel tubes and tubesheet and channel cladding. For fresh water, the author prefers to start the tube material selection with carbon steel and investigate the water characteristics to confirm the initial carbon steel selection, or select a more suitable material such as stainless steel. Fresh-water channels and tubesheets can be of carbon steel and the corrosion tendencies controlled by the addition of an adequate corrosion allowance. The main point to be emphasized is that if the tubeside is filled with a natural water, tubeside material selection must be based on a thorough investigation of the characteristics and corrosion effects of the water.

Typical design conditions for the exchanger are 100 psig and 150 degrees F.

The tube-to-tubesheet joint in this exchanger may be rolled, although a welded joint is recommended if the budget permits. The rolled joint should have at least two grooves, one of which should be in the cladding if the tubesheet is clad. As a maximum requirement, tube-to-baffle clearance should be TEMA "close." In addition, whenever shell-side fluid velocities exceed 4.0 ft per sec, the author prefers a tube-to-baffle clearance not larger than 0.010 inch together with radiused edges on the baffle holes. Note that TEMA "R" requires a ⅛-inch corrosion allowance on carbon steel pressure parts.

13.2.4 *Piping and Valves*—Piping design requirements are the same for both systems: All welded construction with gate type stop valves. All valves which must be operated or may be operated in an emergency must have operators controlled from the control room. Other valves may or may not require operators according to their accessibility during operation.

Closed loop piping should be of carbon steel with an adequate corrosion allowance. Valves should be cast steel with stainless steel trim and weld ends. Valve packing in valves at radioactive heat loads, such as the non-regenerative heat exchanger, should be the same as the packing in the radioactive system. Packing in other locations can be Teflon.

Service water piping material of construction should be developed during the investigation described for the heat exchanger. The most economical choice is usually a carbon steel with adequate corrosion allowance. Many times the new synthetic piping composites, such as fiber glass reinforced epoxy, seem ideal for the service because they will have a sufficiently high pressure rating and excellent corrosion resistance. Before selecting any of these materials a stress analysis must be performed including all Seismic Class I requirements and Code approval obtained. If, after the analysis, the material is still satisfactory, economic, and acceptable to the Code, it can be used.

Service water valving is usually most economic as a cast steel valve with proper trim. Again, this material of construction should be developed during the water investigation.

13.3 Sea Water

As a word of caution in handling sea water, stainless steel should never be used in stagnant sea water and velocities up to 5 ft/sec; its behavior is suspect in velocities between 5 ft/sec and 8 ft/sec. Stainless steel may safely be used in sea water at velocities of 8 ft/sec and higher, provided the equipment can be thoroughly flushed with fresh water immediately after shutdown occurs.

In the design of heat exchangers it is impossible to guarantee the velocity at every point in the shell. Shell side design velocities rarely exceed 5 ft/sec because of undesirable mechanical effects on the tube bundle. Hence, at a great many points in the shell, actual velocities will be under 5 ft/sec. Sea water, then, is always placed in the tube side of these heat exchangers where it can be closely controlled.

CHAPTER 14

Plant Ventilation

14.1 General

The primary purpose of ventilation in a nuclear power station is to maintain a suitable ambient temperature for the operating plant machinery. In certain well-defined areas the comfort of the operator may be an additional design consideration. This situation is logical since most areas of the plant have relatively high radiation levels during plant operation which prevent an operator from entering. These radiation levels decay quickly after the plant shuts down.

In a BWR facility, entrance to the "drywell" is restricted to short periods during reduced load operation. In a PWR containment structure, access to an instrument gallery is permitted for short periods during operation when sufficient shielding is provided. Access is required to the surface of a spent fuel storage pool at all times to permit manipulation and shipping of the spent fuel. Access is required to rooms containing engineered safety features equipment for inspections and maintenance.

In a single building in which there are areas of differing levels of radioactive contamination, the ventilation system is used to control the spread of contamination. The last, and perhaps most important purpose for a ventilation system in a PWR containment structure is to act as a part of the engineered safety features by cooling the structure after the design basis accident.

Design conditions vary—in a drywell or containment structure 120 degrees F is acceptable if it is coordinated with the operating ambient design temperatures of the equipment. In areas of limited personnel access, 105 degrees F can be tolerated. In locations of continuous access, temperatures of 95 or 90 degrees F may be considered, depending on the outside air available. The areas of high design temperatures must have facilities for purge cooling to permit personnel occupation for prolonged periods for maintenance and servicing.

The most widely used system employs "once-through" ventilation with filters in both the supply and exhaust. Air supply systems have a single roughing filter; exhaust systems have a prefilter, a HEPA (absolute) filter, and an activated charcoal filter. Figure 5-14, a BWR ventilation system, illustrates the filtering requirements. The activated charcoal filter is used to remove halogens, principally I_2^{131}, from the air before it is discharged from the facility.

Radioactive or potentially radioactive areas are always operated at presures lower than atmospheric to insure that any leaks are into rather than out of the area. If there are several different areas in a building, with varying degrees of contamination, pressure staging is used to direct the flow of air. The area of least contamination is just below atmospheric pressure—0.05 inch H_2O—with areas of increasing contamination held at the increasingly lower pressures until the area with the highest contamination is at the lowest pressure in the building. It has been found in practice that pressure separations of 0.05 to 0.08 inch H_2O are controllable. Thus, air flow in the building is always from an area of lesser activity to an area of higher activity, and the contamination tends to concentrate rather than to diffuse into larger areas.

Ventilation is directed across the surfaces of spent fuel storage pools and reactor pools during refueling to entrain any gases that may be released from the fuel or coolant. In a similar manner, ventilation air is used to cool control rod drive mechanisms in a PWR and the reactor vessel cavity in both a PWR and a BWR. After an incident, small unit coolers with closed loop cooling water in the coils, are used to extract motor heat from cells with engineered safety feature pumps.

All discharges are channeled to a single exhaust stack where they are monitored for activity and diluted with copious amounts of outside air. The exhaust monitor is arranged so that it shuts down the exhaust system on high activity levels. This same monitor is sometimes used to develop a legal, continuous record of plant gaseous effluent. In this exhaust arrangement, highly suspect areas may have their own individual monitors so that one area may be closed off rather than the complete plant. This type of monitoring can be developed as the plant design progresses, commensurate with need and cost.

14.2 Control Room Systems

A separate air-conditioning and ventilation system must be provided for the control room. This system must cool the control equipment and provide acceptable working conditions for the control room operators in all situations, including the aftermath of a nuclear incident that permits radioactivity to escape to the outdoors. The system must, therefore, have a filtered air inlet system with at least a HEPA filter; where the safety analysis indicates, an activated charcoal filter may even be required. The system must be capable of once-through ventilation, 100 percent recirculation, or any intermediate combination. The system requires an activity monitor that automatically shifts the system to 100 percent recirculation upon high radiation outdoors, as from a second unit on the same site. The installation must be pressure tight (including the building construction). Operating ambient pressure in the control room must be higher than atmospheric so that any leakage is out of the room. This criterion is imposed so that if for any reason the outside air contains active particles, personnel in the control room are secure even when the door is opened. It should be noted that this is an extra precaution in keeping with the overall policy of emphasis on safety. All the equipment and air handling systems must be designed for Seismic Class 1 criteria. Controls and motors must be supplied from the emergency power buses. All equipment will be redundant so that failure of a single fan or refrigeration unit will not compromise operability of the system or occupancy of the control room. All valves (dampers) connecting this system with the outside air must be bubble tight so that when they are closed, the system is isolated from the outside. Figure 14-1 illustrates a typical control room system.

14.3 Containment Systems

The containment ventilating and/or air conditioning systems have many of the same characteristics as a control room system. Valves connecting to the outside air must be pressure tight for containment pressures—45 to 60 psig. In addition, their closing time must be coordinated with air velocity in the ducts so that on closure signal, the valves are closed before the first traces of contaminated air reach the valves. These valves form part of the containment system and so must comply with the same codes, design criteria, and quality levels as other similar parts of the containment system. In a PWR where the system forms a part of the engineered safety features systems, the equipment must conform to Seismic Class I criteria and be powered from the emergency power buses. All equipment and air distribution systems forming a part of the engineered safety features must accept the pressure transient assumed in the design basis accident. In this transient, pressure is assumed to go from atmospheric to containment design (45 to 60 psig) within one- or two-tenths of a second. Several methods of accepting the transient are available, but the most popular is pressure equalization, inside and outside, by means of "gravity dampers" distributed throughout the equipment. These dampers can be seen in Fig. 14-2. In a BWR, the system is not a safety feature so it does not have to conform to Seismic Class I nor withstand the pressure transient.

Containment systems require two more capabilities to fulfill their safety function—they must control

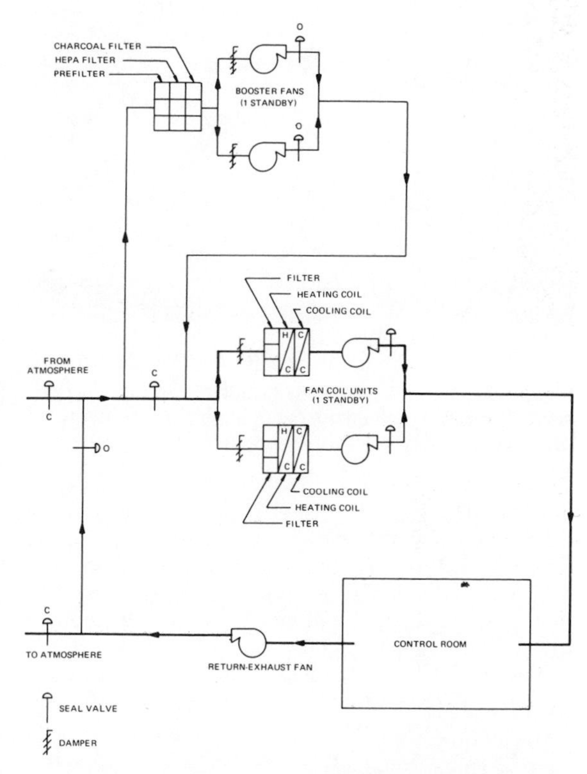

Courtesy, Burns and Roe, Inc.

Fig. 14-1. Air-conditioning system for a control room. The system is capable of any combination from 100 percent down to 0 percent outside air. A radiation monitor outdoors seals off the system from the outside on abnormal radiation level.

iodine and hydrogen concentration after the design basis accident. Iodine would be released from a fuel element if for any reason the element cladding integrity were breached. Hydrogen is evolved from the radiolytic decomposition of water and would continue to evolve slowly after shutdown.

Iodine control is supplied by including in the design a recirculating system with an activated charcoal filter. The filter must be protected from solids and liquids so the designer must furnish roughing and absolute filters as well as a liquid removal device.

The hydrogen evolving, must be prevented from reaching explosive concentrations. In a BWR, the standby, gas treatment system, Fig. 3-24, accomplishes this by controlled purging. In a PWR, this control of concentration must be deliberately exercised. At present, two methods of hydrogen control are available and each has its advocates. The first method is by periodic controlled purging with clean

Courtesy, American Air Filter Co.

Fig. 14-2. Filter housing for installation inside containment. Pressure relief dampers are visible as the openings above the closed access doors.

air—simply reducing the hydrogen concentration by dilution. The second method is by recombination of the hydrogen with oxygen into water. Criterion 41 requires that the I_2 and H_2 control systems have sufficient redundancy of system and equipment so that a single failure will not prevent execution of the systems' safety functions.

The AEC General Design Criteria in effect before May 21, 1971, required that two separate 100-percent capacity cooling systems, based on different principles, be provided to cool and depressurize the containment structure after an incident. In answer to this requirement the containment spray system and this containment cooling system were provided. Criterion 38 now requires that a containment cooling system be provided with adequate redundancy so as to comply with the single failure criterion described in Appendix I. Hence, designers of all future plants are free to examine the economics involved in spray systems, air systems, or combinations of the two and make an intelligent decision. When these economic studies are made it must be remembered that the iodine and hydrogen control systems are still required and these needs should be factored into the containment cooling study. In considering the cooling capability to be provided, it should be pointed out that with two 100-percent systems, if each system loses 50 percent of its capability by failure of a pump or fan, then the remaining two half systems combine into a third full-size system.

A typical PWR containment air cooling system is shown in Fig. 14-3.

14.4 Hydrogen Control

As an active nuclear core is cooled with water, a minute amount of energy is used in the radiolytic decomposition of water into hydrogen and oxygen which dissolve into the coolant. In a PWR, this decomposition is controlled by dissolving an excess of hydrogen in the coolant in the volume control tank

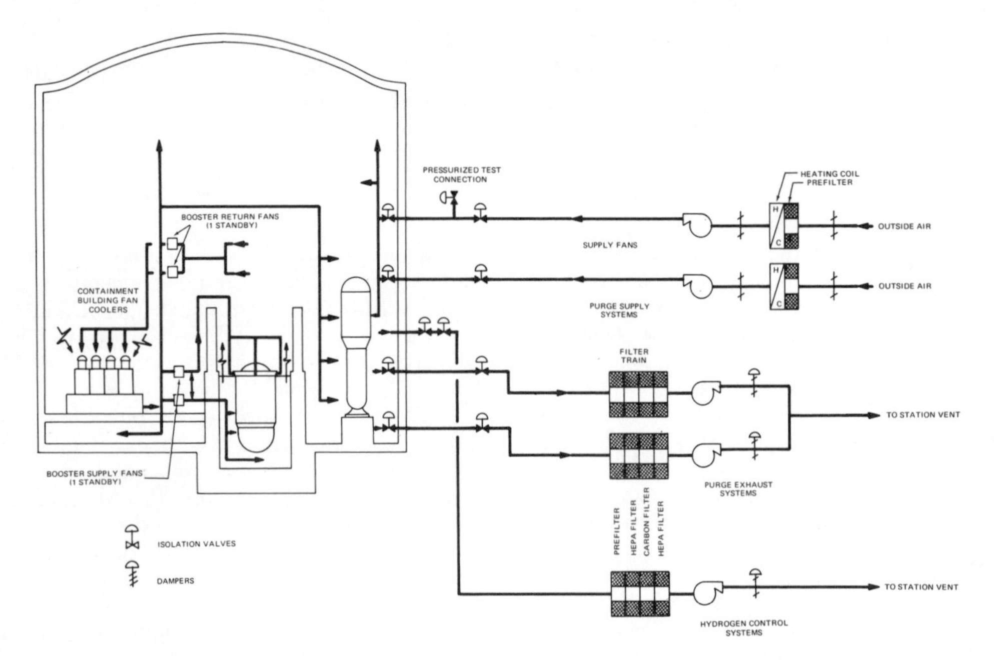

Courtesy, Burns and Roe, Inc.

Fig. 14-3. Air cooling, supply, and purge system for a PWR, showing special requirements for reactor pressure vessel and control rod drive-mechanism cooling.

of the Chemical and Volume Control System. In a BWR, this gas is bled off continuously through the main condenser air ejector; the processing system for the off-gas is shown in Fig. 12-4. When the primary system is breached and rapid depressurization occurs, the hydrogen evolves from the solution in the gaseous state. Radiation from the core, after shutdown and depressurization, continues to decompose water by radiolysis and evolves hydrogen. The concentration of hydrogen in air must be kept below 4.1 percent by volume, which is the lower explosive limit of hydrogen in air. The first step in hydrogen control is to prevent local concentrations. This is accomplished most simply by thorough circulation of air throughout the containment structure. Hydrogen concentration should be measured and continuously recorded. Suitable instruments are available commercially. The best sampling point for this is in the duct where a circulating fan head can be used to propel the samples.

When the hydrogen concentration exceeds 50 percent of the lower explosive limit, hydrogen should be removed from the atmosphere. Some facilities bleed containment air out through a filtering system, replacing the bleed air with filtered outside air. Internal pressure is, of course, kept below atmospheric during this procedure.

Two chemical methods are available for removing the hydrogen by recombination with oxygen to form water; in one method, the hydrogen is burned to water in a flame recombiner. This device has a gas pilot flame which is arranged to burn continuously. Monitoring the burning flame confirms that the recombiner is in operating condition. This device should also have two remotely initiated electric igniters each of which is powered from a separate bus.

The other recombiner is a catalytic type, using a bed of chloride-free alumina pellets impregnated with approximately ½ percent palladium or platinum. This recombiner will not operate with wet catalyst so it must be preceded by moisture removal equipment. Inlet gas temperature must be higher than saturation temperature to guarantee continuous proper operation. The recombination reaction of H_2 with O_2 is exothermic, requiring that the gas mixture be cooled after leaving the recombiner.

Initial design estimates can be based on a face velocity of 50 fpm, a bed depth of 12 inches of catalyst, and a pressure drop of ½ psig. When exact conditions are known—H_2 concentration, O_2 concentration, flow rate, pressure, etc.—exact data must be requested from the manufacturer.

Present trends indicate that the controlled bleed method of hydrogen control is more popular than recombining. There are two principal reasons:

a. Essentially, no additional equipment is needed for plant although some modifications of existing equipment may be required.

b. The sheer volume of air to be handled is such as to make either recombining technique cumbersome and uneconomical.

The flame recombiner is more popular than the catalytic unit although some late BWRs are planning the use of catalytic recombination.

14.5 Equipment

14.5.1 *Air Handlers*—Air handling units in containment are special units with additional requirements over and above air handling. The additional requirements are a decontaminable outer finish, provisions for pressure equalization, inside and out, and mechanical design to accept seismic shock. Fans and motors must be sized to pump air at atmospheric conditions and at accident conditions when the density has greatly increased. Cooling coils in the handlers must be split or duplicated so that in an emergency, sufficient coils can be supplied directly with nuclear service water. In Fig. 14-4, the pressure equalization dampers can be seen on the left side of the unit and in the two recesses above the fan suction opening.

The units must be sized for three operating conditions: the normal situation, the emergency condition, and the containment test condition. The physical size of the equipment is controlled by the "normal" situation because of the lower temperature difference between the air and the cooling water and the low air density at 120 degrees F and atmospheric pressure. The load is usually split among three or four large units because of the size required. When these units are checked for their total capacity under emergency conditions, it is found that they are oversized by a factor of 2 to 4, due mainly to very high temperature differences (inlet air can be as high as 280 degrees

Courtesy, American Air Filter Co.

Fig. 14-4. One air handling unit for Three Mile Island, Unit No. 1. The unit has separate coil banks for normal and emergency use. Pressure reliefs can be seen on the front and left sides.

F) and high air density. At the start of emergency operation, the coils act as condensers because containment is at 100 percent relative humidity; coil operation—heat transfer and drainage—must be investigated and accommodated at these conditions. It is not uncommon to run the fans half-speed, under emergency conditions, to limit the horsepower requirements. Coils may be split simply because of the high temperature differences available. When a coil is split, the section in emergency use must be suitable for use with the raw nuclear service water. If this requires a change in materials of construction, the system must be examined for the possibility of galvanic corrosion. It may be cheaper and easier to install an extra, separate coil for nuclear service water than to properly protect the system from galvanic corrosion. The most economic solution can only be determined by an examination of the individual system. With the exception mentioned above, normal HVAC materials of construction may be used. Coil construction must be brazed with a braze metal having an adequate melting point margin (not less than 100 degrees F) above the highest temperature anticipated during emergency operation.

Fans should be directly mounted on motor shafts to avoid the maintenance problems inherent in a belt-driven unit. Vaneaxial fans are preferred to centrifugal fans. Fans and motors must be designed to operate for a minimum of one year without maintenance. Design temperatures should be about 325 degrees F to permit extended operation during the cooldown period after an incident. Horsepower should be based on unit operation during containment testing as this represents the heaviest air-density handled. Lubrication must be suitable for 10^6R so as to accept normal operating dosage as well as emergency dosage. Lubrication must be changed every year to conserve radiation resistance for emergency duty.

Filters are supplied separately, from the air handlers.

14.5.2 *Isolation Valves*—All ductwork that penetrates the containment envelope must have two isolation valves in tandem, one outside containment and one inside containment. The two valves and the duct connecting them form part of containment and must conform to the same codes and design criteria as the containment system. Valves are, therefore, Class 2, of Section III of the ASME Code. Valve sizes vary with plant capacity but may be as large as 42 or 48 inches, and because of their size are butterfly pattern with a soft elastomer seat. Valve bodies and discs are of cast steel or other ductile material. Cast iron is not used because it cannot accept the mechanical shock associated with Seismic Class I design.

Disc shafts may be of stainless steel or a type of hardenable tool steel. Shaft material must be coordinated with the bearings. Bearings may be either anti-friction or bushing type. Lubricant may be graphite, molybdenum disulfide or a radiation resistant grease. When specifying the lubricant, an estimate must be made of the radiation dosage to be withstood by the lubricant; if desired, the dosage may be based on the assumption that the lubricant is flushed out and replaced during refueling. The dosage estimate must include a reasonable amount to be received after an accident.

Seats should be of ethylene-propylene rubber or neoprene for maximum radiation resistance. Valve operators must fail closed; spring loaded cylinders using 100 psig compressed air are the most common. It should be noted that these valve operators usually determine the instrument air pressure needed in the facility. The operator must be equipped with some type of shock absorber so that even though the valve closes quickly, it does not bottom on the soft seat.

Valve ambient conditions should be checked to determine if a nil ductility transition temperature need be specified for the valve outside containment.

14.5.3 *Particle Filters*—Two particle filters in tandem are always supplied. The first, or prefilter, is provided to remove all the common larger dusts in order to reduce the load on the second, relatively expensive, high-efficiency filter. The high-efficiency filter is commonly known as the HEPA filter.

Prefilter. The prefilter is sometimes also known as the roughing filter. It is a fiber glass filter unit and should be Underwriters Laboratory, Class I and have a 90-percent minimum discoloration rating in accordance with the NBS dust spot procedure on atmospheric dust.

HEPA Filter. The HEPA filter is sometimes known as an absolute filter; it has an efficiency of 99.95 percent in removal of particles down to 0.3 microns. A cell is designed for a clean pressure drop of 1 inch H_2O and run to a dirty pressure drop of 2 inches H_2O. The filter is made in individual cells 24 × 24 inches × 11½ inches deep. The filter media conforms to Specification MIL-F-51079. The media is an outgrowth of work done at Edgewood Arsenal during World War II. The media is cut to length and folded into pleats with a separator between each pleat. The entire assembly is then inserted into a metal frame and the two faces covered with metal cloth. The entire assembly is described in Specification MIL-F-51068. The unit must also comply with Underwriters Laboratory UL586 and be UL Class I. The edges must be sealed with adhesive to prevent by-passing within the cell. Normal separators in a commercial or bacteriological unit are aluminum. In a nuclear application when, as in a PWR, the possibility exists of a chemical reaction between aluminum and caustic from containment sprays releasing hydrogen, waterproof asbestos separators should be substituted for aluminum. The frame should be welded steel, 14 or 16 gauge, with the gasket face ground flat after welding. Then the frame should be galvanized or cadmium plated. The gasket should be neoprene sponge-rubber, cemented to the cell; gasket corners should be keyed to prevent air leakage paths. Figures 14-5 and 14-6 show a HEPA cell being installed in a filter bank frame. Cross bracing, as shown in Fig. 14-6, should be specified for strength against racking.

Filter Frame. The filter frame must be a rugged, welded Seismic Class I structure. Gasket surfaces should be ground flat and smooth after welding. Figure 14-7 shows a filter frame with the additional bracing of a structural steel channel border around the entire assembly. The assembly must be rigid enough to prohibit deflection under gasket loads and

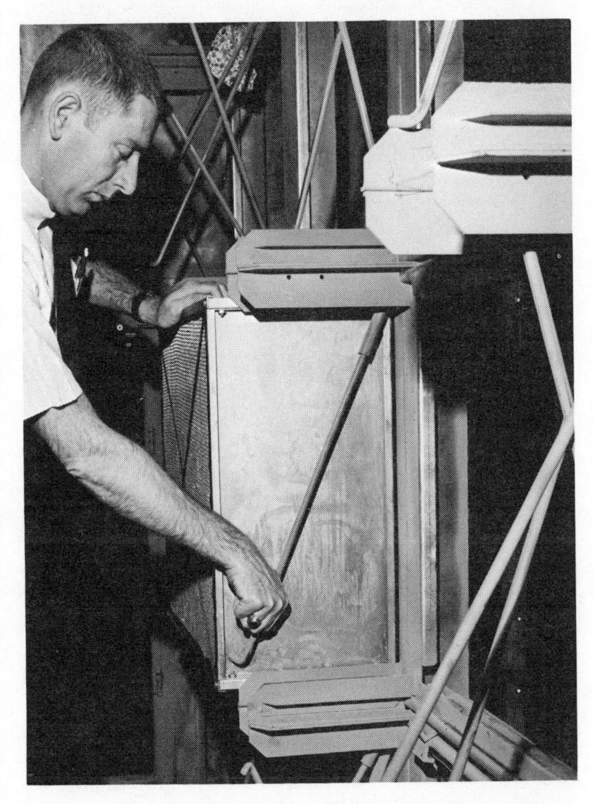

Courtesy, American Air Filter Co.

Fig. 14-5. HEPA cell in a frame system. Installation of the cell takes about 20 seconds.

strong enough to withstand the pressure differential loads as the filters accumulate dirt. Design pressure differentials are of the order of 2 to 4-inch water gauge. Means must be provided to remove a dirty filter and install a new one in a minimum of time. The crossed rods taped together in Fig. 14-7, and the cam lock in Fig. 14-6 seal the cell into place quickly and efficiently.

14.5.4 *Vapor Filter—Charcoal Cell.* The vapor filter is provided to remove halogens, principally Iodine-131 (I_2^{131}), from the air after an incident. The filter media is an activated charcoal. Vapor re-

Courtesy, American Air Filter Co.

Fig. 14-6. HEPA cell installation clamp (arrow), partially closed. Weight of cell is carried by the support brackets.

Courtesy, American Air Filter Co.

Fig. 14-7. HEPA cell bank mounting frame. Taped cross bars are operating handles of quick acting cell clamps. Note the welded structural channel reinforcement around the entire frame.

moval is accomplished by adsorption of gas molecules into the pores of the charcoal. Theoretically, the units can be regenerated by the application of heat after they have become saturated. With the present-day requirements for "zero release," research is investigating the adsorption properties of activated carbon for zenon and krypton, at the cryogenic temperatures of liquid nitrogen. Activated charcoal, at the present time, is made from coconut shells because it produces a hard, structurally strong charcoal. Manufacturers in the United States are trying to develop other sources in the event their supply of coconut shells is cut off. No industry standards exist, but the charcoal is generally provided in a cell as two 2-inch-thick flat beds of charcoal in an assembly about 26 inches long, 24 inches wide, and 6 inches high. One end of the assembly has a "facing" extending out past the dimensions given above for a gasket surface. Gasketing material is the same as for HEPA filters. This can be seen in Fig. 14-8 which also shows the cell mounted in a horizontal position. Each bed is made with solid sides and ends and perforated screen, top and bottom. In the unit pictured in Figs. 14-8 and 14-9, air flow is into the center of the cell and out through the top and bottom of the cell. Construction is stainless steel; usually 14-gauge side plates and 12-gauge face plates. A single cell weighs about 80 pounds, so the carrying handles, seen in Fig. 14-8, are furnished.

Charcoal is not selective in its adsorption. As a result, any stray vapors passing through will be adsorbed and occupy adsorption sites; this is particularly true of solvent cleaning vapors left in containment by cleaning operations. Therefore, the structure must be thoroughly purged of all vapors before installation of the charcoal cells in order to preserve adsorption capacity.

Courtesy, American Air Filter Co.

Fig. 14-8. Activated charcoal cell installed in frame, ready for clamping. Threaded studs are used for fastening. Narrow rectangular opening in cell is the air inlet.

Courtesy, American Air Filter Co.

Fig. 14-9. Rear view of the cell shown in Fig. 14-8. The blanks above and below the cell are used to isolate the cell during an individual cell test.

There is no device that can measure vapors adsorbed continuously and read out the capacity remaining in the charcoal bed. However, capacity remaining can be determined from charcoal samples. In specifying the cells, charcoal should be identified by individual batch and only one batch permitted per cell. The adsorption capability of each batch should be determined by a radio-iodide or methyl-iodide test. Sample cannisters should be prepared and installed where they will see the same tramp gas as the charcoal cells. Usually, these cannisters are 2 inches in diameter and 2 inches deep. Then at intervals the cannisters are removed, the adsorption capacity determined, and compared with the capacity of the original material. The cannisters may be installed in the filter bank or in a valved by-pass that permits removal of the cannister at any time. Six samples of each batch are recommended, although four are sometimes used. The procurement specification should require both the cannisters and complete mounting to be supplied by the cell manufacturer.

Frame. The frame is all welded stainless steel as pictured in Fig. 14-10 which shows the rear side of the filter frame. The stainless steel vertical separators between columns of cells are provided to limit the spread of fire if the charcoal is accidentally ig-

Courtesy, American Air Filter Co.

Fig. 14-10. Rear view of an assembled frame for a bank of charcoal filter cells. The vertical separators are used to limit spread of fire. The large opening between the two frame assemblies will contain a charcoal bed by-pass damper.

nited. Gasketing surfaces must be smooth and flat, similar to the HEPA filter frame. In Fig. 14-8, the gasket surfaces have been reinforced by bending the metal over in a 90-degree angle. Unlike the HEPA filters, a charcoal cell has a life of several years; hence, the requirement for rapid replacement is not as great and they are fastened in place with screwed fasteners, as shown in Fig. 14-8, rather than with quick-opening fasteners.

14.5.5 *Duct Work*—Distribution systems are designed and arranged by conventional methods and calculations must be made for all operating conditions. Thus, inside containment, flow calculations must be performed at normal conditions and at emergency conditions.

Duct fabrication is different than conventional in certain parts of the facility. The ductwork section between the isolation valves, which forms part of containment, must be designed to the same temperature and pressure as the containment structure for both internal and external pressure. It must be Seismic Class I and conform to the rules for Class 2 piping, in Section III of the ASME Code. The external pressure will be developed in operation if an accident should occur and the isolation system actuates as a result of the accident. During facility testing the external pressure will be imposed on the ductwork. Note that because of the requirement to conform to Class 2 of Section III, the piping material used for this ductwork must be acceptable to the code and appear in code material lists. The outside of this ductwork and all other ductwork in radioactive or potentially radioactive areas should be covered with a polyvinyl or epoxy protective coating system to facilitate future washdown and decontamination.

Other portions of ductwork inside containment, required for use after the design basis accident, must accept the accident pressure transient and be designed for Seismic Class I. The best procedure to be followed in designing this ductwork is to arrange and size the ductwork, next, to do the preliminary seismic design of both duct and supporting systems. With this completed, check the ability of the ductwork to withstand the external pressure transient. Adjust the ductwork, as necessary, to accept the pressure transient by either equalizing pressure with gravity-type dampers or increasing the duct wall thickness. Each alternate should be investigated and priced; the lower priced alternate then would be selected. The increased load on the supporting system, after ductwork adjustment, should be included in the economic comparison so that a true picture results.

Ductwork not required for containment may be fabricated of sheet metal, in conventional fashion, with lock seams. This is a seeming contradiction to previous statements that prohibit crevices because they collect radioactive particles. However, consider the situation. All potentially radioactive areas are finished with a protective coating (epoxy or vinyl) to facilitate future decontamination; these surfaces do not generate dust. The rigid cleanliness stipulations require these plants to be spotlessly clean and dust-free before startup. Air intake systems have filters to remove atmospheric dusts. Finally, only dust particles that are exposed to neutron irradiation can become radioactive; the only location in the plant where this irradiation can take place is in the open volume between the reactor pressure vessel and its surrounding concrete missile shield. Thus, the possibility of a buildup of active dust in a duct during normal operation is so remote as to be regarded as improbable.

For further safety, if the designer wishes, the recirculation system in the reactor pressure vessel area of Fig. 14-3 may be a light gauge, seamless pipe.

When selecting ducting material it must be remembered that aluminum cannot be used in buildings with alkaline sprays for H_2 control.

CHAPTER 15

Instrumentation

15.1 General

All parameters in a nuclear power station are monitored by four different control systems—the reactor protection system, the reactor control system, the process control system, and the radiation monitoring system. In the present state of development all control functions are implemented by high quality solid state, three-mode industrial analog controllers. Modern digital computers are used for data logging, supervision of planned operations and alarms. Present-day regulations require redundancy of equipment as protection against a single failure. This would be interpreted as redundancy of the complete computer system. It is thus simpler and cheaper to divorce the computer system from control functions than to duplicate computers and interconnect them. It must, however, be noted that the Canadian Pressurized Heavy Water Reactor has duplicate computers and uses them for control.

A typical control channel for any variable is composed of a primary detector which develops a signal, an indicating or recording device, a controller, and a bistable for each alarm or limit switch to be activated by the channel. The controller, after comparing the parameter signal with the set point, generates a signal for the final device that is proportional to the error between the parameter and the set point. The final device is the actual mechanism that performs the required control action. In a nuclear plant this device may be a valve, a control rod, a variable-speed drive, a metering pump, or a compressor cylinder unloader.

15.2 Reactor Protection System

The reactor protection system is a "go no-go" system that monitors all the parameters important to safe operation of the reactor. When an unsafe condition is detected the reactor protection system shuts the reactor down. Since a shutdown caused by a spurious reactor trip requires several hours to restart the plant, it is necessary that the protection system not only react to all unsafe conditions but also avoid unnecessary shutdowns that may be caused by component deterioration or failure. The protection system is, therefore, designed with a logic matrix so that it will not react to a single unsafe measurement of a parameter but requires two agreeing measurements to trip the reactor. The various combinations used by each reactor vendor are described in Chapters 2, 3, and 4. The multiple channels operate so that failure of a single component or channel is interpreted by the system as a single unsafe reading of the parameter. The reactor continues operation but now an actual single unsafe reading of the same parameter will shut down the reactor because it is interpreted as the second unsafe reading. The circuits are designed so that they are de-energized in the tripped condition and are, therefore, inherently fail safe.

The equipment is designed to include alarms to signal equipment failure as well as unsafe parameters. The channels are designed so that each component, as well as the complete channel, may be tested at any time. Channels are designed so that they are completely independent except possibly, for the sensor. As an illustration, a single differential pressure producer is used to measure primary flow but three transmitters are used to generate three parallel signals for three parallel flow channels. Contrasted with that are three ion chambers making independent measurements of the neutron flux field to which they are being exposed; each of these channels is completely independent and does not even share the sensor.

The parameters monitored by the protection system are actually chosen by each of the reactor vendors for their own design. However, the following list is typical of the variables chosen:

a. Neutron flux
b. Power level
c. Rate of power increase (Period)
d. Coolant exit temperature
e. Coolant temperature rise
f. Coolant pressure
g. Low water level reactor vessel (BWR)
h. Low water level steam generator (PWR)
i. Manual trip
j. Loss of coolant pump power
k. Isolation of reactor from turbine
l. High radiation level
m. Low coolant flow rate.

Neutron flux is the measurement of the neutrons bombarding the detector per square centimeter per second, and is of the order of 1 to 2×10^{13}. The flux level is a measure of the total neutron population in the reactor and varies according to an exponential function. Since any control system is a fixed design with fixed limitations, some protection systems incorporate the rate-of-power-increase trip which insures that the reactor never gets beyond the ability of the control system to control.

Reactor power level is measured by the neutron instrumentation; the power level channels are initially calibrated using a power level calculated from mass flow-rate and enthalpy rise across the reactor. The reactor protection neutron measuring channels and reactor control neutron measurement channels sometimes share the same detectors but separate immediately into different channels.

Process detectors such as those for flow, temperature, and pressure are high quality industrial devices with time constants rapid enough to meet the requirements of the protection system.

In the early 1960s it was common practice to measure containment ambient pressure and include high containment pressure as another trip parameter in the protection system. High containment pressure results from breach of the primary system. As coolant passes from the primary system into containment, containment ambient pressure rises. This was originally a "back-up" concept for system flow and pressure in the era of vacuum tube instrumentation. With the introduction of solid state control devices and their reliability, the requirement for the inclusion of high containment pressure as a reactor trip has become an "engineering judgement" item. The author prefers to include it for the following reason:

The basic measurement channels are required to furnish information to the operator so he may start his containment cooling systems. Automatic monitoring is needed so that if one of the containment cooling systems fails in operation or fails to start, the second system will start. Actually then, in order to add ambient pressure to the reactor protection system, all that is really necessary are the bistables on each measuring channel, the additional positions in the logic matrix and the installation wiring. The differential cost for this is very small compared to the total overall cost of the protection system.

15.3 Reactor Control System

This system is responsible for maintaining the reactor at a predetermined stable power level. It responds to neutron variation to keep the reactor at the selected power level or to an outside signal (the load dispatcher) to change the power level. In all cases the system responds by varying the reactivity in the core. The system is permitted, by design, to vary the power level according to a predesigned program from 100 percent down to 15 or 20 percent. Beyond this, an alarm is sounded which then requires a manual readjustment of some of the control rods.

The system must be able to detect and control power levels ranging from 10^{-9} to 10^{+2} percent of full power. There is no single measuring channel that can be used over the whole range; the final display component, with appropriate range switches may be used for two adjacent channels. Note that the range of power measurements covers 11 decades. It is most common to split the complete range of measurement into three separate channels with at least a one decade overlap between adjacent ranges. The start-up range would go from 10^{-9} to 10^{-4} percent; the intermediate range from 10^{-5} to 10^{+1} percent and the power range from 10^{0} to 2×10^{2} percent; thus an acceptable reading is reached on the second range before leaving the first range.

As is well known in instrumentation, a signal range is chosen so that a "zero" parameter value gives a definite, measurable, absolute signal value. As an example, for a signal transmission range of 10 to 50 milliamps, the absolute signal value of 10 milliamps corresponds to "zero" parameter value. The same principle is applied to neutron measurement. It is extremely important to know when the neutron population starts to increase so that it may be kept under control at all times. If the instrumentation system had to wait until the core fission neutrons appeared before anything could be measured it is possible that the chain reaction could multiply too rapidly to be controlled. This is especially true the first time the reactor goes critical if the instruments are not properly adjusted. To avoid this possibility, neutron sources with known values are installed in the reactor. This gives a predictable reading on the lowest range of nuclear instrumentation before the first fuel element is installed and while the reactor is shut down.

Redundant neutron sources are always used. A vendor may provide primary and secondary sources; for initial reactor core loading redundant sources are provided to guarantee a signal in the neutron measuring systems. The neutron sources are combinations of antimony-124 and beryllium or polonium-210 and beryllium. The polonium-beryllium source is preferred for the primary source because it has a half-life of 138 days compared to a half-life of 60 days for the antimony-beryllium source. Simply stated, a polonium-beryllium source of the same initial strength as an antimony-beryllium source gives the user twice as much time to load the reactor as the antimony-beryllium source. The initial primary neutron sources are brought to the reactor "live" after being irradiated in another reactor. "Dead" sources are also installed in the reactor and are irradiated by it. After the first year of operation all installed neutron sources contribute to the "shutdown" neutron population of the reactor.

Neutrons are detected by the ionization of gas in a chamber. Detailed descriptions of neutron detection chambers are beyond the scope of this book; suffice it to say that the neutrons generate ions which, aided by an impressed voltage differential across the chamber, are collected and cause a measurable current, of the order of 10^{-6} to 10^{-3} amps, to flow. The value of the current is directly proportional to the neutron population in the reactor. The neutron population, in turn, is directly proportional to the power level. The current value is, therefore, directly proportional to the reactor power level. In the lower end of the range, from 10^{-9} to 10^{-4} percent, the neutron population is so small that the current is measured in pulses per second and the display is in counts per second. In the two upper ranges, the population is large enough to generate a continuous current and the readout is either in megawatts or percent of full power. The current is amplified to an acceptable signal range and used as the input to a controller or to the protection system.

Unlike the protection system, channels for power control are not redundant simply because they have no safety function.

The parameter being controlled is called "reactivity." In an equilibrium situation—stable power level —when the reactivity is 1, as many neutrons are produced per second as are being "destroyed" by absorption or leakage. When the power level is to be raised, the control system adds a small amount of reactivity—less than 1 percent. In a PWR, this is done by withdrawing the "control group" of control rods slightly, from the core. In a BWR, this is done by increasing the recirculation flow rate. These changes are done at very precise speeds so that at no time can the rate of neutron population increase exceed the control capability of the control system. The neutron population increases until a new stabilization level is reached and the reactor continues at the new level. Control margins are of the order of 0.1 to 0.2 percent of reactivity while shutdown margins are of the order of 3 to 4 percent. In all of the industrial American reactors the devices used for shutdown are fixed geometric shapes, because it is easier to manipulate these than solutions.

Control rod drive devices are proprietary mechanisms and may be electric, hydraulic, pneumatic, mechanical, or any convenient combination of them all. They are available in designs responding to impulses to move a small fixed increment, or as devices that move in true modulating fashion. All drive mechanisms have withdrawal speed limitations designed into them so they may never exceed the speed limitation requirement previously described.

15.4 Process Control System

The process control system accomplishes the measurement, indication or recording, and control of all parameters in the plant, except nuclear and electrical. It includes such characteristics as temperature, pressure, flow, level, pH, conductivity, relative humidity, weight, etc.—the characteristics normal to any power station. Instruments are the same type as in fossil fuel plants with a few modifications. Recognizing that the previously described nuclear instrumentation must be electronic, process systems use electronic instrument channels with an electric or pneumatic final control device.

It is almost standard practice today to furnish Teflon gasketing in instruments; this is not permitted in a nuclear plant. Gasketing is specified as asbestos filled spiral metallic or asbestos. Electronic parts of transmitters are separated from the rest of the instrument as much as possible and placed in a shielded area; in some instances where the separation cannot be made, special shielding may be provided for the electronic assembly. Instruments must always be designed to accept a total radiation dosage of 10^7R. Instrument reaction time must always be compared with system reaction time and the two correlated. Of particular importance in this respect are the temperature detectors used to measure coolant temperatures at the inlet and outlet of the reactor. Reaction time of these detectors must be of the order of hundredths-of-a-second so that the full control channel has time to react.

The process systems in a nuclear power station are the primary system, residual heat removal system, purification system, spent fuel cooling system, radioactive waste system, and the boron management system. All of these require measurements of flow, temperature, and pressure. Primary cooling, purification, and spent fuel cooling also require conductivity measurement.

Temperature is measured with a resistance thermometer or a thermocouple. A resistance thermometer is preferred because with a limited range and expanded scale, its accuracy and repeatability can be enhanced. Filled system thermometers are used sparingly and carefully because most systems are filled with organic compounds that deteriorate in radiation fields.

All of the better known instrument manufacturers are now familiar with nuclear plant process instrumentation and can supply the proper components.

15.5 Radiation Monitoring System

The radiation monitoring system is provided to detect and measure radiation outside the reactor. Since neutrons are present only inside the reactor, and alpha and beta radiation is very simply shielded, the monitors measure gamma rays. The system serves the following purposes:

a. It measures and records radiation levels in all spaces of the power station where radiation may be anticipated.

b. It measures and records radiation levels all around the site boundary and such other locations as the jurisdictional authorities may direct.

c. It measures and records the radiation levels of any effluent stream leaving the power station.

d. It detects a failed fuel element.

The radiation monitors for Items "a" and "b" are installed in the buildings where radiation is anticipated, and outside, on the site boundary. The monitors along the boundary always record their readings to provide a legal record. Monitors in the buildings may record or indicate at the designer's option. When the designer elects to indicate, it is common practice to include some trend recorders so that the operator may record an individual monitor if he wishes. All the monitors are mounted in a single panel in the plant control room and have individual alarms with safety actions.

An alarming monitor in the reactor containment building will scram the reactor and seal the building. An alarming monitor at the site boundary or in an auxiliary building will seal the facility to cut-off flow of effluents. All alarms are audible and, in addition, have a visual display that identifies the individual channel alarming. These monitors are usualy ion chambers that produce an electric current that is amplified and used.

Liquid and gas effluents are monitored to prevent the discharge of waste effluents in excess of design objectives. The exhaust stack has a gas sample continuously monitored and if the radiation exceeds allowable limits, the stack inlet is closed off.

The air itself is not radioactive, but some minute airborne particles are. Therefore, the stack gas is passed through a filter and the filter scanned by a Geiger counter. Liquid waste is handled in the same way, but the detector is the most convenient type for the particular application. In both cases the system is arranged so the sample is extracted and counted before the effluent reaches the last shut-off valve.

Failed fuel detection is handled differently. Failed fuel always gives off I_2^{131}, which emits gamma rays at a specific energy level. Reactor coolant (water) cooling a normal reactor core will not have any traces of I_2^{131}. The procedure used is to by-pass a continuously flowing stream in the inlet side of the purification system to a shielded demineralizer which absorbs all the impurities in the coolant. The demineralizer resin bed is monitored by a radiation detector which is electrically biased so as to be sensitive to I_2^{131} gamma radiation energy levels. In the presence of I_2^{131} radiation, the monitor alarms. Although it does not shut down the reactor, it may close off the purification system, however, so as to confine the I_2^{131} to the reactor coolant loop until the extent of the I_2^{131} is determined.

CHAPTER 16

Quality Assurance

16.1 General

In the first half of the 1960s it became evident that there was much to be desired between the plant that was designed and the plant that was constructed. Traditional construction methods and practices did not produce a facility capable of operating for a year in a radioactive environment unobserved and uninspected. As a result, individual engineering organizations started upgrading their specifications in the areas of quality control and documentary proof of compliance to specification. In January, 1968, the ASME issued Appendix IX to Section III, defining nondestructive testing methods to be used for Code examination and the qualifications of the personnel executing and interpreting the examinations. The Code very clearly assigns responsibility for quality control of a vessel to the fabricator. No other official document discussed quality until the AEC issued the Quality Assurance Criteria in 10CFR50 in June, 1970, covering the Quality Assurance Program required for a nuclear power facility. Appendix B of 10CFR50, containing the program requirements, is reproduced in this book as Appendix II, and it also requires a formal quality assurance plan. In July, 1971, the new edition of Section III included requirements for a quality assurance program approximately equal to the requirements of 10CFR50, issued a year earlier.

At this point let us define what is meant by the term "quality assurance." Quality assurance is all of the planned and systemic actions necessary to assure that a nuclear power station has been designed and built in accordance with the commitments made in the Safety Analysis Report. The most important point to note in this definition is the phrase "planned and systemic actions" because it requires a complete written plan covering the entire project from inception of design to commercial operation.

All responsibility for proper design and construction of a nuclear power facility is vested in the Owner (the license applicant) of the facility. The Owner must, therefore, assure himself that the desired quality level, as described in the Preliminary Safety Analysis Report (PSAR), has been achieved and that all records and documents required to prove conformance are complete and filed so as to be readily accessible. The execution of these responsibilities has evolved into what is known as the three level quality assurance program shown in Fig. 16-1. Level 1, the lowest level, is called quality control and is executed by the organization performing the work. Typical actions of this type are an engineer checking a calculation, a designer checking a drawing, and an inspector checking the finish on a gasket surface or the execution and interpretation of a nondestructive examination by qualified personnel.

Level 2 is called quality assurance surveillance and is a regular periodic review of all the actions and records of the Level 1 quality control groups. Typically, a surveillance operation will review a file for organization, completeness, accuracy, and ease of

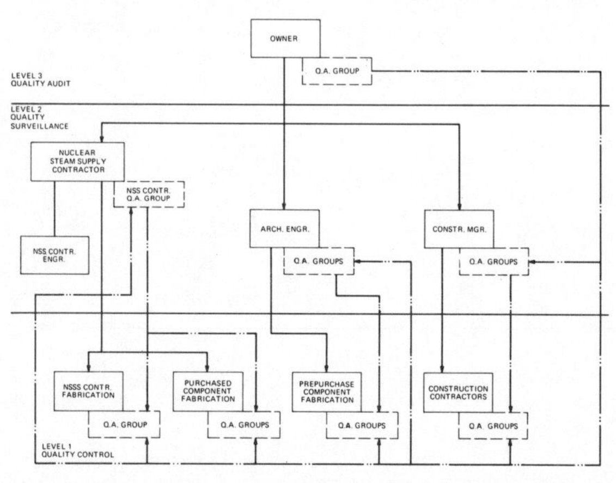

Fig. 16-1. A typical quality assurance organization for an entire project in which the owner has bought a nuclear steam supply and retained an architect engineer and a construction manager.

retrieval. It may then pass on to a random check of any quality control operation to see if the surveillance check agrees with the original control decision as recorded in the files. In the case of field operations at a construction site, surveillance also includes observation of construction and first-level quality control operations to verify conformance to approved procedures. This is usually performed by the three prime contractors—the nuclear steam supply system vendor, architect engineer, and construction manager —in their own areas of responsibility.

Level 3 is the quality assurance audit and is a regular periodic review of all organizations performing quality assurance, design, or construction operations. Its purpose is to assure that all organizations are performing properly, that all commitments of the PSAR are met, and that all documentary proof of conformance is complete and readily retrievable on request. This is performed by the Owner.

Figure 16-1 is a typical project organization in which the facility owner has three prime contractors, one for the nuclear steam supply system (NSSS), one for the overall plant design, and the third for construction management. Each of these three contractors, in turn, is responsible for the fabrication or construction of different parts of the facility. Hence, in the ideal situation each of the organizations in Levels 1 and 2 is checked by a different organization. When this is not the case and the same corporate organization works vertically in Levels 1 and 2, then the Level 2 quality assurance surveillance group must be organizationally independent of the Level 1 operation, such as operating in a separate division, and able to require corrective action from Level 1.

The Atomic Energy Commission always addresses itself to the facility Owner thus making the Owner legally responsible for compliance with all licensing regulations. Hence, the Owner should, whenever possible, assume direct responsibility for execution of all third level auditing as the means by which it will assure itself that all requirements have been met. The architect engineer assumes surveillance responsibilities for all prepurchased components, and monitors shop operations of manufacturers. The construction manager awards construction contracts and assumes surveillance responsibilities for field work. Whenever the construction manager chooses to do force account work then surveillance responsibility for the force account work passes to the engineer. The manufacturing and constructing organizations actually performing the work are responsible for the day-to-day quality control work of approving dimensions, welds, concrete pours, etc.

The NSSS contractor operates in both Levels 1 and 2 when he manufactures his components; in that instance, Level 2 must be organizationally separate from Level 1, as previously pointed out.

Quality assurance, as required by AEC regulations, must be applied to all structures, systems, and components that may affect the health and safety of the public. The design and construction activities subject to the program are: designing, purchasing, fabrication, handling, shipping, storing, cleaning, erecting, installing, inspecting, and testing. Each of these activities must be controlled by a written procedure that defines how the operation will be performed and the documents that will be developed during the operation. Thus, a procedure may require that Engineer A perform a calculation and sign each page; that Engineer B review and check the calculation, signing each page as Checker; and that Discipline Lead Engineer C review the calculation, again signing each sheet as approved. The procedure would further define the format of the calculations, the duties of the Checker, and of the Lead Engineer.

In a similar manner, all other activities are described in precise auditable procedures. In a typical quality assurance audit, the auditor would start with the calculation and check to see that it has been properly performed, checked, and approved as defined in the procedure. He might next see if the results have been properly carried over into a procurement document or the appropriate next design step.

In any specific organization the decision must be made whether or not the Quality Assurance group will assess the quality of the design work or limit the audit to the manner in which the engineering process is executed and the records kept. Quality Assurance usually limits itself to the manner of execution and records produced, while engineering review is performed by personnel skilled in the particular discipline. Thus, the quality assurance audit checks that the engineering design review has actually been performed and that proper records exist to prove it.

This method of operation is followed for all paperwork—design, procurement, changes, deviations, etc.

16.2 Quality Assurance Categories

In the matter of hardware, however, the situation is not as simple. It is certainly possible to build a plant with the highest quality in every item, but is it necessary? Should the same care and precautions be used in manufacturing and installing a service air receiver as is used for a reactor vessel? The answer, of course, is no. There must be a suitable set of quality category definitions that protect the public, and at the same time, permit economic construction. An idealized set of definitions that could be used is the following:

Category 1. Category 1 shall apply to all those systems, components, and structures which prevent or mitigate the consequences of postulated accidents that would cause undue risk to the health and safety of the public. It shall also include those items or components whose failure could cause or increase the severity of a loss of coolant accident or result in a release of excessive amounts of radioactivity. It shall also apply to those components and systems vital to the safe shutdown and isolation of the reactor. All

Category 1 items shall be subject to all requirements of Appendix B, "Quality Assurance Criteria for Nuclear Power Plants" of 10CFR50.

Category 2. Category 2 applies to those systems, components, and structures important to reactor operation but not essential to safe shutdown nor to isolation of the reactor, and whose failure could not result in release of excessive amounts of radioactivity.

Category 3. Category 3 applies to all other systems, components, or structures containing or controlling radioactivity that do not fall either into Category 1 or 2.

Category 4. Category 4 applies to the remainder of the plant and is, essentially, upgraded industrial equipment.

Categories 2 and 3 would require various portions of Appendix B (Appendix II in this book) as defined during project design.

These definitions however are either unworkable or impractical. As an illustration, a diesel generator used for emergency power would fall into Category 1. Imagine the cost of performing a full-dress quality assurance program on one sub-assembly of the diesel engine even if it could be done. Hence, some modifications are in order to recognize what is possible, and also the presence of redundant systems and components.

Category 1 can be modified so that it includes all systems, structures, and components whose failure could cause the design basis accident. It should also include the containment system and all non-redundant portions of engineered safety features systems or systems vital to the safe shutdown of the reactor. This definition now recognizes that cooling or containment of the core and its fission products, is a primary requirement. It also recognizes that while safety systems are of major importance, some lessening of quality can be permitted if the function or component has redundance.

The next, logical, lower quality category should include all systems, structures, and components whose failure could cause an excessive release of radioactivity, and all remaining portions of engineered safety feature systems. This will be Category 2. The "excessive release of radioactivity" can be defined in terms of the allowable isotopic releases to the environment, the allowable radiation dose rates at the site boundary as defined in 10CFR20 combined with Appendix I of 10CFR50, and excessive exposure of plant personnel. One may say that any component failure releasing more than Appendix I design objectives constitutes an "excessive release." With this type of definition, a health physicist, knowing the component size and contents, can determine the quality category of a component and document the selection.

The second portion of Category 2 is controversial. The author prefers the definition given, but many users claim that there is room for a cost reduction and a slightly lower quality category, Category 3, if one considers the accessibility of the parts of the safety features systems. Thus, the portions of redundant engineered safety features inside containment where they cannot be regularly inspected during operation should be a higher quality category, Category 2, than the parts outside containment which can be inspected during operation. The accessible parts of the system can be Category 3.

Category 3 should also include radioactive systems whose failure could not result in excessive releases of radioactivity. The remainder of the systems not yet classified should be Category 4.

The author prefers the four categories that have just been defined because they are logical and can be implemented. Some have used a temperature-pressure relationship to determine quality categories. For instance, high temperature and pressure is Category 2, and low temperature and pressure is Category 3. Reference points can be 150 psig and 220 degrees F. The author feels that temperature and pressure are not logical parameters for selecting quality categories when the lowest quality category in the plant must be that used in the modern, fossil fuel, steam power plant. Hence, the second set of quality definitions is based on the combined concepts of containment of radioactivity, public safety, redundancy, and accessibility during operation.

To recapitulate, a workable set of quality category definitions with a reasonable cost that can be implemented is the following:

Category 1. All systems, structures, and components whose failure can cause the design basis accident and the containment system shall be Quality Category 1. All non-redundant portions of engineered safety systems and systems vital to the safe shutdown of the reactor shall also be Category 1.

Category 2. All systems, structures, and components whose failure could cause an excessive release of radioactivity and all remaining portions of engineered safety features systems shall be Category 2.

Category 3. All radioactive systems or components whose failure cannot result in an excessive release of radioactivity shall be Category 3.

Category 4. All systems, components, or structures not classified as Categories 1, 2, or 3 shall be Category 4.

16.3 Requirements of 10CFR50

The quality assurance criteria of 10CFR50 and its Appendix B (Appendix II in this book) were made a legal requirement in June, 1970. Since the AEC always addresses itself to the applicant, the facility Owner is responsible for execution of an adequate quality assurance program in the project. The requirements of Appendix B are as follows:

I. ORGANIZATION

An organization shall be set up to plan, monitor, and enforce the quality assurance program. This organization must be separate from the design, procurement, and construction organizations in the project and report directly to a corporate level with sufficient authority to direct correction of any nonconforming condition or construction. This organization must be described in writing or on charts. Descriptions must be given of the relationships among all the participating organizations as shown in Fig. 16-1. Additionally each organization must describe its own structure; a hypothetical organization for an Architect Engineer is shown in Fig. 16-2. In the illustration, engineering and procurement for the project are under the Project Manager in the Operations Division. Note that in Fig. 16-2, if the design review function is performed by the Engineering Division, then the reviewing organization is independent of the group that did the engineering. Regardless of whether the reviewing group is a Chief Engineer's staff or a Quality Assurance Manager's staff, it is actually a quality assurance audit that is performed.

II. QUALITY ASSURANCE PROGRAM

This criterion requires that a formal written program or plan be established, delineating all the quality assurance activities that will be performed during the life of the project. The Owner must write his program in which project policies and the responsibilities of each organization are defined. Project quality levels and the requirements of each level are set forth. Each participating organization then prepares its own plan which must conform to the policies and definitions of the Owner's program. At this point, procedures must be written for each activity to be performed—a procedure for inspection or surveillance of a pump, a valve, or a mechanical

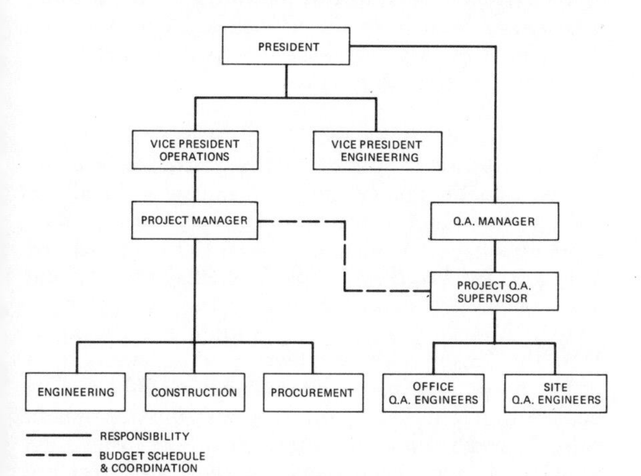

Fig. 16-2. One possible architect engineer organization. In this illustration the detail design work is performed in the Operations Division and the independent engineering review is performed by the Engineering Division. The internal audit of both divisions is performed by the Quality Assurance Department. All QA functions are independent of project pressures.

installation. The procedure describes the steps to be taken and the reports to be prepared and distributed. Nothing is to be done in an impromptu manner; every action is to be documented. Similarly, procedures are required for auditing the design program of each designing entity.

III. DESIGN CONTROL

This criterion provides that all facets of the design activity shall be formalized and performed in an auditable manner. Everything must be documented. For example, the format for performing and checking design calculations must be set forth and rigidly followed so that an independent auditor can verify that the calculation was performed as directed. It must be noted that auditing will take two forms: a design review audit will be performed to check the adequacy of the engineering, and a second audit will also be performed to check the adequacy of documentation that permits tracing a component from installation back through detail design, to design concept, and criteria.

To be completely responsive to this requirement, everything in the engineering process would be formalized. It is difficult to say at this time what the full impact of this item will be; one thing is known, however, there must be complete design traceability from initial concept to final installation of each system and component. Each variation and change that takes place must be recorded and stored in a history file so that the evolution of any item will be easily identifiable.

IV. PROCUREMENT DOCUMENT CONTROL

This criterion is relatively simple. It requires a formal review and sign-off of the procurement documents to insure that all requirements are present. When appropriate, a formal quality assurance program should be imposed on the vendor. The specification should require submittal of the program with the bid. If any portion of the program is unacceptable the differences can be resolved during negotiation before signing the contract, rather than causing disruptions after signing.

The procurement action must impose on the vendor all codes, standards, documentation, traceability, etc., required by the Owner's procurement program.

V. INSTRUCTIONS, PROCEDURES, AND DRAWINGS

This criterion requires that no activity affecting quality be accomplished in an impromptu manner. All construction, inspection, surveillance, auditing, design review, or checking activities shall be done according to procedures that themselves have been reviewed and approved.

To comply with this requirement the engineering organizations are required to state in writing, for each discipline, when they will conduct their design reviews, how these will be executed and in what manner they will be documented. Construction organizations must prepare formal procedures describing their methods of manufacturing and placing concrete, installing wire and cable, or installing mechanical equipment.

VI. DOCUMENT CONTROL

This criterion requires that formal channels be established for the review, approval, reproduction, and distribution of all documents in the project including drawings and specifications. It further requires that revisions and changes pass through the same channels. Finally, it requires an additional procedure that acknowledges receipt of changes, and verification that only the latest revision of documentation is in use at the point of activity execution.

This later requirement is very important as it insures that all organizations involved in the project are always working from the most current documents.

VII. CONTROL OF PURCHASED MATERIAL, EQUIPMENT, AND SERVICES

This criterion requires that prospective bidders be examined and evaluated to ascertain that all the quality asurance requirements of the proposed contract can be met. Any doubts pertaining to the bidder's ability should be resolved during a prebidding survey. Next the contractor (successful bidder) must prepare his detailed procedures required to implement his plan. During execution of the work the contractor must generate all the documentation required to prove conformance to all the requirements of the contract. Additionally the Owner is required to execute a formal program of surveillance and/or audit to insure proper functioning of the contractor's quality assurance program.

VIII. IDENTIFICATION AND CONTROL OF MATERIALS, PARTS, AND COMPONENTS

This criterion requires that all parts, sub-assemblies, and components be identified by tags, etching, low-stress stamps, etc., during all phases of fabrication, installation, and erection. Further, the identification symbology used shall be consistent throughout the project and permit complete traceability at all times to both raw material used and design requirements.

IX. CONTROL OF SPECIAL PROCESSES

This criterion requires that all special processes such as heat treating, welding, electroplating, cleaning, nondestructive testing, etc., be accomplished by qualified personnel according to written procedures which include acceptable standards of performance for the procedure. Requirement VII requires that execution of each of these special processes produces complete documentation attesting to proper performance of the process.

X. INSPECTION

This criterion requires that all work be inspected in process to prove conformance to approved drawings, instructions, and specifications. Inspections must be performed by different organizations from those that did the work. If the requirements include mandatory hold points at which the Owner will perform inspections, then the processing documents must include these hold points.

All inspections must be performed in accordance with detailed written procedures which contain or indicate the source of specific acceptance criteria.

XI. TEST CONTROL

This criterion requires that all testing—component proof tests, pre-operational tests, low power testing, ascent to power testing, and full power testing must be performed in accordance with written procedures. Each procedure should contain step-by-step instructions with acceptable limits for each parameter defined for each step. The instrumentation must be defined as well as the method of reading the instrument and treatment of the reading. As an example, an overall instrument channel may have an accuracy of $\pm 1\frac{1}{2}$ percent; the procedure must define whether or not the reading is to be treated as an absolute value or recognize that a range of values may exist. Special instrumentation and calibration must be precisely specified; methods of calculation should be defined.

XII. CONTROL OF MEASURING AND TEST EQUIPMENT

This criterion requires that all measuring equipment used on the project be serialized, calibrated, and controlled. All devices must be calibrated and adjusted at fixed intervals and records kept for each item. This extends to both company owned and craftsman owned instruments.

XIII. HANDLING, STORAGE, AND SHIPPING

This criterion requires that the procurement documents give the generic requirements for handling, storage, and shipping. Following that, the manufacturer must prepare specific written instructions to be followed, for his component. The final requirement is to insure that the approved instructions are carefully followed during handling, shipping, and storage. Note that this may require heated storage areas, inert gas atmospheres, or air-conditioned chambers. The goal of this criterion is to insure that the component is in as good a condition at the moment of installation as when it was finished in the shop and declared acceptable for shipment.

XIV. INSPECTION, TEST, AND OPERATING STATUS

This criterion requires that all parts and components be marked at all times during fabrication and installation with their status—"Acceptable" or "Unacceptable." Systems and components during the construction and testing must be marked as well with their condition—"Ready for Test," "Repair," "Ready for Operation" or "Do Not Operate." Similarly, after the plant is in commercial operation any component whose operability is in doubt or that cannot operate must be properly identified to avoid inadvertent operation.

XV. NONCONFORMING MATERIALS, PARTS, OR COMPONENTS

This criterion requires that all nonconforming items be identified, marked, and segregated. The nonconformance must be described in a report, repair procedures proposed and a disposition made —repair, rework, scrap, etc.—by a Review Board or specific engineering channels. The nonconforming item or material cannot be removed from the segregated hold area without complete written instructions from the reviewing authority.

XVI. CORRECTIVE ACTION

This criterion requires that all nonconformances be identified and corrected. These corrective actions must also be documented so that the repair or rework can be proven. In cases of significant deviations, the causes must be identified and measures instituted to prevent repetition of the nonconformance. This latter type of nonconformance is followed up by a report to corporate management describing the nonconformance and the changes made to prevent recurrence. It is also customary to increase second level surveillance over the operation until it has been shown that the revised operation can be relied upon to produce an acceptable product.

XVII. QUALITY ASSURANCE RECORDS

These records must include all the records accumulated during the course of the project. They must include personnel certifications, mill test reports, results of equipment inspections, radiographs, nondestructive test reports, survey and audit reports, construction inspection reports, pre-operation test reports, acceptance test reports, etc., to provide a complete history of the design and construction of the plant. All examination data required to be retained by a fabricator to comply with codes must be identified, and certified descriptions provided at the plant site telling where the documentation is and how it may be retrieved.

XVIII. AUDITS

The entire project shall be subjected to periodic auditing of all phases of the quality assurance plan. These audits must be performed to written procedures and documentation generated to substantiate the auditing activity and its results. This auditing must investigate design review, filing, component fabrication, construction, etc. In addition, each organization participating should have some type of self-auditing activity to ensure that it is operating properly.

16.4 Application of Quality Assurance Categories

As can be seen from Appendix II, the quality assurance program must be planned to cover two separate facets. One facet must be directed at the execution of the project's engineering and the other facet at the project's procurement and construction. The engineering organizations must implement within their own project operation, Criteria I, II, III, IV, V, VI, VII, XIII, XVI, XVII, and XVIII. The construction and fabrication operation must implement any or all of the requirements that the specifications impose.

Quality Category 1. Quality Category 1 must impose all eighteen requirements plus two more. First is access at all times for the Owner and his representatives, for auditing, surveillance, or witnessing tests. The second requirement is availability of all documentation, complete or semi-finished, at all quality assurance visits.

Quality Category 2. Quality Category 2 must be based on opinion and engineering judgement. Consequently the requirements that are selected here represent an engineering judgement and the reader may alter them in his own project as he sees fit. The criteria of access and availability of documentation from Category 1 should be retained.

Requirements V, VI, VIII, IX, X, XII, XIII, and XV should be as complete and formal as in Category 1. Requirements I, II, IV, V, VII, and XVII should be present and identifiable in a preaward survey. Any questions that may exist about the latter group of requirements should be resolved before award of the contract because they will simply be identified and not made a contract requirement.

Quality Category 3. Category 3 is also an engineering judgement category. This category should also retain access and availability of documents. Requirements V, VI, VIII, IX, X, and XIII should be as complete and formal as in Category 1. Requirement No. XII should be required for instrumentation procurements. As in Category 2, the presence of the other elements of a quality assurance program should be assessed in a pre-award survey and any questions resolved before award of contract.

Quality Category 4. This quality category consists of the requirements embodied in each individual procurement specification. As a minimum they should include all codes and standards, control of welding and metallurgical processing, control of nondestructive examination procedures, control of cleaning procedures of stainless steel and any special proof tests and performance tests required by the component.

Specific acceptance criteria must be given for each examination or test.

16.5 Specification Requirements

All of the foregoing discussion has been devoted to the planning of a quality assurance program. It must not be forgotten that the procurement specification must include specific quality control requirements, viz: A quality assurance plan may require that nondestructive examinations be executed according to written procedures but unless the specification requires that nondestructive examination be performed, the control imposed by the quality assurance plan cannot be invoked. The procurement documents must have complete requirements for examinations, documentation, proof tests, performance tests, etc. The use of a generic quality assurance specification does not relieve the specification writer of the responsibility for including inspection and test requirements in his specification. Actually, the reverse is true. The writer has decided that he requires a quality assurance program; therefore he must now include in the technical section of his specification all the areas of raw materials, fabrication, inspection, and testing which are the particular factors to be subjected to the quality assurance plan.

Appendix

I. GENERAL DESIGN CRITERIA FOR NUCLEAR POWER PLANTS—Effective Date: May 21, 1971

TABLE OF CONTENTS

VI. Fuel and Radioactivity Control:

INTRODUCTION

Pursuant to the provisions of § 50.34, an application for a construction permit must include the principal design criteria for a proposed facility. The principal design criteria establish the necessary design fabrication, construction, testing and performance requirements for structures, systems, and components important to safety; that is, structures, systems, and components that provide reasonable assurance that the facility can be operated without undue risk to the health and safety of the public.

These General Design Criteria establish minimum requirements for the principal design criteria for water-cooled nuclear power plants similar in design and location to plants for which construction permits have been issued by the Commission. The General Design Criteria are also considered to be generally applicable to other types of nuclear power units and are intended to provide guidance in establishing the principal design criteria for such other units.

The development of these General Design Criteria is not yet complete. For example, some of the definitions need further amplification. Also, some of the specific design requirements for structures, systems, and components important to safety have not as yet been suitably defined. Their omission does not relieve any applicant from considering these matters in the design of a specific facility and satisfying the necessary safety requirements.

These matters include:

(1) Consideration of the need to design against single failures of passive components in fluid systems important to safety. (See Definition of Single Failure.)

(2) Consideration of redundancy and diversity requirements for fluid systems important to safety. A "system" could consist of a number of subsystems each of which is separately capable of performing the specified system safety function. The minimum acceptable redundancy and diversity of subsystems and components within a subsystem, and the required interconnection and independence of the subsystems have not yet been developed or defined. (See Criteria 34, 35, 38, 41, and 44.)

(3) Consideration of the type, size, and orientation of possible breaks in components of the reactor coolant pressure boundary in determining design requirements to suitably protect against postulated loss-of-coolant accidents. (See Definition of Loss-of-Coolant Accidents.)

(4) Consideration of the possibility of systematic, nonrandom, concurrent failures of redundant elements in the design of protection systems and reactivity control systems. (See Criteria 22, 24, 26, and 29.)

It is expected that the criteria will be augmented and changed from time to time as important new requirements for these and other features are developed.

There will be some water-cooled nuclear power plants for which the General Design Criteria are not sufficient and for which additional criteria must be identified and satisfied in the interest of public safety. In particular, it is expected that additional or different criteria will be needed to take into account unusual sites and environmental conditions, and for water-cooled nuclear power units of advanced design. Also, there may be water-cooled nuclear power units for which fulfillment of some of the General Design Criteria may not be necessary or appropriate. For plants such as these, departures from the General Design Criteria must be identified and justified.

DEFINITIONS AND EXPLANATIONS

Nuclear Power Unit

A nuclear power unit means a nuclear power reactor and associated equipment necessary for electric power generation and includes those structures, systems, and components required to provide reasonable assurance that the facility can be operated without undue risk to the health and safety of the public.

Loss-of-Coolant Accidents

Loss-of-coolant accidents means those postulated accidents that result from the loss of reactor coolant at a rate in excess of the capability of the reactor coolant makeup system from breaks in the reactor coolant pressure boundary, up to and including a break equivalent in size to the double-ended rupture of the largest pipe of the reactor coolant system.[1]

Single Failure

A single failure means an occurrence which results in the loss of capability of a component to perform its intended safety functions. Multiple failures resulting from a single occurrence are considered to be a single failure. Fluid and electrical systems are considered to be designed against an assumed single failure if neither (1) a single failure of any active component (assuming passive components function properly), nor (2) a single failure

[1] Further details relating to the type, size, and orientation of postulated breaks in specific components of the reactor coolant pressure boundary are under development.

of a passive component (assuming active components function properly), results in a loss of the capability of the system to perform its safety functions.[2]

Anticipated Operational Occurrences

Anticipated operational occurrences mean those conditions of normal operation which are expected to occur one or more times during the life of the nuclear power unit and include but are not limited to loss of power to all recirculation pumps, tripping of the turbine generator set, isolation of the main condenser, and loss of all off site power.

I. OVERALL REQUIREMENTS

CRITERION 1 — *Quality standards and records.* Structures, systems, and components important to safety shall be designed, fabricated, erected, and tested to quality standards commensurate with the importance of the safety functions to be performed. Where generally recognized codes and standards are used, they shall be identified and evaluated to determine their applicability, adequacy, and sufficiency and shall be supplemented or modified as necessary to assure a quality product in keeping with the required safety function. A quality assurance program shall be established and implemented in order to provide adequate assurance that these structures, systems, and components will satisfactorily perform their safety functions. Appropriate records of the design, fabrication, erection, and testing of structures, systems, and components important to safety shall be maintained by or under the control of the nuclear power unit license throughout the life of the unit.

CRITERION 2 — *Design bases for protection against natural phenomena.* Structures, systems, and components important to safety shall be designed to withstand the effects of natural phenomena such as earthquakes, tornadoes, hurricanes, floods, tsunami, and seiches without loss of capability to perform their safety functions. The design bases for these structures, systems, and components shall reflect:

(1) appropriate consideration of the most severe of the natural phenomena that have been historically reported for the site and surrounding area, with sufficient margin for the limited accuracy, quantity, and period of time in which the historical data have been accumulated;

(2) appropriate combinations of the effects of normal and accident conditions with the effects of the natural phenomena; and

[2] Single failures of passive components in electric systems should be assumed in designing against a single failure. The conditions under which a single failure of a passive component in a fluid system should be considered in designing the system against a single failure are under development.

(3) the importance of the safety functions to be performed.

CRITERION 3 — *Fire protection.* Structures, systems, and components important to safety shall be designed and located to minimize, consistent with other safety requirements, the probability and effect of fires and explosions. Noncombustible and heat resistant materials shall be used wherever practical throughout the unit particularly in locations such as the containment and control room. Fire detection and fighting systems of appropriate capacity and capability shall be provided and designed to minimize the adverse effects of fires on structures, systems, and components important to safety. Firefighting systems shall be designed to assure that their rupture or inadvertent operation does not significantly impair the safety capability of these structures, systems, and components.

CRITERION 4 — *Environmental and missile design bases.* Structures, systems and components important to safety shall be designed to accommodate the effects of and to be compatible with the environmental conditions associated with normal operation, maintenance, testing, and postulated accidents including loss-of-coolant accidents. These structures, systems, and components shall be appropriately protected against dynamic effects, including the effects of missiles, pipe whipping, and discharging fluids, that may result from equipment failures and from events and conditions outside the nuclear power unit.

CRITERION 5 — *Sharing of structures, systems, and components.* Structures, systems, and components important to safety shall not be shared among nuclear power units unless it is shown that such sharing will not significantly impair their ability to perform their safety function, including, in the event of an accident in one unit, an orderly shutdown and cooldown of the remaining units.

II. PROTECTION BY MULTIPLE FISSION PRODUCT BARRIERS

CRITERION 10 — *Reactor design.* The reactor core and associated coolant, control, and protection systems shall be designed with appropriate margin to assure that specified acceptable fuel design limits are not exceeded during any condition of normal operation, including the effects of anticipated operational occurrences.

CRITERION 11 — *Reactor inherent protection.* The reactor core and associated coolant systems shall be designed so that in the power operating range the net effect of the prompt inherent nuclear feedback characteristics tends to compensate for a rapid increase in reactivity.

CRITERION 12 — *Suppression of reactor power oscillations.* The reactor core and associated coolant, control, and protection systems shall be designed to assure that power oscillations which can result in conditions exceeding specified acceptable fuel design limits are not possible or can be reliably and readily detected and suppressed.

Criterion 13 — *Instrumentation and control.* Instrumentation and control shall be provided to monitor variables and systems over their anticipated ranges for normal operation, for anticipated operational occurrences, and for accident conditions as appropriate to assure adequate safety, including those variables and systems that can affect the fission process, the integrity of the reactor core, the reactor coolant pressure boundary, and the containment and its associated systems. Appropriate controls shall be provided to maintain these variables and systems within prescribed operating ranges.

Criterion 14 — *Reactor coolant pressure boundary.* The reactor coolant pressure boundary shall be designed, fabricated, erected, and tested so as to have an extremely low probability of abnormal leakage, of rapidly propagating failure, and particularly of gross rupture.

Criterion 15 — *Reactor coolant system design.* The reactor coolant system and associated auxiliary, control, and protection systems shall be designed with sufficient margin to assure that the design conditions of the reactor coolant pressure boundary are not exceeded during any condition of normal operation, including anticipated operational occurrences.

Criterion 16 — *Containment design.* Reactor containment and associated systems shall be provided to establish an essentially leaktight barrier against the uncontrolled release of radioactivity to the environment and to assure that the containment design conditions important to safety are not exceeded for as long as postulated accident conditions require.

Criterion 17 — *Electric power systems.* An on site electric power system and an off site electric power system shall be provided to permit functioning of structures, systems, and components important to safety. The safety function for each system (assuming the other system is not functioning) shall be to provide sufficient capacity and capability to assure that:

(1) specified acceptable fuel design limits and design conditions of the reactor coolant pressure boundary are not exceeded as a result of anticipated operational occurrences; and

(2) the core is cooled and containment integrity and other vital functions are maintained in the event of postulated accidents.

The on site electric power supplies, including the batteries, and the on site electric distribution system, shall have sufficient independence, redundancy, and testability to perform their safety functions assuming a single failure.

Electric power from the transmission network to the on site electric distribution system shall be supplied by two physically independent circuits (not necessarily on separate rights-of-way) designed and located so as to minimize to the extent practical the likelihood of their simultaneous failure under operating and postulated accident and environmental conditions. A switchyard common to both circuits is acceptable. Each of these circuits shall be designed to be available in sufficient time following a loss of all on site alternating current power supplies and the other off site electric power circuit, to assure that specified acceptable fuel design limits and design conditions of the reactor coolant pressure boundary are not exceeded. One of these circuits shall be designed to be available within a few seconds following a loss-of-coolant accident to assure that core cooling, containment integrity, and other vital safety functions are maintained.

Provisions shall be included to minimize the probability of losing electric power from any of the remaining supplies as a result of, or coincident with, the loss of power generated by the nuclear power unit, the loss of power from the transmission network, or the loss of power from the on site electric power supplies.

Criterion 18 — *Inspection and testing of electric power systems.* Electric power systems important to safety shall be designed to permit appropriate periodic inspection and testing of important areas and features, such as wiring, insulation, connections, and switchboards, to assess the continuity of the systems and the condition of their components. The systems shall be designed with a capability to test periodically:

(1) the operability and functional performance of the components of the systems, such as on site power sources, relays, switches, and buses; and

(2) the operability of the systems as a whole and, under conditions as close to design as practical, the full operation sequence that brings the systems into operation including operation of applicable portions of the protection system, and the transfer of power among the nuclear power unit, the off site power system, and the on site power system.

Criterion 19 — *Control room.* A control room shall be provided from which actions can be taken to operate the nuclear power unit safely under normal conditions and to maintain it in a safe condition under accident conditions, including loss-of-coolant accidents. Adequate radiation protection shall be provided to permit access and occupancy of the control room under accident conditions without personnel receiving radiation exposures in excess of 5 rem whole body, or its equivalent to any part of the body, for the duration of the accident.

Equipment at appropriate locations outside the control room shall be provided:

(1) with a design capability for prompt hot shutdown of the reactor, including necessary instrumentation and controls to maintain the unit in a safe condition during hot shutdown; and

(2) with a potential capability for subsequent cold shutdown of the reactor through the use of suitable procedures.

III. PROTECTION AND REACTIVITY CONTROL SYSTEMS

CRITERION 20 — *Protection system functions.* The protection system shall be designed:

(1) to initiate automatically the operation of appropriate systems including the reactivity control systems, to assure that specified acceptable fuel design limits are not exceeded as a result of anticipated operational occurrences; and

(2) to sense accident conditions and to initiate the operation of systems and components important to safety.

CRITERION 21 — *Protection system reliability and testability.* The protection system shall be designed for high functional reliability and inservice testability commensurate with the safety functions to be performed. Redundancy and independence designed into the protection system shall be sufficient to assure that:

(1) no single failure results in loss of the protection function; and

(2) removal from service of any component or channel does not result in loss of the required minimum redundancy unless the acceptable reliabiltiy of operation of the protection system can be otherwise demonstrated. The protection system shall be designed to permit periodic testing of its functioning when the reactor is in operation including a capability to test channels independently to determine failures and losses of redundancy that may have occurred.

CRITERION 22 — *Protection system independence.* The protection system shall be designed to assure that the effects of natural phenomena, and of normal operating, maintenance, testing, and postulated accident conditions on redundant channels do not result in loss of the protection function, or shall be demonstrated to be acceptable on some other defined basis. Design techniques, such as functional diversity or diversity in component design and principles of operation, shall be used to the extent practical to prevent loss of the protection function.

CRITERION 23 — *Protection system failure modes.* The protection system shall be designed to fall into a safe state or into a state demonstrated to be acceptable on some other defined basis if conditions such as disconnection of the system, loss of energy (e.g., electric power, instrument air), or postulated adverse environments (e.g., extreme heat or cold, fire, pressure, steam, water, and radiation) are experienced.

CRITERION 24 — *Separation of protection and control systems.* The protection system shall be separated from control systems to the extent that failure of any single control system component or channel, or failure or removal from service of any single protection system component or channel which is common to the control and protected systems leaves intact a system satisfying all reliability, redundancy, and independence requirements of the protection system. Interconnection of the protection and control systems shall be limited so as to assure that safety is not significantly impaired.

CRITERION 25 — *Protection system requirements for reactivity control malfunctions.* The protection system shall be designed to assure that specified acceptable fuel design limits are not exceeded for any single malfunction of the reactivity control systems, such as accidental withdrawal (not ejection or dropout) of control rods.

CRITERION 26 — *Reactivity control system redundancy and capability.* Two independent reactivity control systems of different design principles shall be provided. One of the systems shall use control rods, preferably including a positive means for inserting the rods, and shall be capable of reliably controlling reactivity changes to assure that under conditions of normal operation, including anticipated operational occurrences, and with appropriate margin for malfunctions such as stuck rods, specified acceptable fuel design limits are not exceeded. The second reactivity control system shall be capable of reliably controlling the rate of reactivity changes resulting from planned, normal power changes (including xenon burnout) to assure acceptable fuel design limits are not exceeded. One of the systems shall be capable of holding the reactor core subcritical under cold conditions.

CRITERION 27 — *Combined reactivity control systems capability.* The reactivity control systems shall be designed to have a combined capability, in conjunction with poison addition by the emergency core cooling system, of reliably controlling reactivity changes to assure that under postulated accident conditions and with appropriate margin for stuck rods the capability to cool the core is maintained.

CRITERION 28 — *Reactivity limits.* The activity control systems shall be designed with appropriate limits on the potential amount and rate of reactivity increase to assure that the effects of postulated reactivity accidents can neither:

(1) result in damage to the reactor coolant pressure boundary greater than limited local yielding; nor

(2) sufficiently disturb the core, its support structures or other reactor pressure vessel internals to impair significantly the capability to cool the core. These postulated reactivity accidents shall include consideration of rod ejection (unless prevented by positive means), rod dropout, steam-line rupture, changes in reactor coolant temperature and pressure, and cold water addition.

CRITERION 29 — *Protection against anticipated operational occurrences.* The protection and reactivity control systems shall be designed to assure an extremely high probability of accomplishing their safety functions in the event of anticipated operational occurrences.

IV. FLUID SYSTEMS

CRITERION 30 — *Quality of reactor coolant pressure boundary.* Components which are part of the reactor coolant pressure boundary shall be designed, fabricated, erected, and tested to the highest quality standards practical. Means shall be provided for detecting and, to the extent practical, identifying the location of the source of reactor coolant leakage.

CRITERION 31 — *Fracture prevention of reactor coolant pressure boundary.* The reactor coolant pressure boundary shall be designed with sufficient margin to assure that when stressed under operating, maintenance, testing, and postulated accident conditions:

(1) the boundary behaves in a nonbrittle manner; and

(2) the probability of rapidly propagating fracture is minimized. The design shall reflect consideration of service temperatures and other conditions of the boundary material under operating, maintenance, testing, and postulated accident conditions and the uncertainties in determining:

(1) material properties;

(2) the effects of irradiation on material properties;

(3) residual, steady-state and transient stresses; and

(4) size of flaws.

CRITERION 32 — *Inspection of reactor coolant pressure boundary.* Components which are part of the reactor coolant pressure boundary shall be designed to permit:

(1) periodic inspection and testing of important areas and features to assess their structural and leaktight integrity; and

(2) an appropriate material surveillance program for the reactor pressure vessel.

CRITERION 33 — *Reactor coolant makeup.* A system to supply reactor coolant makeup for protection against small breaks in the reactor coolant pressure boundary shall be provided. The system safety function shall be to assure that specified acceptable fuel design limits are not exceeded as a result of reactor coolant loss due to leakage from the reactor coolant pressure boundary and rupture of small piping or other small components which are part of the boundary. The system shall be designed to assure that for on site electric power system operation (assuming off site power is not available) and for off site electric power system operation (assuming on site power is not available) the system safety function can be accomplished using the piping, pumps, and valves used to maintain coolant inventory during normal reactor operation.

CRITERION 34 — *Residual heat removal.* A system to remove residual heat shall be provided. The system safety function shall be to transfer fission product decay heat and other residual heat from the reactor core at a rate such that specified acceptable fuel design limits and the design conditions of the reactor coolant pressure boundary are not exceeded.

Suitable redundancy in components and features, and suitable interconnections, leak detection, and isolation capabilities shall be provided to assure that for on site electric power system operation (assuming off site power is not available) and for off site electric power system operation (assuming on site power is not available) the system safety function can be accomplished, assuming a single failure.

CRITERION 35 — *Emergency core cooling.* A system to provide abundant emergency core cooling shall be provided. The system safety function shall be to transfer heat from the reactor core following any loss of reactor coolant at a rate such that:

(1) fuel and clad damage that could interfere with continued effective core cooling is prevented; and

(2) clad metal-water reaction is limited to negligible amounts.

Suitable redundancy in components and features, and suitable interconnections, leak detection, isolation, and containment capabilities shall be provided to assure that for on site electric power system operation (assuming off site power is not available) and for off site electric power system operation (assuming on site power is not available) the system safety function can be accomplished, assuming a single failure.

CRITERION 36 — *Inspection of emergency core cooling system.* The emergency core cooling system shall be designed to permit appropriate periodic inspection of important components, such as spray rings in the reactor pressure vessel, water injection nozzles, and piping, to assure the integrity and capability of the system.

CRITERION 37 — *Testing of emergency core cooling system.* The emergency core cooling system shall be designed to permit appropriate periodic pressure and functional testing to assure:

(1) the structural and leaktight integrity of its components;

(2) the operability and performance of the active components of the system; and

(3) the operability of the system as a whole and, under conditions as close to design as practical, the performance of the full operational sequence that brings the system into operation, including operation of applicable portions of the protection system, the transfer between normal and emergency power sources, and the operation of the associated cooling water system.

CRITERION 38 — *Containment heat removal.* A system to remove heat from the reactor containment shall be provided. The system safety function shall be to reduce rapidly, consistent with the functioning of other associated systems, the containment pressure and temperature following any loss-of-coolant accident and maintain them at acceptably low levels.

Suitable redundancy in components and features, and suitable interconnections, leak detection, isolation, and containment capabilities shall be provided to assure that for on site electric power system operation (assuming off site power is not available) and for off site electric power system operation (assuming on site power is not available) the system safety function can be accomplished, assuming a single failure.

CRITERION 39 — *Inspection of containment heat removal system.* The containment heat removal system shall be designed to permit appropriate periodic inspection of important components, such as the torus, sumps, spray nozzles, and piping to assure the integrity and capability of the system.

CRITERION 40 — *Testing of containment heat removal system.* The containment heat removal system shall be designed to permit appropriate periodic pressure and functional testing to assure:

(1) the structural and leaktight integrity of its components;

(2) the operability and performance of the active components of the system; and

(3) the operability of the system as a whole, and, under conditions as close to the design as practical, the performance of the full operational sequence that brings the system into operation, including operation of applicable portions of the protection system, the transfer between normal and emergency power sources, and the operation of the associated cooling water system.

CRITERION 41 — *Containment atmosphere cleanup.* Systems to control fission products, hydrogen, oxygen, and other substances which may be released into the reactor containment shall be provided as necessary to reduce, consistent with the functioning of other associated systems, the concentration and quality of fission products released to the environment following postulated accidents, and to control the concentration of hydrogen or oxygen and other substances in the containment atmosphere following postulated accidents to assure that containment integrity is maintained.

Each system shall have suitable redundancy in components and features, and suitable interconnections, leak detection, isolation, and containment capabilities to assure that for on site electric power system operation (assuming off site power is not available) and for off site electric power system operation (assuming on site power is not available) its safety function can be accomplished, assuming a single failure.

CRITERION 42 — *Inspection of containment atmosphere cleanup systems.* The containment atmosphere cleanup systems shall be designed to permit appropriate periodic inspection of important components, such as filter frames, ducts, and piping to assure the integrity and capability of the systems.

CRITERION 43 — *Testing of containment atmosphere cleanup systems.* The containment atmosphere cleanup systems shall be designed to permit appropriate periodic pressure and functional testing to assure:

(1) the structural and leaktight integrity of its components;

(2) the operability and performance of the active components of the systems such as fans, filters, dampers, pumps, and valves; and

(3) the operability of the systems as a whole and, under conditions as close to design as practical, the performance of the full operational sequence that brings the systems into operation, including operation of applicable portions of the protection system, the transfer between normal and emergency power sources, and the operation of associated systems.

CRITERION 44 — *Cooling water.* A system to transfer heat from structures, systems, and components important to safety, to an ultimate heat sink shall be provided. The system safety function shall be to transfer the combined heat load of these structures, systems, and components under normal operating and accident conditions.

Suitable redundancy in components and features, and suitable interconnections, leak detection, and isolation capabilities shall be provided to assure that for on site electric power system operation (assuming off site power is not available) and for off site electric power system operation (assuming on site power is not available) the system safety function can be accomplished, assuming a single failure.

CRITERION 45 — *Inspection of cooling water system.* The cooling water system shall be designed to permit appropriate periodic inspection of important components, such as heat exchangers and piping, to assure the integrity and capability of the system.

CRITERION 46 — *Testing of cooling water system.* The cooling water system shall be designed to permit appropriate periodic pressure and functional testing to assure:

(1) the structural and leaktight integrity of its components;

(2) the operability and the performance of the active components of the system; and

(3) the operability of the system as a whole and, under conditions as close to design as practical, the performance of the full operational sequence that brings the system into operation for reactor shutdown and for loss-of-coolant accidents, including operation of applicable portions of the protection system and the transfer between normal and emergency power sources.

V. REACTOR CONTAINMENT

CRITERION 50 — *Containment design basis.* The reactor containment structure, including access openings, penetrations, and the containment heat removal system shall be designed so that the containment structure and its internal compartments can accommodate, without exceeding the design leakage rate and, with sufficient margin, the calculated pressure and temperature conditions resulting from any loss-of-coolant accident. This margin shall reflect consideration of:

(1) the effects of potential energy sources which have not been included in the determination of the peak conditions, such as energy in steam generators and energy from metal-water and other chemical reactions that may result from degraded emergency core cooling functioning;

(2) the limited experience and experimental data available for defining accident phenomena and containment responses; and

(3) the conservatism of the calculational model and input parameters.

CRITERION 51 — *Fracture prevention of containment pressure boundary.* The reactor containment boundary shall be designed with sufficient margin to assure that under operating maintenance, testing, and postulated accident conditions:

(1) its ferritic materials behave in a nonbrittle manner; and

(2) the probability of rapidly propagating fracture is minimized. The design shall reflect considerations of service temperatures and other conditions of the containment boundary material during operation, maintenance, testing, and postulated accident conditions, and the uncertainties in determining:

(1) material properties;

(2) residual, steady-state, and transient stresses; and

(3) size of flaws.

CRITERION 52 — *Capability for containment leakage rate testing.* The reactor containment and other equipment which may be subjected to containment test conditions shall be designed so that periodic integrated leakage rate testing can be conducted at containment design pressure.

CRITERION 53 — *Provisions for containment testing and inspection.* The reactor containment shall be designed to permit:

(1) appropriate periodic inspection of all important areas, such as penetrations;

(2) an appropriate surveillance program; and

(3) periodic testing at containment design pressure of the leaktightness of penetrations which have resilient seals and expansion bellows.

CRITERION 54 — *Piping systems penetrating containment.* Piping systems penetrating primary reactor containment shall be provided with leak detection, isolation, and containment capabilities having redundancy, reliability, and performance capabilities which reflect the importance to safety of isolating these piping systems. Such piping systems shall be designed with a capability to test periodically the operability of the isolation valves and associated apparatus and to determine if valve leakage is within acceptable limits.

CRITERION 55 — *Reactor coolant pressure boundary penetrating containment.* Each line that is part of the reactor coolant pressure boundary and that penetrates primary reactor containment shall be provided with containment isolation valves as follows, unless it can be demonstrated that the containment isolation provisions for a specific class of lines, such as instrument lines, are acceptable on some other defined basis:

(1) One locked closed isolation valve inside and one locked closed isolation valve outside containment; or

(2) One automatic isolation valve inside and one locked closed isolation valve outside containment; or

(3) One locked closed isolation valve inside and one automatic isolation valve outside containment.

A simple check valve may not be used as the automatic isolation valve outside containment; or

(4) One automatic isolation valve inside and one automatic isolation valve outside containment. A simple check valve may not be used as the automatic isolation valve outside containment.

Isolation valves outside containment shall be located as close to containment as practical and upon loss of actuating power, automatic isolation valves shall be designed to take the position that provides greater safety.

Other appropriate requirements to minimize the probability or consequences of an accidental rupture of these lines or of lines connected to them shall be provided as necessary to assure adequate safety. Determination of the appropriateness of these requirements, such as higher quality in design, fabrication, and testing, additional provisions for inservice inspection, protection against more severe natural phenomena, and additional isolation valves and containment, shall include consideration of the population density, use characteristics, and physical characteristics of the site environs.

CRITERION 56 — *Primary containment isolation.* Each line that connects directly to the containment atmosphere and penetrates primary reactor containment shall be provided with containment isolation valves as follows, unless it can be demonstrated that the containment isolation provisions for a specific class of lines, such as instrument lines, are acceptable on some other defined basis:

(1) One locked closed isolation valve inside and one locked closed isolation valve outside containment; or

(2) One automatic isolation valve inside and one locked closed isolation valve outside containment; or

(3) One locked closed isolation valve inside and one automatic isolation valve outside containment. A simple check valve may not be used as the automatic isolation valve outside containment; or

(4) One automatic isolation valve inside and one automatic isolation valve outside containment. A simple check valve may not be used as the automatic isolation valve outside containment.

Isolation valves outside containment shall be located as close to the containment as practical and upon loss of actuating power, automatic isolation valves shall be designed to take the position that provides greater safety.

CRITERION 57 — *Closed system isolation valves.* Each line that penetrates primary reactor containment and is neither part of the reactor coolant pressure boundary nor connected directly to the containment atmosphere shall have at least one containment isolation valve which shall be either automatic, or locked closed, or capable of remote manual operation. This valve shall be outside containment and located as close to the containment as practical. A simple check valve may not be used as the automatic isolation valve.

VI. FUEL AND RADIOACTIVITY CONTROL

CRITERION 60 — *Control of releases of radioactive materials to the environment.* The nuclear power unit design shall include means to control suitably the release of radioactive materials in gaseous and liquid effluents and to handle radioactive solid wastes produced during normal reactor operation, including anticipated operational occurrences. Sufficient holdup capacity shall be provided for retention of gaseous and liquid effluents containing radioactive materials, particularly where unfavorable site environmental conditions can be expected to impose unusual operational limitations upon the release of such effluents to the environment.

CRITERION 61 — *Fuel storage and handling and radioactivity control.* The fuel storage and handling, radioactive waste, and other systems which may contain radioactivity shall be designed to assure adequate safety under normal and postulated accident conditions. These systems shall be designed:

(1) with a capability to permit appropriate periodic inspection and testing of components important to safety;

(2) with suitable shielding for radiation protection;

(3) with appropriate containment, confinement, and filtering systems;

(4) with a residual heat removal capability having reliability and testability that reflects the importance to safety of decay heat and other residual heat removal; and

(5) to prevent significant reduction in fuel storage coolant inventory under accident conditions.

CRITERION 62 — *Prevention of criticality in fuel storage and handling.* Criticality in the fuel storage and handling system shall be prevented by physical systems or processes, preferably by use of geometrically safe configurations.

CRITERION 63 — *Monitoring fuel and waste storage.* Appropriate systems shall be provided in fuel storage and radioactive waste systems and associated handling areas to:

(1) detect conditions that may result in loss of residual heat removal capability and excessive radiation levels;

(2) initiate appropriate safety actions.

CRITERION 64 — *Monitoring radioactivity releases.* Means shall be provided for monitoring the reactor containment atmosphere, spaces containing components for recirculation of loss-of-coolant accident fluids, effluent discharge paths, and the plant environs for radioactivity that may be released from normal operations, including anticipated operational occurrences, and from postulated accidents.

II. QUALITY ASSURANCE CRITERIA FOR NUCLEAR POWER PLANTS—Effective Date: June 27, 1970

Introduction. Every applicant for a construction permit is required by the provisions of § 50.34 to include in its preliminary safety analysis report a description of the quality assurance program to be applied to the design, fabrication, construction, and testing of the structures, systems, and components of the facility. Every applicant for an operating license is required to include, in its final safety analysis report, information pertaining to the managerial and administrative controls to be used to assure safe operation. Nuclear power plants include structures, systems, and components that prevent or mitigate the consequences of postulated accidents that could cause undue risk to the health and safety of the public. This appendix establishes quality assurance requirements for the design, construction, and operation of those structures, systems, and components. The pertinent requirements of this appendix apply to all activities affecting the safety-related functions of those structures, systems, and components; these activities include designing, purchasing, fabricating, handling, shipping, storing, cleaning, erecting, installing, inspecting, testing, operating, maintaining, repairing, refueling, and modifying.

As used in this appendix, "quality assurance" comprises all those planned and systematic actions necessary to provide adequate confidence that a structure, system, or component will perform satisfactorily in service. Quality assurance includes quality control, which comprises those quality assurance actions related to the physical characteristics of a material, structure, component, or system which provide a means to control the quality of the material, structure, component, or system to predetermined requirements.

I. ORGANIZATION

The applicant[1] shall be responsible for the establishment and execution of the quality assurance program. The applicant may delegate to other organizations the work of establishing and executing the quality assurance program, or any part thereof, but shall retain responsibility therefor. The authority and duties of persons and organizations performing quality assurance functions shall be clearly established and delineated in writing. Such persons and organizations shall have sufficient authority and organizational freedom to identify quality problems; to initiate, recommend, or provide solutions; and to verify implementation of solutions. In general, assurance of quality requires management measures which provide that the individual or group assigned the responsibility for checking, auditing, inspecting, or otherwise verifying that an activity has been correctly performed is independent of the individual or group directly responsible for performing the specific activity.

II. QUALITY ASSURANCE PROGRAM

The applicant shall establish at the earliest practicable time, consistent with the schedule for accomplishing the activities, a quality assurance program which complies with the requirements of this appendix. This program shall be documented by written policies, procedures, or instructions and shall be carried out throughout plant life in accordance with those policies, procedures, or instructions. The applicant shall identify the structures, systems, and components to be covered by the quality assurance program and the major organizations participating in the program, together with the designated functions of these organizations. The quality assurance program shall provide control over activities affecting the quality of the identified structures, systems, and components, to an extent consistent with their importance to safety. Activities affecting quality shall be accomplished under suitably controlled conditions. Controlled conditions include the use of appropriate equipment; suitable environmental conditions for accomplishing the activity, such as adequate cleanness; and assurance that all prerequisites for the given activity have been satisfied. The program shall take into account the need for special controls, processes, test equipment, tools, and skills to attain the required quality, and the need for verification of quality by inspection and test. The program shall provide for indoctrination and training of personnel performing activities affecting quality as necessary to assure that suitable proficiency is achieved and maintained. The applicant shall regularly review the status and adequacy of the quality assurance program. Management of other organizations par-

[1] While the term "applicant" is used in these criteria, the requirements are, of course, applicable after such a person has received a license to construct and operate a nuclear power plant. These criteria will also be used for guidance in evaluating the adequacy of quality assurance programs in use by holders of construction permits and operating licenses.

ticipating in the quality assurance program shall regularly review the status and adequacy of that part of the quality assurance program which they are executing.

III. DESIGN CONTROL

Measures shall be established to assure that applicable regulatory requirements and the design basis, as defined in § 50.2 and as specified in the license application, for those structures, systems, and components to which this appendix applies are correctly translated into specifications, drawings, procedures, and instructions. These measures shall include provisions to assure that appropriate quality standards are specified and included in design documents and that deviations from such standards are controlled. Measures shall also be established for the selection and review for suitability of application of materials, parts, equipment, and processes that are essential to the safety-related functions of the structures, systems and components.

Measures shall be established for the identification and control of design interfaces and for coordination among participating design organizations. These measures shall include the establishment of procedures among participating design organizations for the review, approval, release, distribution, and revision of documents involving design interfaces.

The design control measures shall provide for verifying or checking the adequacy of design, such as by the performance of design reviews, by the use of alternate or simplified calculational methods, or by the performance of a suitable testing program. The verifying or checking process shall be performed by individuals or groups other than those who performed the original design, but who may be from the same organization. Where a test program is used to verify the adequacy of a specific design feature in lieu of other verifying or checking processes, it shall include suitable qualification testing of a prototype unit under the most adverse design conditions. Design control measures shall be applied to items such as the following: reactor physics, stress, thermal, hydraulic, and accident analyses; compatibility of materials; accessibility for inservice inspection, maintenance, and repair; and delineation of acceptance criteria for inspections and tests.

Design changes, including field changes, shall be subject to design control measures commensurate with those applied to the original design and be approved by the organization that performed the original design unless the applicant designates another responsible organization.

IV. PROCUREMENT DOCUMENT CONTROL

Measures shall be established to assure that applicable regulatory requirements, design bases, and other requirements which are necessary to assure adequate quality are suitably included or referenced in the documents for procurement of material, equipment, and services, whether purchased by the applicant or by its contractors or subcontractors. To the extent necessary, procurement documents shall require contractors or subcontractors to provide a quality assurance program consistent with the pertinent provisions of this appendix.

V. INSTRUCTIONS, PROCEDURES, AND DRAWINGS

Activities affecting quality shall be prescribed by documented instructions, procedures, or drawings, or a type appropriate to the circumstances and shall be accomplished in accordance with these instructions, procedures, or drawings. Instructions, procedures, or drawings shall include appropriate quantitative or qualitative acceptance criteria for determining that important activities have been satisfactorily accomplished.

VI. DOCUMENT CONTROL

Measures shall be established to control the issuance of documents, such as instructions, procedures, and drawings, including changes thereto, which prescribe all activities affecting quality. These measures shall assure that documents, including changes, are reviewed for adequacy and approved for release by authorized personnel and are distributed to and used at the location where the prescribed activity is performed. Changes to documents shall be reviewed and approved by the same organizations that performed the original review and approval unless the applicant designates another responsible organization.

VII. CONTROL OF PURCHASED MATERIAL, EQUIPMENT, AND SERVICES

Measures shall be established to assure that purchased material, equipment, and services, whether purchased directly or through contractors and subcontractors, conform to the procurement documents. These measures shall include provisions, as appropriate, for source evaluation and selection, objective evidence of quality furnished by the contractor or subcontractor, inspection at the contractor or subcontractor source, and examination of products upon delivery. Documentary evidence that material and equipment conform to the procurement requirements shall be available at the nuclear power plant site prior to installation or use of such material and equipment. This documentary evidence shall be retained at the nuclear power plant site and shall be sufficient to identify the specific requirements, such as codes, standards, or specifications, met by the purchased material and equipment. The effectiveness of the control of quality by contractors and subcontractors shall be assessed by the applicant or designee at intervals consistent with the importance, complexity, and quantity of the product or services.

VIII. IDENTIFICATION AND CONTROL OF MATERIALS, PARTS, AND COMPONENTS

Measures shall be established for the identification and control of materials, parts, and components, including partially fabricated assemblies. These measures shall assure that identification of the item is maintained by heat number, part number, serial number, or other appropriate means, either on the item or on records traceable to the item, as required throughout fabrication, erection, installation, and use of the item. These identification and control measures shall be designed to prevent the use of incorrect or defective material, parts, and components.

IX. CONTROL OF SPECIAL PROCESSES

Measures shall be established to assure that special processes, including welding, heat treating, and nondestructive testing, are controlled and accomplished by qualified personnel using qualified procedures in accordance with applicable codes, standards, specifications, criteria, and other special requirements.

X. INSPECTION

A program for inspection of activities affecting quality shall be established and executed by or for the organization performing the activity to verify conformance with the documented instructions, procedures, and drawings for accomplishing the activity. Such inspection shall be performed by individuals other than those who performed the activity being inspected. Examinations, measurements, or tests of material or products processed shall be performed for each work operation where necessary to assure quality. If inspection of processed material or products is impossible or disadvantageous, indirect control by monitoring processing methods, equipment, and personnel shall be provided. Both inspection and process monitoring shall be provided when control is inadequate without both. If mandatory inspection hold points, which require witnessing or inspecting by the applicant's designated representative and beyond which work shall not proceed without the consent of its designated representative are required, the specific hold points shall be indicated in appropriate documents.

XI. TEST CONTROL

A test program shall be established to assure that all testing required to demonstrate that structures, systems, and components will perform satisfactorily in service is identified and performed in accordance with written test procedures which incorporate the requirements and acceptance limits contained in applicable design documents. The test program shall include, as appropriate, proof tests prior to installation, preoperational tests, and operational tests during nuclear power plant operation, of structures, systems, and components. Test procedures shall include provisions for assuring that all prerequisites for the given test have been met, that adequate test instrumentation is available and used, and that the test is performed under suitable environmental conditions. Test results shall be documented and evaluated to assure that test requirements have been satisfied.

XII. CONTROL OF MEASURING AND TEST EQUIPMENT

Measures shall be established to assure that tools, gages, instruments, and other measuring and testing devices used in activities affecting quality are properly controlled, calibrated, and adjusted at specified periods to maintain accuracy within necessary limits.

XIII. HANDLING, STORAGE AND SHIPPING

Measures shall be established to control the handling, storage, shipping, cleaning and preservation of material and equipment in accordance with work and inspection instructions to prevent damage or deterioration. When necessary for particular products, special protective environments, such as inert gas atmosphere, specific moisture content levels, and temperature levels, shall be specified and provided.

XIV. INSPECTION, TEST, AND OPERATING STATUS

Measures shall be established to indicate, by the use of markings such as stamps, tags, labels, routing cards, or other suitable means, the status of inspections and tests performed upon individual items of the nuclear power plant. These measures shall provide for the identification of items which have satisfactorily passed required inspections and tests, where necessary to preclude inadvertent by-passing of such inspections and tests. Measures shall also be established for indicating the operating status of structures, systems, and components of the nuclear power plant, such as by tagging valves and switches, to prevent inadvertent operation.

XV. NONCONFORMING MATERIALS, PARTS, OR COMPONENTS

Measures shall be established to control materials, parts, or components which do not conform to requirements in order to prevent their inadvertent use or installation. These measures shall include, as appropriate, procedures for identification, documentation, segregation, disposition, and notification to affected organizations. Nonconforming items shall be reviewed and accepted, rejected, repaired or reworked in accordance with documented procedures.

XVI. CORRECTIVE ACTION

Measures shall be established to assure that conditions adverse to quality, such as failures, malfunctions, deficiencies, deviations, defective material and equipment, and nonconformances are promptly identified and corrected. In the case of significant conditions adverse to quality, the measures shall assure that the cause of the condition is determined and corrective action taken to preclude repetition. The identification of the significant condition adverse to quality, the cause of the condition, and the corrective action taken shall be documented and reported to appropriate levels of management.

XVII. QUALITY ASSURANCE RECORDS

Sufficient records shall be maintained to furnish evidence of activities affecting quality. The records shall include at least the following: Operating logs and the results of reviews, inspections, tests, audits, monitoring of work performance, and materials analyses. The records shall also include closely related data such as qualifications of personnel, procedures, and equipment. Inspection and test records shall, as a minimum, identify the inspector or data recorder, the type of observation, the results, the acceptability, and the action taken in connection with any deficiencies noted. Records shall be identifiable and retrievable. Consistent with applicable regulatory requirements, the applicant shall establish requirements concerning record retention, such as duration, location, and assigned responsibility.

XVIII. AUDITS

A comprehensive system of planned and periodic audits shall be carried out to verify compliance with all aspects of the quality assurance program and to determine the effectiveness of the program. The audits shall be performed in accordance with the written procedures or check lists by appropriately trained personnel not having direct responsibilities in the areas being audited. Audit results shall be documented and reviewed by management having responsibility in the area audited. Followup action, including reaudit of deficient areas, shall be taken where indicated.

III. CODES AND STANDARDS FOR NUCLEAR POWER PLANTS — Effective Date: July 12, 1971

Extract from Chapter 10, Part 50 of the Code of Federal Regulations.

Section 50.2 DEFINITIONS.

As used in this part:

* * *

(v) "Reactor coolant pressure boundary" means all those pressure-containing components of boiling and pressurized water-cooled nuclear power reactors, such as pressure vessels, piping, pumps, and valves, which are:

(1) Part of the reactor coolant system; or

(2) Connected to the reactor coolant system, up to and including any and all of the following:

(i) The outermost containing isolation valve in system piping which penetrates primary reactor containment,

(ii) The second of two valves normally closed during normal reactor operation in system piping which does not penetrate primary reactor containment,

(iii) The reactor coolant system safety and relief valves.

For nuclear power reactors of the direct cycle boiling water type, the reactor coolant system extends to and includes the outermost containment isolation valve in the main steam and feedwater piping.

2. Paragraph (c) of Section 50.55 is amended to read as follows:

Section 50.55 CONDITIONS OF CONSTRUCTION PERMITS

Each construction permit shall be subject to the following terms and conditions:

* * *

(c) Except as modified by this section and Section 50.55a, the construction permit shall be subject to the same conditions to which a license is subject.

3. A new Section 50.55a is added to 10 CFR Part 50 to read as follows:

Section 50.55a CODES AND STANDARDS

Each construction permit for a utilization facility shall be subject to the following conditions, in addition to those specified in Section 50.55:

(a) Structures, systems, and components shall be designated, fabricated, erected, constructed, tested, and inspected to quality standards commensurate with the importance of the safety function to be performed.

(b) As a minimum, the systems and components of boiling and pressurized water-cooled nuclear power reactors specified in paragraphs (c), (d), (e), (f), and (g) of this section shall meet the requirements described in those paragraphs, except that the American Society of Mechanical Engineers (hereinafter referred to as ASME) Code N-symbol need not be applied, and the protection systems of nuclear power reactors of all types shall meet the requirements described in paragraph (h) of this section, except as authorized by the Commission upon demonstration by the applicant for or holder of a construction permit that:

(1) Design, fabrication, installation, testing, or inspection of the specified system or component is, to the maximum extent practical, in accordance with generally recognized codes and standards, and in compliance with the requirements described in paragraphs (c) through (h) of this section or portions thereof would result in hardships or unusual difficulties without a compensating increase in the level of quality and safety; or

(2) Proposed alternatives to the described requirements or portions thereof will provide an acceptable level of quality and safety. For example, the use of inspection or survey systems other than those required by the specified ASME Codes and Addenda may be authorized under this subparagraph provided that an acceptable level of quality and safety in design, fabrication, installation, and testing is achieved.

(c) Pressure vessels:

(1) For construction permits issued before January 1, 1971, for reactors not licensed for operation, pressure vessels which are part of the reactor coolant pressure boundary[1] shall meet the requirements for Class A vessels set forth in Section III of the ASME Boiler and Pressure Vessel Code, applicable Code Cases, and Addenda[2] "in effect"[3] on the date of order[4] of the vessel. The pressure vessels may meet the requirements set forth in editions of this Code, applicable Code Cases, and Addenda which have become effective after the date of vessel order, unless the Commission has published a notice in the FEDERAL REGISTER that compliance with such requirements or any part thereof is unacceptable for such pressure vessels.

(2) For construction permits issued on or after January 1, 1971, pressure vessels which are part of the reactor coolant pressure boundary[1] shall meet the requirements for Class A vessels set forth in Section III of the ASME Boiler and Pressure Vessel Code and Addenda[5] "in effect"[3] on the date of order[4] of the pressure vessel, unless the Commission has published a notice in the FEDERAL REGISTER that compliance with such requirements or any part thereof is unacceptable or unnecessary for such pressure vessels: PROVIDED, however, that if the pressure vessel is ordered more than 18 months prior to the date of issuance of the construction permit, compliance with the requirements for Class A vessels set forth in Section III of the ASME Boiler and Pressure Vessel Code and Addenda in effect 18 months prior to the date of issuance of the construction permit is required. The pressure vessels may meet the requirements set forth in editions of this Code and Addenda which have become effective after the date of vessel order or after 18 months prior to the date of issuance of the construction permit unless the Commission has published a notice in the FEDERAL REGISTER that compliance with such requirements or any part thereof is unacceptable for such pressure vessels.

(d) Piping:

(1) For construction permits issued before January 1, 1971, for reactors not licensed for operation, piping which is part of the reactor coolant pressure boundary[1] shall meet the requirements set forth in:

(i) The American Standard Code for Pressure Piping (ASA B31.1), Addenda and applicable Code Cases[2] or the USA Standard Code for Pressure Piping (USAS B31.1.0), Addenda, and applicable Code Cases[2] or the Class I Section of the USA Standard Code for Pressure Piping (USAS B31.7)[2] "in effect"[3] on the date of order of the piping and

(ii) The nondestructive examination and acceptance standards of ASA B31.1 Code Cases N7, N9 and N10, except that the acceptance standards of Class I piping of the USA Standard Code for Pressure Piping (USAS B31.7) may be applied.

The piping may meet the requirements set forth in editions of ASA B31.1, USAS B31.1.0 and USAS B31.7, Addenda, and Code Cases which became effective after the date of order of the piping unless the Commission has published a notice in the FEDERAL REGISTER that compliance with such re-

[1] Components which are connected to the reactor coolant system and are part of the reactor coolant pressure boundary defined in Section 50.2 (v) need not meet these requirements, provided:

(a) In the event of postulated failure of the component during normal reactor operation, the reactor can be shut down and cooled down in an orderly manner, assuming makeup is provided by the reactor coolant makeup system only, or

(b) the component is or can be isolated from the reactor coolant system by two valves (both closed, both open, or one closed and the other open). Each open valve must be capable of automatic actuation and, assuming the other valve is open, its closure time must be such that, in the event of postulated failure of the component during normal reactor operation, each valve remains operable and the reactor can be shut down and cooled down in an orderly manner, assuming makeup is provided by the reactor coolant makeup system only.

[2] Copies may be obtained from the American Society of Mechanical Engineers, United Engineering Center, 345 East 47th Street, New York, N.Y. 10017. Copies are available for inspection at the Commission's Public Document Room, 1717 H. Street NW, Washington, D.C.

[3] ASME and United States of America Standard Code Addenda are considered "in effect" 6 months after their date of issuance.

[4] The Code issue applicable to a component is governed by the order of contract date for the component, not the contract date for the nuclear energy system.

[5] The use of specific Code Cases may be authorized by the Commission upon request pursuant to Section 50.55a(b) (2).

quirements or any part thereof is unacceptable for such piping.

(2) For construction permits issued on or after January 1, 1971, piping which is part of the reactor coolant pressure boundary[1] shall meet the requirements for Class I piping set forth in the USA Standard Code for Pressure Piping (USAS B31.7) and Addenda[5] "in effect"[3] on the date of order[4] of the piping and the requirements applicable to piping of articles 1 and 8 of Section III of the ASME Boiler and Pressure Vessel Code and Addenda[5] "in effect" on the date of order of the piping unless the Commission has published a notice in the FEDERAL REGISTER that compliance with such requirements or any part thereof is unacceptable or unnecessary for such piping: PROVIDED, however, that if the piping is ordered more than 6 months prior to the date of issuance of the construction permit, compliance with the requirements for Class I piping set forth in USAS B31.7 and Addenda in effect 6 months prior to the date of issuance of the construction permit is required. The piping may meet the requirements set forth in editions of these Codes and Addenda[5] which have become effective after the date of piping order or after 6 months prior to the date of issuance of the construction permit, unless the Commission has published a notice in the FEDERAL REGISTER that compliance with such requirements or any part thereof is unacceptable for such piping.

(e) Pumps:

(1) For construction permits issued before January 1, 1971, for reactors not licensed for operation, pumps which are part of the reactor coolant pressure boundary shall meet —

(i) The requirements for Class I pumps set forth in the Draft ASME Code for Pumps and Valves for Nuclear Power, Addenda, and Code Cases[2] "in effect"[3] on the date of order[4] of the pumps, or

(ii) The nondestructive examination and acceptance standards set forth in ASA B31.1 Code Cases N7, N9, and N10, except that the acceptance standards for Class I pumps set forth in the Draft ASME Code for Pumps and Valves for Nuclear Power and Addenda "in effect" on the date of order of the pumps may be applied.

The pumps may meet the requirements set forth in editions of the Draft ASME Code for Pumps and Valves for Nuclear Power, Addenda, and Code Cases which became effective after the date of order of the pumps, unless the Commission has published a notice in the FEDERAL REGISTER that compliance with such requirements or any part thereof is unacceptable for such pumps.

(2) For construction permits issued on or after January 1, 1971, pumps which are part of the reactor coolant pressure boundary[1] shall meet the requirements for Class I pumps set forth in the Draft ASME Code for Pumps and Valves for Nuclear Power and Addenda[5] "in effect"[3] on the date of order[4] of the pumps and the requirements applicable to pumps set forth in Articles 1 and 8 of Section III of the ASME Boiler and Pressure Vessel Code and Addenda "in effect" on the date of order of the pumps, unless the Commission has published a notice in the FEDERAL REGISTER that compliance with such requirements or any part thereof is unacceptable or unnecessary for such pumps: PROVIDED, however, that if the pumps are ordered more than 12 months prior to the date of issuance of the construction permit, compliance with the requirements for Class I pumps set forth in the Draft ASME Code for Pumps and Valves for Nuclear Power and Addenda[5] and the requirements applicable to pumps set forth in Articles 1 and 8 of Section III of the ASME Boiler and Pressure Vessel Code and Addenda in effect 12 months prior to the date of issuance of the construction permit, unless the Commission has published a notice in the FEDERAL REGISTER that compliance with such requirements of any part thereof is unacceptable for such pumps.

(f) Valves:

(1) For construction permits issued before January 1, 1971, for reactors not licensed for operation, valves which are part of the reactor coolant pressure boundary[1] shall meet the requirements set forth in

(i) The American Standard Code for Pressure Piping (ASA B31.1), Addenda, and applicable Code Cases, or the USA Standard Code for Pressure Piping (USAS B31.1.0), Addenda, and applicable Code Cases, "in effect"[3] on the date of order[4] of the valves or the Class I Section of the Draft ASME Code for Pumps and Valves for Nuclear Power,[2] Addenda, and Code Cases "in effect" on the date of order of the valves or

(ii) The nondestructive examination and acceptance standards of ASA B31.1 Code Cases N2, N7, N9 and N10, except that the acceptance standards for Class I valves set forth in the Draft ASME Code for Pumps and Valves for Nuclear Power and Addenda "in effect" on the date of order of the valves may be applied.

The valves may meet the requirements set forth in editions of ASA B31.1, USAS B31.1.0, and the Draft ASME Code for Pumps and Valves for Nuclear Power, Addenda, and Code Cases, which became effective after the date of order of the valves unless the Commission has published a notice in the FEDERAL REGISTER that compliance with such requirements or any part thereof is unacceptable for such valves.

(2) For construction permits issued on or after January 1, 1971, valves which are part of the reactor coolant pressure boundary[1] shall meet the requirements for Class I valves set forth in the Draft ASME Code for Pumps and Valves for Nuclear Power and Addenda[5] "in effect"[3] on the date of order[4] of the

valves and the requirements applicable to valves set forth in Articles 1 and 8 of Section III of the ASME Boiler and Pressure Vessel Code and "Addenda,"[5] in effect on the date of order of the valves, unless the Commission has published a notice in the FEDERAL REGISTER that compliance with such requirements or any part thereof is unacceptable or unnecessary for such valves: PROVIDED, however, that if the valves are ordered more than 12 months prior to the date of issuance of the construction permit, compliance with the requirements for Class I valves set forth in the Draft ASME Code for Pumps and Valves for Nuclear Power and Addenda[5] and the requirements applicable to valves set forth in Articles 1 and 8 of Section III of the ASME Boiler and Pressure Vessel Code and Addenda in effect 12 months prior to the date of issuance of the construction permit is required. The valves may meet the requirements set forth in editions of these Codes or Addenda which have become effective after the date of valve order or after 12 months prior to the date of issuance of the construction permit, unless the Commission has published a notice in the FEDERAL REGISTER that compliance with such requirements or any part thereof is unacceptable for such valves.

(g) Inservice inspection requirements: For construction permits issued on or after January 1, 1971, systems and components shall meet the requirements set forth in Section XI of the ASME Boiler and Pressure Vessel Code and Addenda[2, 5] in effect 6 months prior to the date of issuance of the construction permit, unless the Commission has published a notice in the FEDERAL REGISTER that compliance with such requirements or any part thereof is unacceptable or unnecessary for such systems and components. Systems and components may meet the requirements set forth in editions of this Code and Addenda which have become effective 6 months prior to the date of issuance of the construction permit, unless the Commission has published a notice in the FEDERAL REGISTER that compliance with such requirements or any part thereof is unacceptable for such systems and components.

(h) Protection systems: For construction permits issued after January 1, 1971, protection systems shall meet the requirements set forth in the Institute of Electrical and Electronics Engineers Criteria for Nuclear Power Plant Protection Systems (IEEE 279) in effect[6] 12 months prior to the date of issuance of the construction permit, unless the Commission has published a notice in the FEDERAL REGISTER that compliance with such requirements or any part thereof is unacceptable or unnecessary for such protection systems. Protection systems may meet the requirements set forth in later editions or revisions of IEEE 279 which have become effective after 12 months prior to the date of issuance of the construction permit, unless the Commission has published a notice in the FEDERAL REGISTER that compliance with such requirements or any part thereof is unacceptable for such protection systems.

(i) Power reactors for which a notice of hearing on an application for a provisional construction permit has been published on or before December 31, 1970, may meet the requirements of paragraphs (c) (1), (d) (1), (e) (1), and (f) (1) of this section instead of paragraphs (c) (2), (d) (2), (e) (2) and (f) (2) of this section, respectively.

[6] For purposes of this regulation, the proposed IEEE 279 became "in effect" on Aug. 30, 1968, and future IEEE 279 editions or revisions will become "in effect" on the effective date printed on the document. Copies may be obtained from the Institute of Electrical and Electronics Engineers, United Engineering Center, 345 East 47th Street, New York, N.Y. 10017. A copy is available for inspection at the Commission's Public Document Room, 1717 H. Street, N.W., Washington, D.C.

IV. AREAS REQUIRING SERVICE INSPECTION IN NUCLEAR REACTOR COOLANT SYSTEMS*

EXAMINATION CATEGORIES

AREAS SUBJECT TO EXAMINATIONS — EXTENT AND FREQUENCY OF EXAMINATIONS

A

PRESSURE-CONTAINING WELDS IN REACTOR VESSEL BELT-LINE REGION

The areas subject to examination shall include the shell longitudinal and circumferential welds in the reactor vessel wall opposite the length of the reactor vessel thermal shield, where used, or opposite the effective length of reactor fuel where thermal shield is not used.

The area to be examined shall include the weld metal and base metal for one plate thickness beyond the edge of weld.

In addition, material which has been repaired by welding shall be subject to examination when the repair depth exceeds 10 percent of the nominal wall thickness and when the repair area is located in the vessel wall opposite the length of the reactor vessel thermal shield, where used, or opposite the effective length of reactor fuel where the thermal shield is not used. The examination must completely cover the repair; if the location of the repair is not positively and accurately known, then the individual shell plate, forging, or shell course positively known to contain the repair shall be subject to examination.

The individual examination of the shell longitudinal and circumferential welds may be performed at or near the end of each inspection interval, and shall cover at least 10 percent of the length of each longitudinal weld, and 5 percent of the length of each circumferential weld, with the minimum length of weld equal to one wall thickness.

When the longitudinal and circumferential weld have received an exposure to neutron fluence in excess of 10^{19} nvt (E_n of 1 Mev or above), the length of weld in the high fluence region to be examined shall be increased to, at least, 50 percent.

At the end of the first inspection interval when the repaired areas have received an exposure to neutron flux in excess of 10^{19} nvt (E_n of 1 Mev or above) during the first inspection interval, in any case no later than at the end of the second inspection interval; the extent of the examination shall be increased to at least 50 percent of the repaired areas. Examination shall be continued in the succeeding intervals in accordance with IS-243.

B

PRESSURE-CONTAINING WELDS IN VESSELS

The areas subject to examination shall include the longitudinal and circumferential welds in the vessel shell and meridional and circumferential welds in vessel heads.

The area to be examined shall include weld metal and base metal for one place thickness beyond the edge of weld.

The examinations performed during each inspection interval shall cover at least 10 percent of the length of each longitudinal shell weld and meridional head weld, and 5 percent of the length of each circumferential shell weld and head weld.

For welds on the reactor vessel, the individual examinations may be performed at or near the end of each inspection interval.

C

PRESSURE-CONTAINING WELDS, VESSEL-TO-FLANGE, AND HEAD-TO-FLANGE

The areas subject to examination shall include the vessel-to-flange and head-to-flange welds.

The examination of the vessel-to-flange weld may be conducted from the bolting face of the flange.

The individual examinations performed during each inspection interval shall cumulatively cover 100 percent of each circumferential weld.

The number and extent of areas examined during each individual examination of a circumferential weld shall provide a representative sampling of the entire weld.

* This appendix is a reprint of Table IS-251 from Section XI of the ASME Code Inservice Inspection, 1971 edition, and it includes the Summer 1971 addendum.

EXAMINATION CATEGORIES

AREAS SUBJECT TO EXAMINATIONS — EXTENT AND FREQUENCY OF EXAMINATIONS

D

PRESSURE-CONTAINING NOZZLES IN VESSELS

The areas subject to examination shall include the nozzle-attachment weld to vessel and the integral extension of nozzle inside vessel. Excluded from this category are the nozzle-to-connecting pipe circumferential welds and pressure-containing seal welds.

The extent of examination of each nozzle shall cover 100 percent of the nozzle-to-vessel weld, and 100 percent of the inner radius section of the nozzle-to-vessel juncture. All the nozzles of the vessel shall be examined during each inspection interval.

The number of nozzles examined during each inspection interval shall be the total number in a vessel subject to examination divided approximately equally among the expected number of examinations during the inspection interval. The resulting number shall include a representative sampling among each group of nozzles of comparable size.

E-1

PRESSURE-CONTAINING WELDS IN VESSEL PENETRATIONS

The areas subject to examination shall include those pressure-containing welds of reactor control rod drive penetrations[1] in reactor vessel heads, in the control rod drive housings, at vessel instrumentation connections and at heater connections in pressurizer vessels, among which a weld failure in any single penetration results in conditions that fail to meet the exclusion criterion of IS-121.

The examinations performed during each inspection interval shall cumulatively cover at least 25 percent of the vessel penetrations, in each group of penetrations of comparable size.

E-2

PRESSURE-CONTAINING WELDS IN VESSEL PENETRATIONS

Penetrations excluded from the examination by IS-121 shall be given a visual examination for evidence of leaking at the time of the system hydrostatic test as required by IS-520.

The extent of examination shall include twenty-five percent of the penetrations in the vessel during the inspection interval.

F

PRESSURE-CONTAINING DISSIMILAR METAL WELDS

The areas subject to examination shall include dissimilar metal welds (safe-end welds) between combinations of ferritic steels (carbon, low alloy and high tensile steels) and austenitic stainless steels, nickel-chromium-iron alloys, nickel-iron-chromium alloys and nickel-copper alloys.

The area to be examined shall include the base material for, at least, one wall thickness beyond the edge of weld.

The individual examinations performed during each inspection shall cover 100 percent of the circumference of the safe-end welds. All of the safe-end welds shall be examined during the inspection interval.

[1] The pressure containing welds of reactor control rod drive penetrations that meet the exclusion criterion of IS-121 are required to be visually examined, rather than volumetrically examined, provided analysis of the rod ejection accident complies with the AEC General Design Criteria (Appendix I).

EXAMINATION CATEGORIES

AREAS SUBJECT TO EXAMINATIONS	EXTENT AND FREQUENCY OF EXAMINATIONS

G-1
PRESSURE-RETAINING BOLTING

Bolting 2 inches and larger in diameter shall be examined either in place under tension, or when the bolting is removed, or when the bolted connection is disassembled.

The bolting and areas to be examined shall include the studs, nuts, bushings, threads in base material and the flange ligaments between threaded stud holes.

The examinations performed during each inspection interval shall cumulatively cover the entire number of bolts, studs and nuts.

Examinations of bushings, threads and ligaments in base material of flanges may be performed from a face of the flange and are required to be examined only when the connection is disassembled.

G-2
PRESSURE-RETAINING BOLTING

Bolting below 2 inches in diameter shall be visually examined either in place, if the bolted connection is not disassembled during the inspection interval, or whenever the bolted connection is disassembled.

The bolting to be examined shall include studs and nuts.

Excluded from examination is bolting of a single connection whose failure results in conditions that satisfy the exclusion criterion of IS-121.

The extent of visual examination performed during each inspection interval shall cumulatively cover the entire number of bolts and studs and nuts.

H
VESSEL EXTERNAL SUPPORTS

The areas subject to examination shall include the integrally welded vessel external support attachment (vessel support skirts) which includes the welds to the vessel and the base metal beneath the weld zone and along the support attachment member for a distance of two base-metal thicknesses.

In the case of nozzle type supports consisting of pads integral with the nozzle, the area to be examined shall be the weld connection between the nozzle and the vessel shell as described in Examination Category *D*.

The examination performed during each inspection interval in the case of vessel support skirts shall be, at least, 10 percent of the lineal feet of welding to the vessel.

In the case of nozzles with supports, all of the nozzles with supports shall be examined in accordance with Category *D*.

I-1
INTERIOR CLAD SURFACES OF REACTOR VESSELS

In the case of the reactor vessel, the areas to be examined shall include at least 6 patches (each 36 square inches) evenly distributed, in the closure head, and 6 patches (each 36 square inches), evenly distributed in accessible sections of vessel shell.

Cladding examination patches shall be prepared to facilitate examination, and shall be located in areas with identifiable indications, insofar as practical.

The examinations performed during each inspection interval shall include the patch areas subject to examination in the reactor vessel shell and in the closure head.

EXAMINATION CATEGORIES

AREAS SUBJECT TO EXAMINATIONS — EXTENT AND FREQUENCY OF EXAMINATIONS

I-2

INTERIOR-CLAD SURFACES OF VESSELS OTHER THAN REACTOR VESSELS

In the case of other than reactor vessels, the areas to be examined shall include at least 1 patch (36 square inches) near each manway in the primary side of the vessel.

Cladding examination patches shall be prepared to facilitate examination, and shall be located in areas with identifiable indications, insofar as practical.

The examination of the patches may be performed at or near the end of the inspection interval.

J-1

PRESSURE-CONTAINING WELDS IN PIPING

The areas subject to examination shall include longitudinal and circumferential welds in piping and the base metal for one wall thickness on both edges of these welds. Longitudinal welds shall be examined for at least a length of one-foot as measured from the intersection with the circumferential weld selected for examination.

Pipe branch connections shall be subject to examinations which shall include all the weld metal, at least the base metal for one pipe wall thickness beyond the edge of the weld (on the main pipe run), and at least the base metal in the branch pipe within the limit of compensation defined in the Nuclear Power Piping Code.

The examinations performed during each inspection interval shall cumulatively cover: (a) 25 percent of the total number of circumferential joints (and the adjoining sections of longitudinal weld) selectively distributed within the system boundary, and (b) 25 percent of the pipe branch connection welded joints.

J-2

PRESSURE-CONTAINING WELDS IN PIPING

Welds in piping (including pipe branch connections) which are excluded from the examinations by IS-121 shall be visually examined whenever the system boundary is subjected to a hydrostatic test, prior to each plant startup following refueling, and at the time of the system hydrostatic test as required by IS-520.

The extent of examination shall cumulatively cover 100 percent of the welds in piping (including pipe branch connections) during the inspection interval.

K-1

SUPPORT MEMBERS AND STRUCTURES FOR PIPING, VALVES AND PUMPS

The support members and structures subject to examination shall include the integrally welded external support attachments which include the welds to the pressure-containing boundary and the base metal beneath the weld zone and along the support attachment member for a distance of two base metal thicknesses.

Excluded from examination are supports whose failure results in conditions that satisfy the exclusion criteria of IS-121.

The examinations required to be performed during each inspection interval shall cumulatively cover 25 percent of the total number of integrally welded supports within the system boundary.

EXAMINATION CATEGORIES

AREAS SUBJECT TO EXAMINATIONS — EXTENT AND FREQUENCY OF EXAMINATIONS

K-2

SUPPORT MEMBERS AND STRUCTURES FOR PIPING, VALVES AND PUMPS

The support members and structures subject to examination shall include those supports for piping, valves and pumps within the system boundary, whose structural integrity is relied upon to withstand the design loads and seismic-induced displacements.

The support settings of constant and variable spring type hangers, snubbers and shock absorbers shall be inspected to verify proper distribution of design loads among the associated support components.

The examination performed during each inspection interval shall cumulatively cover all support members and structures.

L-1

PRESSURE-CONTAINING WELDS IN PUMP CASINGS

The areas subject to examination on pump casings shall include the welds in the pressure-containing boundary.

The area to be examined shall include the weld metal and the base metal for one wall thickness beyond the edge of the weld.

The examinations performed during each inspection interval shall include 100 percent of the pressure-containing welds in, at least, one pump (with pressure-containing welds) in each group of pumps performing similar functions in the system (e.g., recirculating coolant pumps).

The examinations of pump casings may be performed at or near the end of the inspection interval.

L-2

PUMP CASINGS

The areas subject to examination on pump casings shall include the internal pressure boundary surfaces.

The internal surfaces of one disassembled pump (with or without pressure-containing welds) in each of the group of pumps performing similar functions in the system shall be visually examined during each inspection interval. The internal examinations may be performed on the same pump selected for the volumetric examinations of pressure-containing welds.

The examinations of pump casings may be performed at or near the end of the inspection interval.

M-1

PRESSURE-CONTAINING WELDS IN VALVE BODIES

The areas subject to examination shall include the pressure-containing welds in valve bodies three inches and over, in nominal pipe size.

The area to be examined shall include the weld metal and the base metal for one wall thickness beyond the edge of the weld.

The examination performed during each inspection interval shall include 100 percent of the pressure-containing welds in at least one valve (with pressure-containing welds) within each group of valves which are of the same constructional design, (e.g., globe, gate, or check valve), manufacturing method, and manufacturer and which are performing similar functions in the system (e.g., containment isolation, system overpressure protection, etc.).

The examination on valve bodies may be performed at or near the end of each inspection interval.

EXAMINATION CATEGORIES

AREAS SUBJECT TO EXAMINATIONS / EXTENT AND FREQUENCY OF EXAMINATIONS

M-2

VALVE BODIES

The areas subject to examination shall include the internal pressure boundary surfaces, on valves three inches and over, in nominal pipe size.

The internal surfaces of one disassembled valve (with or without pressure-containing welds) in each of these groups of valves of the same constructional design, manufacturing method, manufacturer and performing similar functions in the system, shall be visually examined during each inspection interval. The internal examination may be performed on the same valve selected for the volumetric examination of pressure-containing welds.

The examination on valve bodies may be performed at or near the end of each inspection interval.

N

INTERIOR SURFACES AND INTERNAL COMPONENTS OF REACTOR VESSELS

The areas subject to examination shall include those areas which are made accessible for examination by the removal of components during normal refueling outages, such as the interior surfaces of the reactor vessel, the internal components, the internal support attachments welded to the vessel wall, and the space below the reactor core and above the bottom head.

The examinations of the interior surfaces of the vessel, the internal components and the space below the reactor core shall be performed at the first refueling outage and during subsequent refueling outages at approximately three year intervals.

The examination of internal support attachments welded to the vessel wall whose failure may adversely affect core integrity shall be examined at least once during each inspection interval.

Visual examination shall cover additional selected points throughout the vessel to provide a reasonably representative sampling of the condition of the cladding.

Where access to the space below the reactor core during normal refueling outages precludes inspection of this space, at least one examination, at or near the end of each inspection interval, shall be conducted under conditions which enable inspection. Such examinations may require removal of reactor internal components beyond those removed during normal refueling outages.

List of Abbreviations of Terms and Organizations

AC alternating current
ACI American Concrete Institute
ACRS Advisory Committee on Reactor Safeguards
AEC Atomic Energy Commission
ANS American Nuclear Society
ANSI American National Standards Institute
API American Petroleum Institute
ASA American Standards Association (now ANSI)
ASCE American Society of Civil Engineers
ASME American Society of Mechanical Engineers
ASME Code ASME Boiler and Pressure Vessel Code
ASTM American Society for Testing and Materials
AWWA American Water Works Association
Btu British thermal unit
BWR boiling water reactor
CCS component cooling system
CFR Code of Federal Regulations
COND condensate
CVCS chemical and volume control system
DBE design basis earthquake
DC direct current power
DCH drain collection header
D of C Division of Compliance of the AEC
DM demineralized water
DRL Division of Reactor Licensing of the AEC
FAI fail as is
FC flow controller
FDW feedwater
FIT flow indicator transmitter
FRA flow recorder and alarm
FRC flow recorder controller
F.S. far side
FSAR Final Safety Analysis Report
FT flow transmitter
GA gas analysis
GCH gas collection header
HEPA high efficiency particulate
HP high pressure
HPCS high pressure core spray
HS hand switch
HTGR high temperature gas cooled reactor
HVAC heating, ventilating, and air conditioning
LC level controller
LI level indicator
LIC level indicator controller
LP low pressure
LPCS low pressure core spray
MSS Manufacturers Standardization Society
MWD megawatt day
MWe megawatt electric
MWt megawatt thermal
N_2 nitrogen
NC neutron controller
NDT nil ductility transition temperature
NPSH net positive suction head
N.S. near side
NSSS nuclear steam supply system
NT neutron transmitter
OBE operating basis earthquake
O.D. outside diameter
PA pressure alarm
PAHL pressure alarm, high limit
PC pressure controller
PCRV prestressed concrete reactor vessel
PI pressure indicator
PIC pressure indicator controller
ppm parts per million
PS pressure switch
PSAR Preliminary Safety Analysis Report
PSH pressure switch, high pressure
psi pounds per square inch
psia pounds per square inch absolute
psig pounds per square inch gage
PT pressure transmitter
PWR pressurized water reactor
QA quality assurance
QC quality control
R Roentgen
RCIC reactor core isolation cooling
RCS reactor coolant system
RDH radioactive drain header
RHR residual heat removal
RI radiation indicator
RO restriction orifice
RRC radiation recorder controller
RWCU reactor water cleanup
S solenoid or strainer
SC speed controller
SIAS safety injection actuation signal
ST speed transmitter
TC temperature controller
TEMA Tubular Exchanger Manufacturers Association
TFE tetrafluoroethylene, a form of Teflon
TSH temperature switch, high temperature
TT temperature transmitter
U-233 isotope 233 of uranium
U-235 isotope 235 of uranium
USAS United States of America Standards Institute (now ANSI)

Glossary

absolute filter—An air filter with an efficiency of 99.3 percent for removal of particles down to 0.3 microns.

accumulator—A pressurized vessel that supplies fluid for driving a control rod drive in an emergency.

aggregate—The stone or gravel constituent of concrete.

air lock—A component used for personnel access to a containment structure with two sets of doors interlocked so that there can never be a through path from inside containment to the outside atmosphere.

ambient temperature—Room temperature.

anion—A negative ion such as Cl^-. It travels to the anode.

background activity level—The natural activity level anywhere on earth due to cosmic rays and radioactive elements in the earth.

baseline inspection—Inspection performed just before plant startup to determine the initial condition of the primary coolant pressure boundary. It is used as a reference of comparison for future inspections.

base loaded—Keeping a power station continuously loaded at the maximum load because it is one of the lowest cost power producers on the system.

bed—The term used to describe the mass of material in a component needed to accomplish an absorption or adsorption process such as a "resin bed," "charcoal bed," or "dryer bed."

bistable—An electric, electronic, or mechanical device with two stable positions, that changes from one position to the other when a control signal passes through a predetermined absolute value. In a nuclear power plant safety system the position change of interest occurs with decreasing signal value to provide a "fail safe" instrument channel. The position change actuates alarms and initiates protective actions.

buttering technique—The process of coating one metal with a second metal by welding so that a weld end preparation can be machined in the second metal.

cation—A positive ion such as Na^+. It travels to the cathode.

chain reaction—The process by which at least one neutron released by a fissioning nucleus strikes and fissions a second nucleus so that the fissioning reaction continues.

cladding, fuel—The term used to describe any material that encloses nuclear fuel. In a water cooled power reactor this is the fuel rod tube.

closed loop cooling system—A closed loop of deionized water interposed between radioactive systems and the environment as an added protection in the event of a leak of the radioactive fluid.

collet piston—The name used for the piston on the end of the BWR control rod drive latch. (See Fig. 3-9.)

component—Any machine, vessel, valve, etc., used in the power station.

composite pressure vessel—A vessel whose pressure retaining walls are comprised of two or more materials, as concrete lined with steel plate.

condensate—Condensed water vapor. The water vapor originates in a nuclear plant in a boric acid or radioactive waste evaporator. The same term is used for the condensed steam leaving the turbine condenser.

containment—The structure used to control and confine dispersion of radioactive fluids after an incident involving release of activity from the core.

containment barrier—The boundary that defines the volume used for containment. It usually has a design temperature and pressure associated with it.

containment building—See "containment."

control rod cluster—A group of individual control rods assembled into a single assembly and controlled by one drive package.

control rod drive package—The device used to provide the power for moving control rods. One package may drive a single control rod as in a BWR, or a cluster as in a PWR.

control rod, reactor—A mechanical device with a fixed geometric shape.

control rod, regulating rods—The control rod or group of rods that moves under the control of the automatic power control system to follow the variations of the load demand on the power station.

control rod, safety rods—The groups of rods that shut down the reactor in an emergency.

control rod, stepping rod drive—A drive package that moves a control rod in fixed incremental distances such as one or two centimeters.

core—The complete assemblage of fuel elements inside the reactor vessel that produces thermal energy.

core shroud—A cylindrical structural member inside the reactor vessel that surrounds the core. It adds structural stability to the reactor internals and assists in control of coolant flow paths. It is sometimes called the "core barrel." (See Figs. 2-4 and 3-2.)

cryogenic distillation—Distillation processes carried out at cryogenic temperatures to separate such fluids as krypton and xenon.

cryogenic temperature—Subzero temperature commonly assumed to be 100 or more degrees below zero.

curie—3.7 x 10^{10} nucleus disintegrations per second.

cylinder bank—The group of cylinders in a multicylinder reciprocating charging pump.

damping factor—The factor by which seismic loads are reduced by passing through the structure up to the component in question.

demineralizer—A component in which positive or negative ions in solution are removed by adsorption on a synthetic resin and an H^+ or OH^- ion is added to the solution to replace the ion removed.

design accident—A hypothetical accident, defined during plant design, which is used to develop design criteria.

design basis earthquake—The seismic shock (force and acceleration) at which reactor power production is permitted to fail, but the primary coolant system and the containment structure retain their pressure envelope and all safety systems remain operational.

differential pressure producer—A flow element inserted in a pipeline to produce a differential pressure proportional to the rate of fluid flow.

double-ended pipe break—The pipe has sheared off and an unrestricted stream of water at full system pressure head is jetting out of each side of the break.

drive package—See "control rod drive package."

drive seal—The seal where a control rod package is attached to the reactor pressure vessel. (See Fig. 3-9.)

drive water—Water used as the hydraulic fluid in the control rod drives of the GE BWR.

end caps—Solid metal plugs welded into each end of a fuel tube to seal it.

energy load—Load demand imposed on the power station by the power system.

fail as is—A requirement of a control valve such that upon power or control signal failure the valve does not change position.

filter train—A series of filters in tandem.

fissioning core—The core during the actual fissioning process.

fission product decay—The process by which unstable radioactive fission products lose their radioactive energy after shutdown of a reactor core. The energy is dissipated in the form of gamma rays and heat.

flux monitor—A device used to measure neutron flux in a reactor.

follow load—Varying the power level of the power plant to follow the demands of the power distribution system to which it is connected.

forced outage—An unscheduled plant shutdown resulting in a loss of power production time.

fossil fuel—Coal, oil, or gas fuel.

fossil plant—A power plant fueled by coal, oil, or gas.

freeboard—The free volume in a demineralizer vessel above the resin bed.

fuel—A solid, liquid, or gas containing fissile material. Fissile materials are the uneven (233, 235) isotopes of uranium and plutonium. A liquid is a uranium compound dissolved in an organic liquid or a molten salt such as beryllium fluoride. A solid is an oxide or a carbide.

fuel enrichment—Process of increasing the U-235 content of uranium above the natural concentration of 0.718 percent.

fuel rod—An assembly consisting of a capped zircalloy or stainless steel tube filled with fuel pellets.

going critical—The process of increasing the neutron population in a reactor to the point where there are sufficient neutrons present to sustain a chain reaction.

gravity damper—A damper in an air distribution system, opened by the velocity pressure of the air and closed by the force of gravity on the damper blade.

grid—The name given to the machined metal components used to locate fuel elements in a reactor core.

half life—The period required for an unstable radioactive element to decay to one-half of its initial mass.

heat exchanger, let down—The heat exchanger between the regenerative heat exchanger and the purification demineralizers, that removes the excess heat in the reactor coolant bleed stream and brings the coolant temperature down to approximately 120°F—a temperature acceptable to the demineralizer resins.

heat exchanger, regenerative—A heat exchanger used to recover heat from the coolant stream bled off to the purification system, by transferring the heat to the incoming makeup water.

heat sink—The fluid, water, or air that accepts all the waste heat from the power station condenser and auxiliary systems.

heat transfer surface—Area in a heat exchanger used for the transfer of heat.

heat tracing—Wrapping a pipeline with a heat source to keep the pipe and contents hot.

heater bank—The group of electric heaters in a pressurizer.

HEPA filter—A high efficiency particulate filter, sometimes known as an absolute filter with an efficiency of 99.95 percent in removal of particles down to 0.3 microns.

ice bank—The mass of ice cubes in the Westinghouse ice condenser concept provided to condense the steam resulting from an incident.

impeller—The part in a centrifugal pump or fan that imparts velocity to the fluid being pumped.

incore instrumentation—Instrumentation attached to individual fuel elements to measure fuel performance. Typical parameters are temperature and neutron flux.

inlet conditions—The characteristics of temperature, pressure, pH, etc., at the point of a fluid inlet to a process.

iodine scavenger—A device whose purpose is to collect the iodine released from core into the containment atmosphere.

irradiation—The process of bombarding a material with neutrons.

isotopes—Chemically identical variations of a single element with different atomic weights. The weight variations are said to be due to different numbers of neutrons in the atomic nucleus of the element.

isotopic heat—Radioactive heat evolving from an unstable isotope as it decays to a stable form.

latch—The locking device on a BWR control rod drive package that holds the control rod in a fixed position. (See Fig. 3-9.)

load, dead—The stresses imposed on a component by the weight of the component, insulation, and the contents of the component.

load, pressure—The stresses imposed on a component by the pressure of the fluid in the component.

load, seismic—The stresses imposed on a component by a seismic shock.

load, thermal—The stresses imposed on a component due to restriction of thermal growth caused by temperature changes.

makeup water—Demineralized water used to replace water losses from any demineralized water system. The term used by Babcock and Wilcox to describe the water returning to the reactor coolant system after passing through the purification system.

manifold—A pipe with several lateral outlets.

missile shield—The concrete shield surrounding the reactor pressure vessel. Its primary purpose is to block any missile originating in or from the reactor pressure vessel.

mixed bed column—Term used to describe a resin bed containing both anion and cation resins.

mode—Term used to describe the duty being performed by a system when the system has more than one duty—as "injection mode," decay heat "removal mode," etc.

module, steam generator—One portion of a steam generator in a high temperature gas cooled reactor. The portion includes an economizer, evaporator, superheater, and reheater. Several modules are manifolded together to produce a steam generator.

monitor—The act of measuring a parameter; a device used to measure the radiation field in the areas adjacent to the device.

MWD—Megawatt day, the energy produced in 24 hours when operating at 1 megawatt; 24,000 kilowatt hours.

MWe—Megawatts of electrical energy.

MWt—Megawatts of thermal energy.

NAUTA mixer—Trade name for a type of mixer manufactured by the J. H. Day Co. of Cincinnati, Ohio.

NDT—1. Nondestructive testing of materials by radiography, eddy current, magnetic particle or dye penetrant examination; or 2. Nil ductility transition temperature. The temperature at which carbon or low alloy steels lose their ductility and below which they behave as brittle materials.

neutron—A neutral particle contained in the nucleus of an atom.

neutron flux—The number of neutrons passing through one square centimeter in one second. A typical number in a power reactor is 1×10^{13}.

neutron poison—The general name given to materials that absorb neutrons. These materials either interfere with the fissioning process or are used to control it.

neutron, thermal—A neutron whose energy level has been lowered sufficiently so that upon collision with another atom it will cause the atom to split and release energy. Neutron energy levels can be lowered by recoil off moderating atoms.

nondestructive test—Sometimes called "nondestructive examination"—a method of examination of material that does not change the condition of the material. Methods in use today are radiography, ultrasonics, eddy current, magnetic particle and dye penetrant.

nuclear fission—Splitting of an atomic nucleus as a result of the impact of an external neutron.

nuclear incident—A generic phrase used to refer to an uncontrolled release of radioactivity. As used in this book it refers to incidents involving neutrons rather than gamma radiation.

nuclear island—The buildings and equipment that comprise the reactor and all its emergency and auxiliary systems.

operating basis earthquake—The maximum seismic shock (force and acceleration) included in normal design stresses in addition to any other loads.

parameter transducer—A device used to measure a variable characteristic and produce a controlled transmittable signal.

particulate removal—Removal of solid particles from a fluid stream of filtration.

penetration—A component which becomes part of the containment structure and is used for passage of a pipe or electric cable. Piping penetrations are further subdivided into hot (temperature) and cold penetrations according to the fluid in them. The penetration assemblies are built in accordance with Class 2 or MC of Section III of the ASME Code.

pH control—The process of controlling hydrogen ion concentration by the addition of acid or alkali to primary coolant. pH is normally kept above 7.0 by the addition of lithium hydroxide.

plant shutdown—The act of stopping plant operation for any reason.

plenum—A chamber used as a distribution point for several parallel flow paths.

polar crane—A bridge crane installed in a circular

building whose bridge spans the building diameter. The bridge rotates on a single circumferential rail in the outside wall of the building.

precoating cartridges—The act of applying a coating of ground resin to a cartridge in a ground resin filter/demineralizer.

precolumn—Term used to describe a special purpose resin bed installed upstream of a mixed-bed demineralizer.

pressurizer—A pressure vessel in which pressure is developed by boiling water with electric heaters. The pressure developed is imposed on the reactor coolant system of a pressurized water reactor system.

pressure excursion—An uncontrolled pressure rise.

pressure transient—A rapid pressure change either up or down.

primary coolant—The fluid used to cool the fuel elements. It may be liquid or gas.

pump, canned—A horizontal centrifugal pump without dynamic seals in which the drive motor is inside the pressure boundary; a vertical multistage pump in which the pump casings are mounted inside a shell. The annular space between the shell and the casings serves as part of the suction path of the pump.

pump, charging—The name given to pumps used to return reactor coolant to the primary system after purification. They are sometimes known as makeup pumps.

pump, holding—The recirculating pump used to hold the resin bed in place on the cores in a ground resin filter/demineralizer during periods of low or no flow.

pump, metering—A pump with a very accurate volumetric delivery used to inject measured amounts of chemicals into a system.

pump shutoff head—The pressure head developed by a centrifugal pump when operating against a closed discharge line at zero flow.

radioactive waste—Waste materials, solid, liquid, or gas, that are produced in any type of nuclear facility.

radwaste—Contraction for "radioactive waste."

reactor containment boundary—The pressure envelope in which a reactor and its primary cooling system are located.

reactor coolant pressure boundary—The system of vessels, valves, pumps, and piping that contains the reactor coolant.

reactor subcritical—A reactor with insufficient neutrons to sustain a chain reaction.

regeneration—The process of restoring working capacity to an adsorption bed after it has been exhausted—as in regenerating a dryer bed by heating it to drive off the adsorbed water.

relay rack room—The room adjacent to the control room in which the relays and control circuitry are installed.

relief valve effluent—The fluid discharged from a relief valve when it opens to relieve pressure.

resin—A synthetic organic compound with the ability to absorb a cation ion such as Ca^{++}, and release two hydrogen ions H^{+}, to restore electrical balance. In a different form the resin can do the same thing with anions while releasing OH^{-} ions.

restriction orifice—An orifice installed in a pipeline to restrict flow in the line instead of measuring the flow.

safety system—A mechanical, electrical, or instrumentation system or any combination of these, whose purpose is the safety of the reactor or of the public.

scrammed—Rapid unscheduled shutdown of the reactor.

scupper—An opening cut in the side of a pool at the water surface to allow water to overflow over the opening wall. Scuppers are used both for inlet and outlet purposes.

seal water—Clean water injected into a primary pump seal to insure that exiting water from the seal is clean and nonradioactive.

seiche—An oscillation of a land-locked lake or sea. It can be of interest in intake structure design, containment design, or emergency water supply design.

Seismic Classes I, II, *and* III—An arbitrary set of classifications set up to define the different seismic requirements of equipment.

seismic design—The design criteria associated with a particular seismic class. The criteria include a combination of vertical and horizontal "g" loading, vibration spectra and damping.

seismic duty—The performance required of a component during and after a seismic shock. It ranges from complete failure to no loss of operating characteristics.

service water—Raw water from the heat sink used to cool all waste heat loads in a power plant except the turbine condenser.

shell and tube exchangers—A heat exchanger comprised of a bundle of tubes inside a cylindrical shell. See TEMA Standards published by the Tubular Exchanger Manufacturers Association, New York, N.Y.

shell pass—A flow path through one length of the heat exchanger shell. If a flow path goes through two lengths of the shell, it would be designated as two shell passes.

shielding cask—A transfer cask for radioactive material whose principal requirement is adequate biological shielding for the material inside. The usual shielding material employed is a combination of lead and steel.

shutdown margin—The excess negative reactivity available to shut down the fission process in a reactor.

slurry—A dispersion of a solid in a liquid. The term is associated with transportation of the solid through a pipeline.

sparger—A perforated submerged pipe in a tank or vessel used for introducing a fluid into the vessel below the liquid surface.

spent fuel storage—Storage of fuel elements that

have been removed from the reactor after completion of rated time in the reactor on load.

spent fuel storage pool—The pool of demineralized water in which spent fuel elements are stored pending their shipment from the facility.

steam generator—The main heat exchangers in a pressurized water or gas cooled reactor power plant that generates the steam that drives the turbine generator.

steam generator, once through—A steam generator with straight single-pass tubes. (Used by Babcock and Wilcox Co.)

steam generator, U-tube—A steam generator with two-pass U-shaped tubes. (Used by Westinghouse Electric Corp. and Combustion Engineering Corp.)

stepping switch—A switch with a multiplicity of circuits that controls a series of operations by opening and closing contacts as it indexes from one position to another.

supplier—A manufacturing organization supplying material used in the construction of a nuclear power station.

threshold, damage—The start of damage to an organic material due to bombardment by neutrons and/or gamma rays.

top head—The removable bolted head on a reactor pressure vessel or a BWR drywell.

torous—The mathematical shape initially used by the General Electric Co. for their pressure suppression pool. It has now been supplanted by a cylinder.

tripped—See "scrammed."

tsunami—An ocean wave caused by a submerged earth movement as in an earthquake or volcanic eruption.

turbine generator—The assembled steam turbine coupled to an electric generator that produces the electric power in a power plant.

vendor—Same as "supplier."

volumetric examination—A nondestructive examination of a weld or base material by radiographic or ultrasonic techniques that investigate the complete thickness of the material.

water hammer—The shock load imposed on a flowing pipeline by the rapid closure of a shutoff valve.

Index